GAMING-SIMULATION:
RATIONALE, DESIGN, AND APPLICATIONS

Gaming-Simulation:

RATIONALE, DESIGN, AND APPLICATIONS

A Text with Parallel Readings
for Social Scientists, Educators,
and Community Workers

Cathy S. Greenblat
Douglass College
Rutgers University

and

Richard D. Duke
University of Michigan

SAGE Publications

Halsted Press Division
JOHN WILEY & SONS
New York—London—Sydney—Toronto

Distributed by Halsted Press, a Division of
John Wiley & Sons, Inc., New York

Printed in the United States of America

Library of Congress Cataloging in Publication Data

Greenblat, Cathy S   1940–
   Gaming-simulation—rationale, design, and applications.

   Bibliography: p.
   1. Game theory—Addresses, essays, lectures.
   2. Social sciences—Simulation methods—Addresses, essays, lectures.
   I. Duke, Richard D.   II. Title.
H61.G67   300'.1'84   74-75864
ISBN 0-470-32500-3

FIRST PRINTING

CONTENTS

PART IV: OTHER ARENAS AND APPLICATIONS

APPENDICES

INTRODUCTION

The impetus to put this book together emerged from the frustrations we both had experienced in trying to learn about gaming-simulation, to write about it, and to teach about it to our graduate students and others who had developed an interest in the field. Often we needed, or others requested, reading materials, and it was difficult to make recommendations. We found several very good books dealing with selected aspects of the field, but no comprehensive, in-depth treatments. There existed, of course, a growing body of literature in the form of articles; these, however, were widely scattered and inaccessible to potential readers unless they were willing to expend considerable time and effort. Finally, there were a few topics on which very little had been written at all. The potential of gaming-simulation for research applications and in the area of public policy, for example, have scarcely been addressed in the literature despite increasing numbers of applications in these arenas. Therefore few readings could be found in any source on these topics.

We saw our task, then, as twofold. First, we wanted to develop and present in one volume a set of succinct overview essays on the three major themes of rationale, design, and applications. The "text" part of this volume presents, therefore, integrative essays that could be read separately by the person desiring a comprehensive overview of the field.

Second, we wished to provide the more serious reader the opportunity for in-depth coverage of topics. Therefore we have supplemented our overview with a set of articles, some reprinted and some written especially for this volume. These have been used either to *expand upon ideas* presented in the lead articles or to *offer examples* of points made there. Thus, for example, the selection by Miller and Duke in Part IV amplifies ideas about one stream of future public policy applications that were suggested in the lead article. William Gamson's description of the development of SIMSOC offers a case-study example of some of the design-construction considerations described in our lead articles in Part II.

The new text material by the authors is broken into four parts dealing with the character, design, and present and potential applications of gaming-simulation. Each of these sections contains the related readings in smaller print. In toto, we believe this volume offers to both the newcomer to gaming and to the practicing professional an integrated, comprehensive set of materials which express both a theoretical base and pragmatic guides for day-to-day usage.

Cathy S. Greenblat
Douglass College, Rutgers University

Richard D. Duke
University of Michigan

Part I

THE NATURE AND RATIONALE OF GAMING-SIMULATION

SEEING FORESTS AND TREES:
Gaming-Simulation and Contemporary
Problems of Learning and Communication

CATHY S. GREENBLAT

One spring day last year, I took a visiting friend from another part of the country into New York City. We spent a marvelous afternoon roaming through Greenwich Village and, later, Chinatown. At the conclusion of the day, she commented, "I can't imagine why people speak so disparagingly of Manhattan. It's really very charming."

My friend is an intelligent woman, and she knew that Greenwich Village and Chinatown were not typical of New York, but having seen only these two neighborhoods, she had no idea of where they "fit" into the larger whole called "Manhattan." Were the winding streets and small stores found there unique to those neighborhoods, or could one find many examples of this ambience in the city?

Another few days were spent taking a less intense look at a large number of neighborhoods, getting the feel of one and then going on to another. At the end of the week, confusion still reigned; now there was too much information, and the problem was not knowing how it all fit together. My friend felt a bit like the inchworm who crawled over the Mona Lisa and then tried to understand the painting from the pieces of information it could put together. (See the reading in this section by Rhyne.) Again, I was left with the question of how to give someone a good sense of the city.

The answer came to me the other day through a serendipitous occurrence. A plane strike and a busy schedule led me to take a helicopter from Newark Airport to LaGuardia. For the first time I saw Manhattan from the air at a low altitude. As we hovered at about the level of the top of the Empire State Building, moving across the city, I realized I was able to "see" its major dimensions clearly. Of course, much was missing in such a view: I couldn't smell the fresh-cooked foods in the ethnic neighborhoods, or see the results of differential garbage collection facilities, or note the different paces of life on Broadway, Fifth Avenue, the garment district, and 125 St. These flavors were lacking up there. But what I *was* provided with was an *overview*—a framework that gave me a sense of how big the "beast" was, what its major characteristics were, and how they fit together; how typical skyscrapers were as compared to brownstones, and what kind of transportation routes connected the different neighborhoods, making them variously accessible to members of other sections. In short, from this vantage point, the city as a *geographic system* could be seen and comprehended. Of course, I couldn't see or know of the subways, but then any system has an "underground" or "network" or "black box" not easily found, but there to be discovered by the enterprising researcher.

From now on, then, I have a plan for the best of all possible introductory tours: I would first take friends for a helicopter ride over the city, giving them the framework. Then, bit by bit, I'd expose them to the neighborhoods, confident that they could appreciate the charms and horrors of each place, but also that they could place them in the proper context of the city as a whole. And hopefully someone would do the same for me in another city.

If all this seems discursive and irrelevant to the topic of gaming, I suggest that it's not, for my tour plan deals directly with the central thesis of this book regarding the importance of gaming: we need ways of providing comprehension of wholes to overcome provincialism, to counteract the narrow perspectives that derive from specialization, and to provide a mode of retaining detail. As we shall try to show, gaming represents a mode of developing this holistic understanding, and thus is an important communication and learning tool.

COMMUNICATION-LEARNING NEEDS IN TECHNOLOGICALLY ADVANCED SOCIETIES

In those parts of the world in which the "technological revolution" has taken place, men have achieved extraordinary feats, but have also encountered new kinds of problems. The complex of traits associated with the "civilized" end of the folk-civilization continuum includes larger aggregates, multiple communications channels among varied kinds of people, a complex division of labor, deliberate design in personal and collective life, more impersonality in interpersonal relations, and a world outlook that differs significantly from that of typical members of more traditional societies (cf. Redfield, 1962; Tonnies, 1940; Becker, 1950; Lerner, 1958). Each of these characteristics may be looked upon as positive or negative, as a benefit or a cost. For example, weaker kinship ties are related to both greater mobility of the labor force and to rapid change in information creating "generation gaps." Is it thus to be applauded or bemoaned?

It makes little sense to issue lamentations for the passing of a former idyllic state; while the folk society may have provided more positive, warm, meaningful relationships, it also entailed high disease and death rates, considerable illiteracy, and the maintenance of fear and superstition. The transition has thus been complex, and the interpretation of whether the end result is, on balance, "good" or "bad," is dependent upon values (cf. Tumin, 1958).

These changes have brought about a major set of problems and challenges in the communication-learning domain. No longer can the information needed to sustain a rich and meaningful life be carried through oral transmission from members of one generation to members of another. Older people, rather than serving as repositories of learning and skills and being venerated for their wisdom, find that they are quickly "outdated," replaced by younger colleagues who have had more recent exposure to scientific discoveries and developments, and put "out to pasture."

Knowledge has been increasing at a geometric rate, so that it is estimated that, by the mid-sixties, the output of books approached 1,000 per day, the number of scientific journals and articles was doubling about every fifteen years, and computers were raising the rate of knowledge production to extraordinary speeds (Toffler, 1971:30-32). This acceleration was both a product of technological advance and a creator of it.

The "knowledge explosion," combined with bureaucratization and the increasing complexity of the division of labor have led to the development of cadres of highly trained specialists. Speaking of this crisis in the scientific community, Kenneth Boulding (1968: 4) cautions:

> One wonders sometimes if science will not grind to a stop in an assemblage of walled-in hermits, each mumbling to himself words in a private language that only he can understand. In these days the arts may have beaten the sciences to this desert of mutual unintelligibility, but that may be merely because the swift intuitions of art reach the future faster than the plodding leg work of the scientists. The more science breaks into sub-groups, and the less communication is possible among the disciplines, however, the greater chance there is that the total growth of knowledge is being slowed down by the loss of relevant communications. The spread of specialized deafness means that someone who ought to know something that someone else knows isn't able to find it out for lack of generalized ears.

Within their fields, individuals often must absorb vast quantities of information to develop and maintain their positions. Simultaneously they are urged, as citizens, to develop at least a general comprehension of other aspects of the world in which they live and must operate. If the society is to take advantage of the technological expertise and to flourish, it must find ways to combat the narrow perspectives born of pressures of specialization. *Integrative* modes of learning must be found.

There are, then, two types of communication-learning needs in contemporary societies such as ours: (1) the need for modes of learning large quantities of information; and (2) the need for modes of developing general comprehensions of some domains rather than detailed information about them. In both cases, the need is for an awareness and understanding of elements and relationships—a "systems" awareness. As the world becomes more and more complex, most of us encounter enormous difficulty comprehending those systems in which we operate in our everyday lives. Without time and space being collapsed, elements being reduced to manageable size, and some simplification of the number of variables being effectuated, comprehension of these systems is rendered extremely difficult, if not impossible (cf. Goffman, 1961).

Furthermore, decision makers, be they executives or voters, must develop appreciations of "wholes" in order to judge the general thrust of policy (Michael, 1968). The business executive, the general citizen, and the policy maker must integrate reports with other half-sensed appreciations, for "they require gestalt appreciations rather than explicit knowledge of bits of data" (Rhyne, 1972: 95).

In addition, the need for modes of developing systems awareness is critical because it is a necessary precursor to the development of understanding of details. Without a sense of the whole, detail is rejected. A framework within which ideas can be placed must be developed before those ideas can be integrated. Thus, to effectively approach a new book or article, one should read the introduction, table of contents, skim the chapters, looking at the nature of the presentation, etc. Likewise, to maximize comprehension of a chapter, one should first read the lead paragraph, then read all topic sentences through the chapter, and finally read the concluding paragraph. With this overview established, one is then—and only then—ready to begin the ordinary process of sequential reading of the chapter sentences. This approach is not usually stressed, however.

> For *any* excursion into understanding, we should start first with sweeping comprehensions and then seek to learn, or teach component facts. The route of science, prosaic exposition, and academic specialty has normally been the opposite [Rhyne, 1972: 97].

An approach that stresses bit-by-bit acquisition of information without prior acquisition of a holistic sense carries severe threats of conveying miscomprehensions of the sort that beset the king's blind emissaries who stationed themselves at different parts of an elephant and offered descriptions of the creature they felt. Holistic learning is needed as a first step; hence, my decision concerning the helicopter ride.

There is yet another problem. Many skills, roles, and norms that must be understood cannot be learned by direct experience; *vicarious* learning is required. Attainment of educational goals such as the following, suggested by Edgar Friedenberg (1963: 221-222), must depend to some extent on alternatives to direct experience:

> The highest function, I would maintain, is to help people understand the meaning of their lives, and become more sensitive to the meaning of other peoples' lives and relate to them more fully. Education increases the range and complexity of relationships that make sense to us, to which we can contribute, and on which we can bring to bear competent ethical judgement. If we are to transcend our own immediate environment, we must have access to the record of past and present, learn the skills needed to interpret it, and learn to tell good data from poor, whether it be the empirical data of the sciences or the moral and aesthetic data of the humanities. We must be able to read, and to know where what we read fits into the structure of human experience; and to write with enough subtlety and complexity to convey the

special quality of our mind to others. We must explore, and we must have the privacy and authority necessary to protect ourselves from intrusion if we are to use energy for exploration rather than defense.

At this point, two more needs must be added to those offered above. In a society such as ours, in which change is constant and rapid, we must not only teach people to *see* the forests and the trees, but we must show them *how to find the woods* and *motivate them to want to make the search.* In non-metaphoric terms, we must develop ways of building motivation to learn, and then modes of developing people who know how to learn—to explore, conceptualize, inquire, experiment, and critically analyze. This conception of the problem focuses not only the content of communication-learning needs, but upon processes.

MEETING THE CHALLENGES

How, then, can we meet these challenges? That is, what are the characteristics of the environments and media by means of which the required learning is most likely to transpire?

Educators, of course, have fought over the answers to these questions with no resolution, but many suggest that critical elements include the following: (1) We need to find modes of creating motivation prior to transmitting information; (2) the learner must be an active participant in the learning process, rather than a passive recipient of information transmitted to him; (3) instruction must be individualized such that learning is at the appropriate pace for each learner; (4) there must be prompt feedback on success and error.

In their article in the readings section, Moore and Anderson (1969) develop these ideas into four heuristic principles for the design of learning environments. In summary, they argue the following:

First, the learner should be given the opportunity to operate from various *perspectives.* He should not just be a recipient of information, but should at times be an agent, a referee, and a reciprocator. Second, activities should contain their own goals and sources of motivation, not just represent means to some end (such as grades). That is, in an effective learning environment, activities are *autotelic.* Third, the learner should be freed from a dependence on authority and allowed to reason for himself; he is thus made more *productive* in the learning process. And fourth, the environment should be responsive to the learner's activity. Not only should he be given feedback, but he should be helped to be *reflexive,* evaluating his own progress.

Those familiar with gaming-simulations will recognize that these principles parallel the major arguments given for the effectiveness of gaming in teaching. Games entail the active involvement of learners with the subject matter in autotelic activities that free them from dependence on authority and offer them feedback and ways of measuring their progress toward a goal.

But there is more to it than that. When we argue that gaming-simulation may be a way to deal with the challenges delineated earlier, it is partly because we see games as *communications* devices, rather than holding to the more narrow conception of them as pedagogic devices. The problems outlined in the beginning of this essay are basically problems of communications: the complexity of the world is something that has to be understood more or less all at once. What is required is a mode of simultaneous rather than sequential presentation of parts of the overall message. Rhyne argues, in the reading offered here, that prose is a poor tool for creating pattern comprehension. It is "sequential, working point-to-point along a chain of assertions or questions. It treats one molecule of meaning at a time" (1972: 97).

The systems awareness we urge as the contemporary need also requires a holistic language, one able to convey gestalt. That, we believe, is what gaming-simulation is. The paper by Duke in this section of readings demonstrates the distinctions between gaming-simulation and other forms of communications, and provides the basis for our argument that gaming-simulation is a language for transmitting both specific facts and general principles of social theories. Watson's paper discusses the role of games and play in the socialization process, and Moore and Anderson provide a general schema for understanding the nature of a "learning environment."

REFERENCES

Becker, Howard
    1950 "Sacred and Secular Societies." Social Forces 28: 361-376.

Boulding, Kenneth E.
    1968 "General Systems Theory: The Skeleton of Science." In Walter Buckley (ed.) Modern Systems Research for the Behavioral Scientist. Chicago: Aldine.

Friedenberg, Edgar
    1963 Coming of Age in America. New York: Random House.

Goffman, Erving
    1961 Encounters: Two Studies in the Sociology of Interaction. Indianapolis: Bobbs-Merrill.

Lerner, Daniel
    1958 The Passing of Traditional Societies. New York: Free Press.

Michael, Donald
    1968 "On Coping with Complexity: Planning and Politics." Daedalus (Fall): 97-104.

Moore, Omar Khayyam and Alan Ross Anderson
    1969 "Some Principles for the Design of Clarifying Educational Environments." Pp. 582-594 in David Goslin (ed.) Handbook of Socialization Theory. Chicago. Rand McNally.

Redfield, Robert
    1962 "Civilization as Things Thought About." Pp. 364-391 in Margaret Park Red-
        field (ed.) Human Nature and the Study of Society: The Papers of Robert
        Redfield, Volume 1. Chicago: University of Chicago Press.

Rhyne, R. F.
    1972 "Communicating Holistic Insights." Fields within Fields . . . within Fields 5
        (1).

Toffler, Alvin
    1971 Future Shock. New York: Bantam Books.

Tonnies, Ferdinand
    1940 "Gemeinschaft and Gesellschaft." In Charles Loomis (ed.) Fundamental Con-
        cepts of Society. New York: American Book Company.

Tumin, Melvin
    1958 "Some Social Requirements for Effective Community Development." Com-
        munity Development Review, International Cooperation Administration, 11
        (December).

BASIC CONCEPTS AND LINKAGES

CATHY S. GREENBLAT

Further elaboration of our thesis that games are effective communication devices requires that we spell out more thoroughly what gaming-simulations are and how they relate to social theory.

THEORIES, SYSTEMS, AND TYPES OF MODELS

A theory is a set of logically related propositions about some aspect of reality. Where classical science was concerned mostly with problems involving only a few variables and one-way chains of causality, most of the problems addressed in contemporary social science theory are multivariate. They are problems of "organized complexity"—that is, of the interaction of a large but not infinite number of variables (cf. von Bertalanffy, 1968: 11).

The variables in such theories are organized into a *system*; that is, the conceptualization includes identifiable parts which are mutually interdependent and fit together to make a whole.

> We define a system in general as a complex of elements or components directly or indirectly related in a causal network, such that at least some of the components are related to some others in a more or less stable way *at any one time*. The interrelations may be mutual or unidirectional, linear, non-linear or intermittent, and varying in degrees of causal efficacy or priority. The particular kinds of more or less stable inter-relationships of components that become established at any time constitute the particular *structure* of the system at that time [Buckley, 1968: 493].

The basic problems and questions, then, for the social scientist, are problems of interrelationships.

As the scientists works, he often formulates a *model* or models—a representation of his view of reality, showing the major elements and their relationships. These are statements of theory; they indicate definitions, assumptions, and propositions from the larger body of theory, for even where they are tentative and exploratory, they should be built, insofar as possible, "not from ad hoc fragments, but from the available general theories that seem germane and from appropriate portions of the larger body of sociological knowledge" (Riley, 1963:

9). The model, then, is a way of construing and representing a particular set of social phenomena.

But what do such models look like? The most familiar form to most of us is the *verbal model.* These are linear in presentation; we confront one component at a time. For example, we have all read such statements as "Social stratification systems function to distribute favorable self-images unequally throughout a population." Most teaching and publication in the social sciences depends upon verbal models; in lecture halls, small classrooms, and the library we encounter them by the thousands.

Graphic models are used with increasing frequency, for the presentation of complexity is aided by the simultaneity of expression. Note, for example, the difference in the verbal and the visual model of the mutual causal relationships offered in the example in Figure 1. Once the graphic symbols are understood, the figure gives a concise summary of the elements and their relationships. Where there are systems of simultaneous changes, graphic models seem to convey more than linear verbal models.

Mutual causal relationships may be defined between more than two elements. Let us look at the . . . diagram (Figure [1]). The arrows indicate the direction of influences. + indicates that the changes occur in the same direction, but not necessarily positively. For example, the + between G and B indicates that an increase in the amount of garbage per area causes an increase in the number of bacteria per area. But, at the same time, it indicates that a decrease in the amount of garbage per area causes a decrease in the number of bacteria per area. The−between S and B indicates that an increase in sanitation facilities causes a decrease in the number of bacteria per

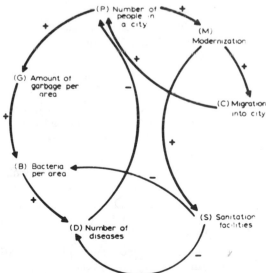

FIGURE 1.   Mutual Causal Relationships Shown in a Verbal and a Graphic Model

area. But, at the same time, it indicates that a decrease in sanitation facilities causes an increase in the number of bacteria per area.

As may be noticed, some of the arrows form loops. For example, there is a loop of arrows from P to M, M to C, and C back to M. A loop indicates mutual causal relationships. In a loop, the influence of an element comes back to itself through other elements. For example, in the loop of P-M-C-P, an increase in the number of people causes an increase in modernization, which in turn increases migration to the city, which in turn increases the number of people in the city. In short, an increase in population causes a further increase in population through modernization and migration. On the other hand, a decrease in population causes a decrease in modernization, which in turn causes a decrease in migration, which in turn decreases population. In short, a decrease in population causes a further decrease in population through decreased modernization and decreased migration. Whatever the change, either an increase or a decrease, amplifies itself. This is so when we take population as our criterion. But the same is true if we take modernization as a criterion: an increase in modernization causes a further increase in modernization through migration and population increase; and a decrease in modernization causes a further decrease in modernization through decreased migration and decreased population. The same holds true if we take the migration as the criterion.

In a loop, therefore, each element has an influence on all other elements either directly or indirectly, and each element influences itself through other elements. There is no hierarchical causal priority in any of the elements. It is in this sense that we understand the mutual causal relationships [Maruyama, 1968: 311-312].

Some statements of theory are made in a special type of graphic form: these are *mathematical models*. We learn in the natural sciences that the circumference of a circle is derivable from the formula: $C = r^2$. So, too, we find that desegregation patterns, urban growth, and population expansion are only some of the social phenomena to have been conceptualized in mathematical form. Generally it is argued that the more formal presentation involved in the formulae are clearer than the purely verbal statements. They are neater, and, once having learned the symbols, the learner can "see" all the elements and their relationships.

A fourth, related kind of model is the *physical model*. Watson, describing the DNA molecule as a double helix, reports that he realized this through "playing around" with a tinker-toy model—a physical representation. Some of you may have seen plastic models of the human anatomy created to show students the parts of the human body and how they fit together. By viewing the model, one can gain a more systemic view of relationships of size, place, and form than by listening to a lecture on the neck bone being connected to the shin bone! And the model is three-dimensional, so it can be turned and viewed from many angles, unlike a diagram of the body.

All the above types of models share one characteristic limitation: they show the structure of the referent system, but do not satisfactorily display the functions or dynamic processes. In other words, they're static models. It's on the basis of this characteristic that we distinguish the fifth type of model: the *simulation*. Included in this category are such things as wind tunnels for testing aircraft under varying conditions; representations of automobile panels for teaching driving; and Zorba the Greek's operating model or simulation of the device he would later construct for bringing logs down the mountain.

The critical factor differentiating the simulation from the other types of models is that the simulation is an *operating model*. That is, it demonstrates not simply the state of the system at some given time, but also the way the system changes. Think, for example, of the difference between the type of model a junior high school science student might make to show the sun and the planets, and contrast it with the model found in a planetarium. The latter shows not only the number and relative size of the planets, but also their velocities. Thus, the observer can see the different patterns of relationships which create eclipses, etc. It is dynamic.

Simulation, then, entails abstraction and representation from a larger system. Central features must be identified and simplified, while less important elements are omitted from the model. It is the very process of highlighting some elements and eliminating others that makes the model useful. A city street map is easier for a driver to use and more helpful to him if it doesn't show the differences in altitude in different parts of the city; but a map showing the route from Denver to Los Angeles is more helpful if it *does* show mountain ranges and if it eliminates most local streets. So, too, with a simulation, the utility derives from correct selection of elements to emphasize and elements to eliminate. The selection of such features depends upon the purpose the model is to serve.

Most of us are familiar with simulations of technological systems, having seen simulated space maneuvers on TV while real spacecraft were travelling to, orbitting, landing on, or departing from the moon. Likewise, many have seen simulations of the control panels of aircraft used to train future pilots to deal with flying conditions and problems at less expense than would be involved in putting trainees in real planes on missions. *Social systems,* too, can be simulated: marital dyads, organizations, neighborhoods, cities, nations, or groups of nations. The principle is the same: central features are identified and put together such that they operate in a manner similar to the real world system.

SIMULATIONS, GAMES, AND GAMING-SIMULATIONS

Simulation models can be made to operate in one of three ways. First, a computer can be used to make decisions and take actions. Other actions and the consequences of the decisions then are yielded by the computer model. Second, a combination of computer and human players can make the model operate. The computer may serve simply as a high-speed calculator, or it may contain a model or set of models within it which are triggered by the actions of the players. Third, all operations may be generated by human players and the consequences calculated by humans. The first of these is a "computer simulation"; the second and third are referred to as "gaming-simulations," "game-simulations," or "simulation games." Some would further distinguish between the two by prefacing them "man-machine" and "all-man," respectively.

The term "game" is applied to those simulations which work wholly or partly on the basis of players' decisions, because the environment and activities of participants have the characteristics of games: players have goals, sets of activities to perform; constraints on what can be done; and payoffs (good and bad) as consequences of the actions. The elements in a gaming-simulation are patterned from real life—that is, the roles, goals, activities, constraints, and consequences and the linkages among them simulate those elements of the real-world system. In the gaming-simulation GHETTO, for example, players are given educational, occupational, and family profiles similar to those found in most ghettos. The choices they are offered in terms of how they will spend their time are similar in character and availability to those offered ghetto dwellers, and the payoffs (e.g., money for jobs and for illegal enterprises, and possibilities of arrest for the latter) are also realistic.

"Gaming-simulation," then, is a hybrid form, involving the performance of game activities in simulated contexts. (This position, it should be noted, differs somewhat from that taken by some others, notably John Raser [1969], who treats games as models not sufficiently developed to warrant the label simulation.)

Three of the readings in this section have been selected to amplify these ideas. James Coleman's "Introduction: In Defense of Games" explores in succinct fashion the reasons social scientists are interested in looking at games and game play. Garry Shirts draws some of the distinctions and notes overlaps between 'games,' 'simulations,' and 'contests.' Finally, Robert Armstrong and Margaret Hobson elaborate upon the discussion here of gaming-simulation components and operation.

REFERENCES

Buckley, Walter
   1968 "Society as a Complex Adaptive System." In Walter Buckley (ed.) Modern Systems Research for the Behavioral Scientist. Chicago: Aldine

Maruyama, Magoroh
   1968 "The Second Cybernetics: Deviation-Amplifying Mutual Causal Processes." In Walter Buckley (ed.) Modern Systems Research for the Behavioral Scientist. Chicago: Aldine.

Raser, John
   1969 Simulation and Society. Boston: Allyn & Bacon.

Riley, Mathilda
   1963 Sociological Research: A Case Approach. New York: Harcourt, Brace & World.

von Bertalanffy, Ludwig
   1968 "General Systems Theory: A Critical Review." In Walter Buckley (ed.) Modern Systems Research for the Behavioral Scientist. Chicago: Aldine.

COMMUNICATING HOLISTIC INSIGHTS

R. F. RHYNE

Patterns and Systems, International

INTRODUCTION

The intention here is to stimulate explorations of the means whereby appreciations of complex wholes may be more quickly and more reliably "told" to
others. The paper seeks to engage the interest of those who are proud enough of
their ingenuity to test themselves against tough problems, experienced in the arts
of seeing and describing how things fit (or clash) together, and practiced
somehow in the deliberate application of intelligence in the surmounting of
difficulties.

The argument stands on four legs.

- There is a great and growing need for the kinds of powers of communication that
  help a person gain, vicariously, a feeling for the natures of fields too extensive and
  diverse to be directly experienced. This need is an objective one, an ineluctable
  concommitant to decision within a highly interconnected biosphere that is beginning
  to fill up.

- Prose and its archetype, the mathematical equation, do not suffice. They offer more
  specificity within a sharply limited region of discourse than is safe, since the clearly
  explicit can be so easily mistaken for truth, and the difference can be large when
  context is slighted. Also, prosaic description has a natural affinity for speciality,
  which is nearly the opposite of the mood within which wholes are to be "felt."

- If these two issues were all that needed to be met, the problem still would be serious.
  The prognosis is even darker in view of trends that may adversely prejudice just those
  analysts, staffers, and decision makers who should help the most. In the intellectual
  climate of the times, intuition and supralogical insight tend to be classed with
  alchemy and magic; efforts to invent an alternative to prosaic exposition will have to
  be justified to skeptics before first trials will be permitted.

- With all of these urgencies and obstacles, there still are signs that spectacular success
  may lie just beyond the first serious trials. Many isolated achievements have shown
  themselves in recent decades—most untouted, some coincidental, some seen as being
  trivial by their makers. Follow-up was rare; the field is open and inviting.

Reprinted from "Fields within Fields . . . within Fields" 5, 1 (1972) with the permission of
the World Institute Council.

THE SPECIAL, CURRENT NEED FOR HOLISTIC COMMUNICATION

There *is* a macro-problem, an interweaving of adverse conditions that is more extensive, more richly structured by interior lines of interaction, and more threatening than any circumstance faced before by all mankind. It is almost as though the universe had turned malevolent again (as it was seen to be by early man) after having seemed neutral for centuries—or even favorable to human aspirations. For decades to come, it seems wisest to expect that surprises will most frequently be unpleasant, in a kind of reverse serendipity. This prospect may be caused by some deep underlying cause, or it may be just a coincidental cresting of mutually exacerbating trends, but in either case it will desperately stress that freedom within order that is the crown of mankind's "Great Ascent."

The signs suggest that really massive social currents will somehow be blocked and shifted to new channels during just the next few years, and the forces at work are unlikely to be gentle. Failure, therefore, could be deadly.

- For instance, it seems nearly sure that population growth (at least in hungry lands) *must* change very soon. The circumstances invite preemptive actions in pursuit of short supplies, tending to damage productivity and distribution long before the theoretic maximum of almost starving poor can have been reached. What is the character of our rich part of such a world, and if it is not famine that steps in to cancel exponential growth, what will it be instead?

- Or think of fuel exhaustion. Technology *may* bail us out, by giving us unlimited atomic power. We'd better hope it does. Supplies of things to burn for power would still be scant—in fifty years, if not in ten—if current users merely were to live no worse, but only wide catastrophe seems apt to interdict the drive toward ease-through-energy by billions more. Fuels will be burned more swiftly, or else the world will face the wrath of those condemned to muscle-power while satellites deliver news of affluence. Of course, this issue ties to that of population.

- To pick one more facet, adversely linked with both of those above, human communication hinges on a sharing of implicit underpinnings of words—patterned traits (never verbalized by those who teach) passed on to neophytes (who do not know they learn). If old and young live together so sketchily that this cultural baton is not passed between them, there tends to be no second chance—confusion and mistrust are fed as "silent languages" diverge and thus, in turn, force more divergence. This alone would threaten the fragile rules and trusts that are prerequisite to openness and democratic rule; when mixed with all the other, tangible causes for anxiety and rage that permeate the macro-problem, the path of continuity within change seems desperately narrow.

The flavor of this meld of difficulties differs greatly from one region to another, especially as glut or scarcity may rule. In every case, however, one fact seems to stand out. *The arenas within which partial solutions are illusory have grown remarkably in subject scope, spacial extent, and time.* Subjects recognized as "hard" and "soft" now interweave, neither to be seen as dominant above the other. Continents are just as apt to be the stage as counties are. Knowing the present and the recent past no longer will suffice, since change demands equivalent concern for future, unknown patterns. In almost every step to meet the macro-problem, the decision maker (whether some executive or the voter who must judge how well he likes the actions taken in his name) is forced to act as though he comprehended fields that are so wide he cannot even fairly sample them.

The other branch of argument at this stage concerns the actual nature of decisions relevant to complex fields. A *field* is any variable space within which a disturbance of one element tends to be significantly felt at every other point. A *complex field* is analogous to the complex variable of mathematics in that its elements are incommensurate, a characteristic shared by almost all component fields within the macro-problem. Such problem areas (of which family life and most business interplays also are examples) are ones in which no wise person thinks of calculation or even of logic as the determinant of decision. Rather, he seeks to use such very powerful but narrow tools to amplify his broader, human powers, readying himself by those and other "drills" for decisions that finally must flow—by intuition, insight, hunch—from inner comprehensions. In this light, research is self-defeatingly presumptuous when it purports to offer answers, and researchers are ridiculous when they express surprise or shock when their results are not adopted. Research produces heuristic cartoons, to be used (if at all) by decision makers in teaching themselves more of what they ought to understand. The function of most staff work is the same—the care and nurturing of those few scattered folk tasked not to reason but to reach decisions on demand. *Decision is a gestalt event; it is not a logically determinable process.*

The piecing together of myriad bits of data fails to serve many kinds of users. This evidently is so regarding voter comprehension. Not only does the citizen lack time for searching. *His* responsibility is general, a breadth of needed cognizance shared in the United States only by his president. All lesser powers may focus their attentions somewhat; the citizen must judge at least the common sense of every topic of concern—which now means looking into every cranny of the macro-problem. Even if rapid building of structures out of fragments of information (working from details toward wholes) were feasible, it still would be unnecessary. The citizen needs only to appreciate the whole in order to judge the general thrust of policy. If he lacks that overview, he and we are vulnerable to demagoguery, as narrow fools or sly bastards hawk simple answers to desperately complex problems.

Even though the formal, professional decision maker is able to devote more time to his work and often has support from staffers and research, his problem finally is similar to that of voters. He must gain a gut awareness—a feeling—for whatever social field his responsibilities may cover. Reports on a shelf, or relations understood by someone else are worthless to him until he integrates them in himself with all the other buried, half-sensed appreciations that combine to found his intuition.

When fields were small and stable, one could (as it were) rub against a situation, observing the faint signs that followed each contact and gradually learning to sense unthinkingly what complex response might flow from any intervention. Such are the appreciations of athletes, freeway drivers, or long-familiar lovers; they are finer, richer, and more competent than reasoned inference in all but very simple fields. That "long, slow, natural way," however, only works when the field obligingly stays still for years on end. When patterns change kaleidoscopically, the lessons learned last year were in a different field than now exists, and they may confuse as much as they complement today's experience. Also, when the field's scope defies direct observation, that too tends to interdict the normal means of building common sense.

The policy researcher needs quick, gestalt appreciations, too, though his need is different from that of official policy makers or citizens. When it comes time in studying a complex field to decide which terms may be worth measuring or to create within the group (and thus convert it to a team) a common sense of the patterned interaction of hard and soft components, then the issue becomes more one of feeling than of reason. And if findings are to be the base for practical design or group response, then comprehensive understanding is required. Man does not move to creative invention out of dry fact; ideas are born, not manufactured in some routine way. Finally, if insights at last are gained concerning policy alternatives, whether by working upward through a sea of scattered facts or in some swifter way, they still must be transmitted to a user. It is not fair to tell one's client, "Come with me to Macedonia," to relive experiences of search and comprehension, nor is it good enough to only write the factual results. The flavor is the point—the product that the client needs—and if reports are dry, as almost always they must be, that point is lost.

The insufficiency of factual reports is rarely known unless knowledge is sought in order to be used creatively. The need to move ahead to appreciation, accordingly, has been most often felt in purposive research, where findings are tested not by scholarly criteria but by their fruitfulness in follow-up decision or design. Because the test of consequence is crisper there, the need for gut appreciations surfaced most decisively in "harder" scientific fields and in military operations. In the former, it was found necessary to *feel* the character of fully calculable answers before insightful design could be pursued. In the latter, commanders long have known that they must steadily appreciate the *whole* situation, woven of opposing strengths and purposes and opportunities. We argue here that if gestalt appreciation is essential in these relatively calculable fields, surely it is still more necessary in problem areas where rules are vague and inferences frail.

In summary of this section, we have argued that the interweaving of problems in this era has forced attention to wider and more complex fields by each decision maker and by staff or research efforts set to aid him. The mode of understanding that is needed is one of gestalt appreciation rather than explicit knowledge of bits of data. This is true whether one views the current macroproblem as a citizen, a responsible executive agent in government or business, or a researcher. The extent of the fields to be appreciated and the contraction of the time available for doing so interdicts the normal, experiential way of gaining deep appreciations, so vicarious routes are needed.

THE STRENGTHS AND WEAKNESSES OF PROSE

Much that is said in this paper "talks down" the worth of logic and scientific rigor, but such judgments are tied to the particular subject at hand. There are many topics that do not call for pattern comprehension that are both important and exciting, and the languages of prose and mathematics are wonderfully adapted to their exploration.

A noun is a powerless thing, and any definition can be imposed on it. Prose sometimes has been identified as lacking rhythm, rhyme, or beauty; such definitions fail in normal usage, but semantic bases of distinction prove to be more serviceable. We here take two factors to be definitive. First, prose consists of propositions having only one meaning and composed of symbols, each with just one definition; each prosaic proposition should be testable for truth on generally positivist grounds, and an equation (with its terms defined) is an archetypical example. Second, prose is sequential, working point-to-point along a chain of assertions or questions. It treats one molecule of meaning at a time.

Prose has been to logic in the broad what in finer detail the calculus has been to the exploration of Newton's laws of motion. In particular, it fits the essential steps of scientific method—the unequivocal designation of a hypothesis, deduction of concomitants that must be true if the hypothesis is, and testing of such predictions against evidence. Crispness of definition and meaning are crucially important at each stage, and prosaic exposition has proved to meet that need.

There has been no lapse of need for scientific exploration. Indeed, the culminating pressures within the macro-problem call for better, quicker science of the sort that underwrote so much of mankind's move toward ease in recent centuries; there is little chance that virtue, humanism, or a "ghost dance" syndrome will see us out of these dark circumstances. But science *is* here, and so is a fully evolved language—prose—to carry it. On the other hand, it seems as clear that bit-wise progress toward appreciation also will prove too little by itself, and there is no adequate vehicle for the holistic communication needed to knit together a complement to science.

Any field comprehension must be at least as sweeping as that offered by a map—two crisply defined dimensions, one of depth less clearly registered, and several others (such as political patterning, routes of movement, population, etc.) indicated at the cartographer's discretion. The problem, then, can be likened to that of describing a picture, provided that one realizes that two-dimensional pictures are at best threadbare vehicles for describing any complex field.

Prose certainly is ill-fitted even to describe a picture. It is as though a near-sighted inchworm sought to randomly traverse the Mona Lisa and to then produce from summed facts a comprehension of the whole. The prognosis of such an effort would be very poor, for many reasons. The linear traverses may miss some crucially important bits; the worm (or prosaic researcher) may remember selectively, bound by preconceptions as to the overall pattern or hampered by forgetfulness within some ten-year scanning; with all the bits at hand, the *picture* still may not be seen.

As another analogy, prosaic communication offers a very poor "impedance match" with either the holistic subjects to be "told" or the receptive powers of the human organism. It is as though a roomful of rich information were to be extruded through a small knothole to another great expanse of eager receptivity. We *know* (from watching athletes, hearing symphonies, driving autos, making love) that humans readily can sense subtle differences within half-sensed, infinitely complex situations. The capacity to receive is there, as it has been for millions of years of unfolding human evolution. The messages to be sent—the fields to be described and the differences to be sensed—have growth more and more complex. The prosaic link between them still is only as wide as one small bit of data.

There is one further limitation of prose that needs discussion here. When we learn a nominally isolated fact, we tend to invent some plausible context—some cognitive structure—within which it *might* fit. (Such first constructions rarely are the best, but they serve as filters through which subsequent perceptions must be pressed; disparities very often cause ill-fitting but true facts to be rejected until, under repeated stress, the initial construct fails abruptly and some new one must be invented. This is an abnormal, inefficient mode of learning, alien to all humanity except the academic, prose-taught student. One should learn wholes *first* as one learns first the shape of distant mountains; afterward is soon enough—and the right time—for learning separate facts or separate trees and streams. There is therefore, a need for holistic communication that predates and supersedes the special urgency deriving from the macro-problem of these times. For *any* new excursion into understanding, we should start first with sweeping comprehensions and then seek to learn, or teach, component facts. The route of science, prosaic exposition, and academic speciality has normally been the opposite.

THE NADIR OF INTUITION

It is not news that the fame and viability of intuition now are at an ebb, but some of the reasons for that waning of reputation are peculiar to these times. Indeed, the temporarily confluent impulses toward an undue currency of the prosaic seem to make up one aspect of the macro-problem, appearing as they do just at the time when prose seems least capable of dealing with the sweeping issues that confront us. Two dominating forces in this evolution have been the seemingly limitless achievements of science during recent centuries and a shorter line of change in which the inner workings of bureaucracy have been the key. These by themselves would not have caused the present circumstance; that called for some apparently coincidental developments since World War II and even since the Eisenhower Calm.

Science has succeeded in relieving men from toil (and, more recently, from hand calculation and some forms of remembering) as nothing ever did before, and it is not surprising that such success should enhance the reputation of those collateral acitivites identified with it. Diamond clarity of definition, logical exactitude, and speciality have been among those hangers on. Ironically, the insistence on uncertainty which is the actual core of science has not fared so well; non-scientists would rather think in terms of white-smocked certainty. While this clutch at peace of mind is understandable within the unending doubt that is the hallmark of democracy, it still set up the later aberrations that beset us now. It is especially important to us here to try to see what caused the sound repute of science to spill over into "softer" fields where it fits poorly.

Perhaps the chief impulse came from successful analyses of man-machine systems during and after World War II. It was understood at first that such analyses were like instructive cartoons, adjunctive to command appreciations. With the era of "massive retaliation," however, the problems deemed most crucial were seen by all to be more nearly calculable, and analytic answers often superseded insight. The pride that this condition bred in systems analysts ought to have eroded during the Kennedy shift toward nation building and attempted manipulation of complex, uncomprehended cultures, but such was not the case. Cost/effectiveness became king in the Pentagon. That fad was brief, and analysis has been graduallly subordinated to military judgment since the mid-1960s, but while military decision makers were shaking off the claims of calculation, two other things were happening.

The American space program, an activity perfectly matched both to the methods and accumulated gains of science, gained momentum and began to show its almost magical results. Incredibly difficult tasks attempted for the first time proceeded just as planned. The predictive power of scientific generalizations was proven so well that when disaster almost struck Apollo 13, the public thought in terms of malfeasance rather than of normal uncertainty. Americans began to think that science could do anything.

At much the same time, civilian branches of the federal bureaucracy belatedly picked up the forms of cost/effectiveness, seeking to use them on especially complex problem fields to which they least applied. And then, to cap this whole ironic jest, the space program was abandoned in the moment of its triumph, releasing crowds of talented and confident systems analysts just in time to feed the hopes in civil agencies that such skills would relieve the burden of social policy decision.

These evolutions were naturally complemented by conditions within the academic and governmental bureaucracies. In the former, the specialization that had been so powerful in science also proved efficient in creating personal security. Interests of several kinds therefore combined to reinforce a pattern featuring subdivision of knowledge rather than its integration, and speciality has become increasingly the route toward academic honor.

The dynamic within governmental staffs has been almost the opposite. As the emerging macro-problem has stretched almost intolerably the fields to be comprehended by decision makers, such persons naturally have looked for ways of lightening their loads. At the same time, those tasked with staff support saw the chance of raising their functions to new levels of autonomy. If analysis could do as it promised (and as the record in defense and space seemed to bear out), the staffer could use it, at long last, to "tell the boss what he ought to do" and thereby preempt the power of decision without bearing the concomitant responsibility. To admit now that analysis has no such powers would burden decision makers (who then would have to seek to spread their commonsense appreciations rapidly and far) and would cut into new claims to pride and indispensability by staff. Something indeed will have to give, as calculations prove inept, but that time is not yet. These two lines of development—prosaic speciality in academia and the preempting of decision by researchers and staff within large agencies—come into confluence in the field of policy analysis as it has come to be defined in recent years. Decision is visualized as a logically determinant process (which it is not), and the models flowing from that idea purport to show how separate bits of knowledge drawn from different experts may be combined to show which decisions would be optimum.

For all these reasons, intuition and gestalt appreciations now are in disrepute, both in government and among professional users of intelligence to solve problems. Efforts to suggest that scientific probings into separate holes must be outflanked before the macro-problem (as they always have been before the simpler problems of family life) must "swim upstream" against tradition, fad, and private interests.

PROSPECTS FOR NEW, HOLISTIC MODES OF COMMUNICATION

If the need were less and the time left for fumbling were longer, there would be no need for such dramatic presentation of this case. Of course, it is true that there now are ways by which gestalt appreciations of complex fields may be indicated to others. There always have been. Unfortunately, the most powerful of these, the arts, seems too imprecise to do that which is needed, and very little effort is being spent upon improvement of the others. Still less attention is addressed to seeking new approaches. What is called for is some way (or much more likely, ways) of capturing directly the almost infinite capacity that we all have for swiftly sensing and discriminating among wholes, a suprarational process that gains little and may lose much (the theme of Hamlet) from intellect.

An illustration here may help. Cartography stands almost alone as an example of holistic "language." A good map lets us see at once the character of the field that it represents—its shape and all the relative locations of its parts—and yet it still can offer crisp details to him who would go beyond such structure. Unfortunately maps are static things; they only point one possible way toward that which we now need.

A good military operations room is another mode of nonprosaic transfer of ideas. In it, as with a map, vast amounts of information bearing on the nature of the situation are disposed in parallel so that a viewer may look at many or few and in whichever order he may choose. Such a room becomes the most important interface between a good commander and his staff, the point at which those facts or pictures that seem most representative of the composite field to which they pertain are offered for his consideration. Each commander adapts his war room to his own modes of appreciation, having those items posted that he thinks will tell him most surely when the field changes so as to call for some related change in operations. The design of such rooms is a highly polished art, quite possibly one starting point for work and adaptation. Most recently, however, analytic talents have been addressed less to the improvement of gestalt communications within such a room than to the ways in which computers might be used to calculate the consequences of action. This is far from being wasted effort (as long as there is no attempt to substitute the computer for the man in overall appreciation), but it is not the topic here, nor will it go far toward meeting the demands of the macro-problem. When we think of ways in which computers and other aids may be used to process data in advance so it may be so presented as to trick our primordial skills of pattern recognition, then we come nearer to the mark.

Several different examples of such trickery have been uncovered, all of them provocative.

The human ability to focus attention upon one conversation in a room humming with many threads of talk is well known. That aptitude can be stimulated synthetically if one takes, say, nine voices talking on different topics and mixes them on tape so that five voices are on each of two tapes, with just one voice shared on both. If one listens to either tape, or to both of them concurrently, monaurally, the common voice is indistinguishable. If heard binaurally, however, with one tape played into each ear, the diverse voices mask each other and the common one emerges clearly (and disconcertingly) just above the listener's head. To what extent might one train himself to interpret statistical information that had been converted to a sonic analogue and then attend to differences by stimulating one or another aspect of binaural hearing?

A somewhat similar use for pseudo-stereo vision was exposed, again through one lonely, unexploited experiment. Those who have watched over the shoulder of a radar operator as he reads an early warning scope understand how the track of an aircraft may be masked by spurious signals or "snow." When a picture so clouded as to defy the perceptions of any but the most perfectly experienced and talented observers was treated binocularly, a startling result emerged. The radar scope was photographed, and the resulting motion picture film was presented so the right eye of an observer saw a picture just one sweep earlier than that offered to his left. Such a combination of views never could be seen in nature, but the brain took it, noted that the "snow" was the same for both eyes while the aircraft track was slightly displaced, interpreted the result as being caused by parallax, and "saw" a stereo picture in which the masking signals were drawn forward; the real track came out clearly, like a moving object seen through haze. The brain is very good at disregarding noise and spotting the faint but relevant changes in a picture—it has had a racial lifetime in which to practice, and the failures during 20,000,000 years or more tended not to mix their genes within the human pool. The discriminations to be made within the macro-problem, however, usually are artificial ones, composed of data bits that never would have concerned a cave man and to which our primordial talents therefore are not tuned. How might production statistics for our economy or for some less intricate, developing one be processed so that daily or seasonal changes might be made to stand out in some kind of pseudo-stereo display?

Tempo also is important, if each of us is to get the magician within to ply his trade less indolently. In one experiment some years ago, a display was made of simulated signals from a field of instruments intended to detect the noise of submarines. Each instrument was indicated by a glow lamp, and each lamp flickered when it would have been triggered by a given, simulated passage of a sub. Each lamp also lit, at random, as it might in answer to wave noises. At real time, with a second of simulated time equal to a second of viewing time, the board was just a sea of flickering lights; no pattern was at all apparent. When, however, the presentation was speeded so a simulated passage took only a few seconds—when the subs moved across the display at something like the speed with which a rat might run across a floor—the random noise became a barely noticeable background of haze, and four subs could be seen taking a complicated, weaving course across the field.

This last effect forms the core of an especially promising experiment in holistic communication, featuring a combination of maps and time-lapse motion picture presentation; it has been nicknamed Tempo-Adjusted, Animated Display, or TAAD. It was first developed by analysts in the hard sciences who indeed believed their calculations but whose analytic models had grown so intricate that each question asked of the computer produced a tubful of print-out, leaving the user almost as uninformed as he was before the answer had been given him. When mapped on any suitable schematic base, speeded or slowed appropriately, and reproduced as color motion picture film, the calculated process changes could be seen and "felt" at once. The appreciation prerequisite either to the posing of more pointed questions or the design application of the existing answers had been provided. Similar exposures of the patterned character of interdiction campaigns have had the same remarkable effect, bringing observers quickly to a common, overall appreciation of the field in question. Application to many kinds of historical (or projected future) change seems evidently promising.

Other points of departure depend less on mechanism or instruments. If one can get himself to function in an appreciative mode (as contrasted with an analytic, intellectual one), that in itself can be important. Gaming of all kinds, from the pseudo-combat of the playing field to more realistic war and business games, has its chief utility in the way in which it prompts its players to behave for a while as though the fictional situation were real and therefore to engage those visceral powers of discrimination that actual competition calls into effect. It is quite wonderful to see how persons engaged in a war game cooperate as though the problem posed were real and bring their common sense to bear upon it, when those same individuals would have haggled over doctrinal differences if the questions had been asked them in an intellectual setting at their office desks.

A less combative way of doing something similiar was developed in the course of research into alternative, holistic future patterns, the general approach of which would serve in any investigation of equilibrium conditions within a complex field. The method, called Field Anomaly Relaxation (FAR) to indicate its debt both to social field theory and to some of the iterative methods used in engineering, need not be treated here in detail. Some steps within it are almost purely holistic, and no methodical procedures have been found to deal with them. Other steps are more analytical in nature and can be described as processes, but even they are fundamentally heuristic explorations of the "feeling" of a field; they can serve us here as illustrations of the way in which nominally routine analytic processes can be made to augment holistic appreciations.

The accompanying chart displays a matrix used in an exploration of alternative patterns of education for this country in coming decades. The factors arrayed along each side of this square matrix are intended to cover the plausible alternative conditions that might obtain within each of five dimensions of description. The regular institutions of public education (marked E, for Establishment) might take on several forms, as might those of ordinary private schools (P), training institutions intended to create or enhance particular skills (S), instructional facilities in military or other national service institutions (N), or broadcast education such as that exposed by Sesame Street (B). The question to be asked for each square of the matrix is, "Would these two factors fit together within the designated national context?" The choices were constrained by assuming a particular embracing condition for the other institutions of this country each time the matrix was filled out. The job then was done again for another illustrative future pattern for the United States, and so on until a representative set of conditions had been explored.

The problem exposed as one tries to fill in such a matrix is almost archetypical of any exploration of an unknown field. One learns about the potentials of the field by working within it in a tentative way, using such lessons to make more insightful judgments as comprehension grows. The important thing, relevant to this discussion, was the way in which one looked at each square of the matrix. The scoring shown was done under the assumption that an over-commitment to welfare during the 1970s had brought the nation close to bureaucratic stalemate by 1985. In such a condition and after such a line of evolution (since a holistic future pattern would be incomplete without its own distinctive history behind it), which sorts of educational components might fit together? The members of the research group were asked to feel as though they

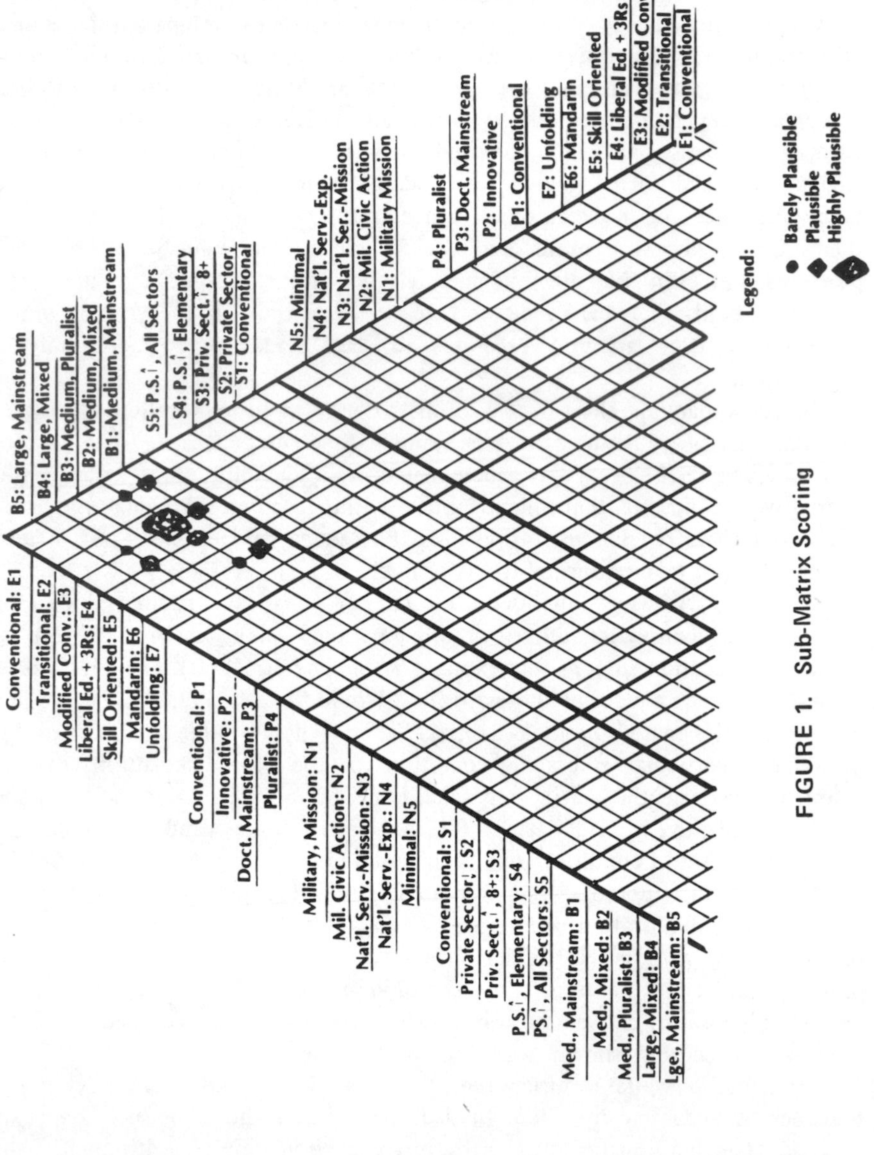

FIGURE 1. Sub-Matrix Scoring

Legend:
● Barely Plausible
◆ Plausible
◈ Highly Plausible

B5: Large, Mainstream
B4: Large, Mixed
B3: Medium, Pluralist
B2: Medium, Mixed
B1: Medium, Mainstream
S5: P.S.↑, All Sectors
S4: P.S.↑, Elementary
S3: Priv. Sect.↑, 8+
S2: Private Sector,
S1: Conventional
N5: Minimal
N4: Nat'l. Serv.-Exp.
N3: Nat'l. Ser.-Mission
N2: Mil. Civic Action
N1: Military Mission
P4: Pluralist
P3: Doct. Mainstream
P2: Innovative
P1: Conventional
E7: Unfolding
E6: Mandarin
E5: Skill Oriented
E4: Liberal Ed. + 3Rs
E3: Modified Conv.
E2: Transitional
E1: Conventional

Conventional: E1
Transitional: E2
Modified Conv.: E3
Liberal Ed. + 3Rs: E4
Skill Oriented: E5
Mandarin: E6
Unfolding: E7
Conventional: P1
Innovative: P2
Doct. Mainstream: P3
Pluralist: P4
Military, Mission: N1
Mil. Civic Action: N2
Nat'l. Serv.-Mission: N3
Nat'l. Serv.-Exp.: N4
Minimal: N5
Conventional: S1
Private Sector: S2
Priv. Sect.↑, 8+: S3
P.S.↑, Elementary: S4
PS.↑, All Sectors: S5
Med., Mainstream: B1
Med., Mixed: B2
Med., Pluralist: B3
Large, Mixed: B4
Lge., Mainstream: B5

lived in that specified condition and then to judge the educational questions posed. The answer whether one alternative for the public school system and another for broadcast education might fit together was no more important to the project than was the increasing depth of comprehension of the projected situation as one tried to place himself functionally within it.

It may be clear even from this brief description of the matrix stage of FAR analysis that it facilitates a mode of thought in which members of the research group profit by each others' judgments while at the same time communicating richly with each other. This raises a topic that should close our analysis. Which kinds of users stand to use most fruitfully each kind of holistic communication method suggested here?

- Analytic methods such as the last one discussed are probably of greatest use to researchers, since at best they are so time-consuming that few others can have access to them. A planning staff, however, might very fruitfully use similar approaches.

- Gaming of various kinds clearly could be used by voters, executives, or staff/research groups. So far, however, except for a few insightful educational applications, most games have had their usefulness vitiated by a belief that they were means of finding answers; they are generally worthless for that purpose, but they are very powerful tools in building field application. To serve the latter purpose best, they must be played by real decision makers—not by staff, role-playing in their stead.

- The use of parallel display in something like a war room, to breach the sequential linearity of prose and at least give to him who wants to learn a situation many kinds of data concerning it all at one time, seems especially adapted to any and all executives, who are (by definiton) tasked to respond upon demand with decisions concerning the field of their responsibility. It also can be useful to research and staff work, however, since insight is needed in deciding which of many tasks deserve emphasis, and such insight is enhanced by the same sort of feeling for the situation that is needed by the executive. And it could be used much more widely than it is to enhance lay comprehensions, as the TV weather map helps all of us to comprehend the weather fields in which we live. A similar display of posted ecological concerns might help get rid of simplistic ideas on that important, complex topic.

- The tricks, like TAAD, that cause us to engage our supralogical capacities for feeling and interpreting wholes will, insofar as they work, serve everyone. They already have been proved, in isolated instances, by researchers communicating first within themselves and then to each other; it seems evident that the executive who faces up to his need for widened gestalt comprehensions would profit from them mightily; they seem perfect for the generation of voter comprehension, when combined with systems such as cable television, since they by-pass the learning of bits of data almost altogether and start with the transmission of whole patterns. They, alone among the items mentioned here, offer prospects of a breakthrough.

In conclusion, then, it has been argued here that the angry interconnectivity of problems in this era demands of almost all of us that we expand our commonsense awarenesses in several dimensions. The nature of the questions to be faced and their identification in most instances with truly complex fields requires that the comprehension be of the gestalt variety, to which scientific method, specialized scholarship, and the prose communication tuned to both are generally inappropriate. Prose (and logic) being insufficient, though still necessary in many kinds of implementing inquiry after broader comprehensions point the way, what then may be done?

The entire problem of finding modes of communication of insights into complex wholes, sharing some of the precision of prose and something like the breadth of coverage of the arts, has been neglected sadly, and the fashions of mind that govern most members of the present knowledge community will militate against any concerted effort that might be mounted now. Such an effort seems, nevertheless, to rate a high priority if we in this decade are to "secure the blessings of liberty to ourselves and our posterity." Some well-known methods could be improved and broadened, and there are hints of vast potentials among the scattered, unexploited discoveries of recent years.

MANAGING COMPLEXITY:
Gaming—A Future's Language

RICHARD D. DUKE

THE PROBLEM

Humankind is a little harried of late. The naked ape barely blinked, only to discover that his animal being has moved from the cave to the moon with precious little time for adjustment, measured in evolutionary terms. It is difficult to derive a valid "alienation index" for a society, perhaps impossible for different points in historical time. Nonetheless, evidence abounds that all is not right with Western civilization as it is currently structured; further, the situation has deteriorated markedly in the past quarter of a century. The tune-out, drop-out, cop-out syndrome is ever apparent although the recurrent waves of enthusiasm (e.g., populist activity in environmental concerns) give reason to believe all is not lost. But even in such cases where enthusiasms are high and a general sense of urgency and responsibility exists at the level of the individual, there is pervasive frustration. The individual needs to be part of those processes affecting his life, but is currently devoid of any effective means to alter things or join the dialogue about potential change. This situation reveals some dimensions of life today that were not previously true:

—The problems of today are infinitely more complex, involving systems and interacting subsystems that go beyond normal human ken and which do not yield to conventional jargon or traditional forms of communication.

—The sheer quantity of individuals who want to be effectively part of the dialogue is large and growing rapidly.

—There is a growing personal urgency because the solutions pursued today constitute a more pervasive intrusion in the individual's life. (In earlier times, the king's men may have come periodically for the taxes, but in the interim period, life was constrained only by the elements and by whatever circumstances might exist within a personalized clan; today, the Internal Revenue Service comes every week and unknown Big Brother, in a thousand ways, constrains the daily actions of our lives.)

This situation, of course, is not new. Without too much difficulty, man's struggle for the personalized control of his life can be traced through the Magna Carta and the decline of the king's power to the Parliament to the Declaration of

Independence and resulting constitutional governmental forms (whose painfully won gains are now threatened by a technical aristocracy, the high priests of 1984). Even the great urban political bosses performed a valid personalizing influence buffering the citizen from the emerging systems and technologies that must control his world. (Sadly, only vestiges of this humanizing function remain; witness the light years between the individual and national politics—can it be other than Alice in Wonderland with spy versus spy, body counts from a constitutionally nonexistent war, complexity of domestic programs that boggle the expert mind while dominating, in strict inverse relation, the lives of the least able citizenry?)

At the very moment when man seemed to have garnered the power to control his personal destiny by his own hands, he has been caught unaware in the grinding pincers movement of the complexity of societal survival in modern times and the inevitable technological response. This crunch has been on its way since the industrial revolution, but its very rapid progression was precipitated by World War II, in particular by the spinoffs in computer technology and the resultant elaboration of the concept of "systems" and related, evolving technologies. Now the high priests of technology speak only to the high priests of technology; God is dead, and the citizen, no matter how strongly motivated, can hardly get a word in edgewise!

Problems of the management of modern Western society (and in a particular sense, the great urban centers) have generated the modern equivalent of the biblical Tower of Babel. To unravel the present "want structure" in human terms, to harness appropriate technologies, and to manage a successful and continuous response in an ongoing societal context generates a communication net (non-net?) that is truly unimaginable and certainly unmanageable. Society's failure to respond to individual need is, in large part, a communication problem.

The naked ape waved and grunted, and we do little better. He lived in a relatively simple world, and over many centuries developed what have been viewed as sophisticated languages but which in reality are only involved extensions of sequential form (including not only a written and spoken English, but also the sophisticated artificial languages of mathematics, computer programming, musical notation, etc.) to deal with this noncomplex environment. The naked (now harried) ape of today still employs these simple sequential tongues, but, in a world several magnitudes more complex, it leaves him speechless. The highly constrained and sequential languages of the past and their related technologies (even in their highest forms) fail at conveying gestalt, and so the complexity of today cannot be comprehended or communicated except with the greatest of effort, and then only by a new elite. For example, consider our great urban centers as they exist today, multisystems within multisystems; alternative upon alternative, presenting an incomprehensible, many-futured state(s)—the tongues of many men, some unborn, in-migrant and out, being daily rearticulated as perceptions of the possible change; a great multifaceted sphere of complexity that cannot be managed, but must be.

Societal response, predictably, has been of four concurrent dimensions: false dichotomies, professional elitism, increasing dependency on technology, and gigantism. The inevitable but false dichotomies appear first: pare out of the total fabric of society some element of great urgency; if we can neither understand nor solve the totality, we can solve some definable part, *no matter that other evils are encountered, the least of which may be inefficiency, the most dangerous irrelevancy* (witness "education" as a system currently practiced in the inner city).

As the bureaucracy (education, transportation, health, housing, ad nauseam) transforms life into disconnected cells, society loses not only in the more obvious negative harvest (the "solution" of urban renewal, originally conceived as a simple-minded clearance problem, now yields to more sophisticated approaches at a cost of two decades, only to be replaced by the "solution" of an "interstate highway system" rather than an "interstate transportation system"); but also in a positive sense, since such dichotomy leaves little room for subtlety of solution.

And with the dichotomy came the armies of the professional elite and with them their empires, and the resulting gigantism dwarfs and smothers the citizen. The fiction of alternatives (witness State Highway Engineering) may seduce the generalized citizen, but the poor beggar in front of the bulldozer soon discovers the moment of truth. And as he turns into the jaws of this giant beast trying to locate some isolable component unit that will be responsive, he is put down by the elites—the professionals who quite literally speak another language—and he is put off by a technology that appears antihuman; his only alternative is to join together with fragmentary bands to throw slingshots at the giant. And, occasionally, his aim is true! First one, then another public scheme is beat back not to be replaced with positive alternatives, but to a frustrated stand-off where the great urban administrations survive through nonaction, and the great creaking structure grinds through time, the moans of unresolved needs and of endless counter-productive conflict emanating from the incongruous mass.

And who is to comprehend it all? Or who is to speak of it all? And to whom? Is there any remote possibility of establishing a real dialogue about this multi-faceted, dynamic gargantua, even among the elite, substituting future time-frame for future time-frame in advance of reality, permitting positive management to replace a negative reactionary reality? And is there any way to enlarge the dialogue to include the activist citizen or someone who might conceivably be called his "representative" in that he transmits a personal translation of ideas for his limited and personally known constituency?

Of course not, not if we insist on restricting ourselves to the languages of the caveman.

But there is hope that the possibility for a quantum jump exists; that communication can move from its rigid and limiting sequentiality to a gestalt mode, and that this supra-language can be used as a simultaneous translation for our modern Tower of Babel.

ESTABLISHING THE NEED FOR FUTURE'S LANGUAGE

"Future Shock" has become part of the popular lexicon. Alvin Toffler in 1970 introduced the concept in a book by that name in which he stresses the death of permanence and the coming of the age, not of Aquarius, but of transience. The book documents in detail his thesis that the world of tomorrow will be significantly different from the world of yesterday along many dimensions. Toffler quotes from Kenneth Boulding:

> As far as many statistical series related to activities of mankind are concerned the date that divides human history into two equal parts is well within living memory.... The world of today ... is as different from the world in which I was born as that world was from Julius Caesar's. I was born in the middle of human history, to date, roughly. Almost as much has happened since I was born as happened before.

The temptation to document here the proof of this accelerated change and its impact on our lives is strong, but the thesis will either be accepted or rejected without elaboration, since it is a familiar one. To place its significance to gaming in proper perspective, I would like to allude to Figure 1, a simple graph. The horizontal axis would represent centuries starting perhaps with the year 1 in our current system of counting; the vertical column would represent an index, however obtained, which would attempt to convey complexity, transience, and rate of change confronting the typical citizen. Using a logarithmic scale, a curve is plotted which attempts to illustrate this change (perhaps the number of new things which must be assimilated in the lifetime of a given citizen). The curve would start in the extreme lower lefthand corner and be virtually a straight line with a slight incline upward, barely perceptible, until perhaps 1900; incremental

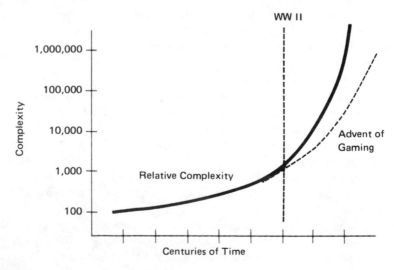

FIGURE 1.

jumps might be noted at the time gunpowder was introduced and certainly as the industrial revolution impacted on society. The curve turns vertically during the period 1900-1940, with a sharp increase during the period of World War II. Subsequent to World War II, the curve would be increasing at a near vertical rate, implying change flowing on change at a totally unprecedented rate. Curiously, a number of authors have independently noted World War II as the approximate time of the pivot from the trend line through antiquity to the modern trend line.

Virtually all our language forms have come from antiquity and have sufficed, in spite of their sequentiality, because they rely heavily on analogy, and the analogies employed are predicated on historic circumstance which is not expected to change, by minor adjustments through time. Note that the curve implies that in the post-World War II period a situation far more involved prevails, particularly in the several dimensions of complexity, future orientation, thoughtful consideration of alternatives, and the inevitable recognition of the nature of systems and interlocking subsystems which are affected.

Necessity is the mother of invention, and the post-World War II period has shown many innovations in communication which attempt to deal with this communications problem; each reflects an attempt to convey gestalt, or at least to escape from the harsh burden of strict sequentiality of the written and spoken language forms. If one were to plot the advent of gaming as indicated by the new games which appear in the various cataloging efforts, the curve mimics rather accurately the curve of accelerated change with perhaps a ten- or 15-year lag. This reflects, in my judgment, a spontaneous solution, "gaming," by many people in many problem situations to the problem of developing a gestalt communication form. In short, we have a new language form, a language form which is "future"-oriented. If this premise holds, to date we have seen no general statement or theory which would explain the wide diversity of materials which appear as games, or which might guide the neophyte in efforts to develop effective games for their own communication purpose.

GAMES IN THE COMMUNICATIONS CONTINUUM

The various modes of communication currently in use rest along a continuum ranging from the primitive to the sophisticated; these are combined here into four major divisions: Primitive, advanced, integrated-simulated, and integrated-real. In a sense, the two extremes of the continuum can be viewed as being linked, in that two parties fully sharing a reality need no overt communication or suffice with primitive modes. (The stadium crowd watching a football game, veteran foot soldiers engaged in a fire fight, two lovers in ecstasy—each situation relies on the fact of shared reality as the basis for communication.) When man does resort to any medium, it is to bridge the gap in the perception of reality between two or more parties. The greater the communications gap and the more involved the reality they wish to confront, the more elaborate and sophisticated their language becomes. Carrying this logic forward, a curious phenomenon is encountered. Attempts to convey elaborate systems emerge as a complex language form (gaming/simulation) approaching reality itself, and, in so doing, the process has gone full circle! The fourth category of communication forms (integrated-real) is included here to emphasize this transition. Because of this circular character of communication media, one is not better than the next. Rather, one may be more appropriate than another.

A quick review of the various communication forms is called for. Primitive forms have been divided into informal (grunts and hand signals) and formal (semaphore or light signals). In situations which are simple and transitory, the former will suffice; but as the communications need becomes more important, more involved, or more consistent, these have been formalized. Both forms are characterized by spontaneity, limited message content, and immediacy to experience. They are generally used in face-to-face contact. For example, a cry of warning is almost universally understood by people of all cultures. Its function is to alert someone to a danger; it is effective only insofar as the warned person shares the message sender's perception of current reality—i. e., he is in the same place at the same time. Similarly, the standardized international traffic signals used to direct traffic are an example from the primitive-formal category.

Advanced forms of communication include spoken languages, written languages, emotional forms (art, acting, role-playing), and technical forms (pictures, mathematic notation, musical notation, schematic diagrams, etc.), which are often used as supplements to other advanced forms. It is quite common, of course, to use these in some combination (for example, slides with a lecture), and such uses can be viewed as rudimentary forms of the integrated languages suggested by the final two categories. The first, integrated-simulated, is characterized by deliberate combinations of media (film and television) or by hybrids (gaming/simulation) which employ all prior forms in any combination which best enhances the transmission of some reality. The final category, integrated-real, does not attempt this in an artificial manner, but rather this level of communication inherently recognizes the circular nature of the communications process, and consequently extracts from reality itself. One illustration of this would be apprenticeship programs where the learner (party receiving the message) is placed in a situation of reality but buffered from the consequences of full participation. As he gains "experience" (better perception of reality), these buffers are systematically removed until he becomes fully part of reality.

Figure 2 identifies six major characteristics associated with the various language forms; and includes a brief interpretation of each of these characteristics relative to the four major types of media. The six characteristics include:

(1) Sequential-gestalt constraints: the inherent ability of the language form to convey gestalt or totality.

(2) Specificity-universality constraints: the degree of flexibility inherent to the language form in adapting the new substantive material.

(3) Spontaneity of use: the ease or relative freedom encountered by the user.

(4) Character of conventions employed: the degree of consistency, the extent of complexity, the relative formality and their general versus special use in a given communications attempt.

(5) Coding-decoding: the extent to which the message must be artificially coded by the initiator and reconstructed by the receiver.

(6) Character of the message that can be conveyed: the success with which a number of message characteristics can by conveyed, including but not limited to complexity, analogy, qualitative thought, quantitative thought, subtlety, permanence (ability to reestablish), precision, intangibles, time constraints, and systems characteristics.

Communication needs (message transmission) can be analyzed by these characteristics, and a medium or communication form is selected which yields the most efficient exchange. When no existing forms are successful, other forms (new or modified) are generated.

Gaming/simulation is the most intricate communication form available and as such its response to these six characteristics is unique.

Because gaming/simulation has emerged spontaneously to convey gestalt considerations, it is not surprising that it excels on this first characteristic. Primitive languages fail completely, although advanced forms can be laboriously employed. The four "integrated" forms are most effective at conveying gestalt with the three artificial media (multimedia, gaming/simulation, and experience) often being more efficient than reality itself.

With regard to the second characteristic used in Figure 2 (specificity-universality), the integrated communication forms are the most specific (by reason of their highly particularized and complex construction to convey a single gestalt phenomenon). An example might help: Using a full primitive capability of grunts and hand signals, a traveler might get by temporarily in *any* culture. If he shifts to advanced languages, he is limited to those particular cultures where he is learned (of course, he can participate more fully). If he now selects gaming/simulation, he will be restricted to communication with those who have a particular interest in the single gestalt represented by that gaming/simulation. (This is meant to imply that each gaming/simulation is best viewed as a separate and independent language, at least to the extent that we might view French, Japanese, and English as separate languages.) It is important to examine this premise carefully, for, if it is found acceptable, it then becomes possible to construct general principles to govern the construction of each new gaming/simulation.

The third characteristic of "spontaneity" is intended to convey the relative ease of use of the various modes. Associated with this characteristic are the concepts of "dryness" (defined as the energy required to overcome the resistance to use of a given communication form) and expertise. The advanced forms are often viewed as "dry," and in their more sophisticated use they are almost universally "dry." This inverse relationship becomes a serious constraint

THE COMMUNICATIONS CONTINUUM*

(Examples of Each Communication Form) / Characteristics	PRIMITIVE		ADVANCED				INTEGRATED			
							SIMULATED		REAL	
	INFORMAL	FORMAL	SPOKEN	WRITTEN	EMOTIONAL	TECHNICAL	MULTI-MEDIA	HYBRID	EXPERIENCE	REALITY
(Examples)	Grunts Hand-Signals	Semiphore Lights Flags	Conversation Lecture Seminar Radio	Telegraph Manuscript Books Text	Acting Art Role-Playing	Math-Notation Musical-Notation Schematics Diagrams	Film Television	Gaming/Simulation	Apprentice Job Training	(Any Shared Real Time Perception)
SEQUENTIAL-GESTALT (Degree to which the form is constrained)	Most Constrained Because of Sequential Nature		Basic Character is Sequential but Various Devices Employed to Ease Constraint				Highest Gestalt Ability Short of Reality		Fully Gestalt Because Actual Reality	
SPECIFICITY-UNIVERSALITY (Degree of Flexibility of Use)	Employed for All Situations but Limited in Material Conveyed		Standard (universal) Modes Selectively Employed to Meet Specific Communication Need				Mode Specifically Tailored to Communication need		Specific (It is the Reality Encountered)	
SPONTANEITY OF USE (User Resistance, Skill Required, "Dryness" of Form)	Natural, Easy, Convenient		Special Skills Required. Sophistication Often Accompanied by Dryness, Artificiality of Use Inherent				Very Special Effort to Initiate; Then Spontaneous in Use		Natural "Life" Form, Skill Limits Involvement	

CHARACTER OF CONVENTIONS EMPLOYED (Formality, Complexity)	Relatively few Simple, Simple, Informal	Formal and Informal, Simple and Complex, Highly Structured, Many	Many, *Unique to Each Situation*, Fairly Complex	Many Informal Complex
CHARACTER OF CODING AND DECODING (Inherent)	None Required or Simple Effort	Essential; May be Elaborate and Highly Specialized	Elaborate Coding to Initiate Simple Effort by User	None Required
CHARACTER OF THE MESSAGE THAT CAN BE CONVEYED (Complexity, Analogy, Qualitative or Quantative thought, Subtility, Permanance, Precision, Intangibles, Time Constrained, Systems Characteristics)	Only Rudimentary Message	Sophisticated Messages	Gestalt Substitute for Reality	Reality

*This diagram is only meant to suggest major relationships among the various media to illustrate the character of gaming/simulation there is no suggestion of the comprehensive review of communication forms or their character.

From: "Gaming Simulation—A New Communication Form", by Richard D. Duke, presented at the Third International Conference on Gaming/Simulation, Birmingham, England, July 1972.

FIGURE 2. The Communications Continuum

when attempting to establish communication about serious and involved problems, and generally results in shared comprehension only among an elite group. Returning to the continuum in Figure 2, there is an increasing requirement for special expertise as we move from the primitive through the advanced communications forms. This continues through the construction stages of the integrated forms, but is markedly reduced in their use. (Films often present intricate messages successfully to audiences who would not master the same material presented through the advanced languages.) Spontaneity is often constrained in the early stages of using a gaming/simulation since the players are literally learning a new and highly specialized language. After several rounds of play, they will master the particulars and now communication advances to a level not previously possible (assuming, of course, that the primary rules of construction have been thoughtfully followed).

The fourth characteristic is "conventions employed." The importance of this is generally not understood by developers of gaming/simulations, with the result frequently being an inferior product. To illustrate convention, consider several examples from the advanced form of communication: grammar for both spoken and written languages; consistency of mathematical notation (also musical notation); the rules governing the use of any given computer language; the symbols employed in flow-charting; etc. Contrast this with several gaming-simulations, all of which employ Lego blocks as a construction device, but, unfortunately, to convey different ideas in each case. Because each gaming/simulation is a separate language, the conventions are frequently invented for that particular instance; there are relatively few conventions universally employed by gaming/simulations except those previously adopted for the communication forms from which the gaming/simulation hybrid is constructed.

The fifth characteristic from Figure 2 is "coding," by which we mean the artificial and temporary translation of the message to some intermediate form from which it must be decoded before communication takes place. Perhaps the most clear example is semaphore transmission, which requires coding from a given written language into separate and distinct symbolic representations (the flag positions) and subsequent decoding before the message would be understood by its intended receiver. Most media employ coding to some degree; when it is cleverly employed in gaming/simulation, the technique becomes much more powerful—for example, the use of the graphic display "SYMAP" to encode a large data set which can be visually decoded by the player at a more useful level of abstraction. The secret of success in "coding" messages for gaming/simulation is to employ codes which can be readily interpreted by the participants of that particular gaming/simulation.

The final row in Figure 2 is "message characteristics" and is intended to suggest a large number of particular characteristics, several of which are illustrated (complexity, analogy, etc.). It is essential for the designer of a gaming/simulation to reflect on the character of the message(s) to be conveyed; this will suggest which communications forms are best combined into the hybrid. As a general observation, it seems that there is a direct correlation between the clarity with which the game/simulation designer specifies his communication problem (message, sender[s], receiver[s]) prior to construction, and the quality of the product. (One very good gaming/simulation had as its objective improved communication among a citizen group who were constructing a city ordinance to control aesthetics of builders. The game designer employed art [painting] as one of the component language forms.)

All along the continuum, the purpose of communication remains the transmission of perceptions of reality. But as the scope and detail for the perception to be conveyed increases, becoming more comprehensive, total, or gestalt-like, a price is always paid in spontaneity of use (see Figure 3).

The need for conveying holistic thought, or gestalt, is urgent; the coming decade will increase this urgency considerably. Perhaps the most trenchant statement on this need is by R. F. Rhyne in the previous article. While describing the need for holistic communication, Rhyne states, "There is a macro-problem, an inter-weaving of adverse conditions that is more extensive, more richly structured by interior lines of interaction, and more threatening than any circumstance faced before by all mankind." Rhyne's article was formulated "to stimulate exploration of the means whereby appreciations of complex wholes may be more quickly and more reliably told to others." He, too, rejects our ancestral language forms as being inadequate to the task and argues that new forms must be invented. Arguing that decision is a gestalt event and not a logically determinable process, he believes that the citizen or the policy re-

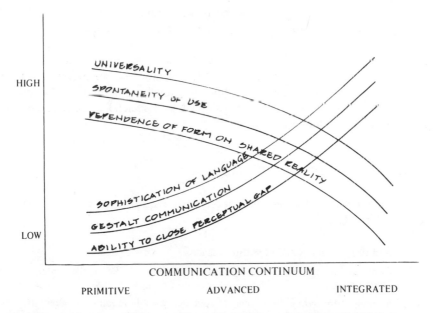

FIGURE 3. Message Characteristics and the Communications Continuum

searcher or other decision maker must first comprehend the whole, the entirety, the gestalt, the system, before the particulars can be dealt with. Rhyne suggests a variety of approaches to this problem and alludes to games as having a particular potential.

We learn through games because, if properly designed, they represent abstract symbolic maps of multidimensional phenomena which serve as a basic reference system for tucking away the bits and pieces of detail which are transmitted and in particular by assisting in the formulation of inquiry from a variety of obtuse angles or perspectives which are meaningful to the individual making the inquiry and which can only be transmitted through an "N"-dimensional, abstract, symbolic-mapping procedure.

If the prior observations on the character of change in the world since World War II are valid, they could perhaps be summarized as follows: prior to World War II, the need for pragmatic information and fact, learned by rote, was imperative; in the new era the need is urgently for the acquisition of heuristics or a flexible set of highly abstract conceptual tools which will let the participant view new and emerging situations, having no precedent, in a way that permits comprehension. We learn through games, then, because it is a relatively safe environment which permits the exploration of many perspectives chosen by the individual, expressed in the jargon of the individual, and subject to fairly prompt feedback in "what-if" contexts. These concepts gain strength when reviewing the work of Moore and Anderson in "Some Principles for the Design of Clarifying Educational Environments" as they conduct research on learning environments. Curiously enough, they pinpoint the time of change in society as being dramatically correlated with the decade of the forties. Properly designed, games have a strong basis in learning theory, which supports their potential as a communication form.

The simultaneous invention of games of a wide diversity of subject matter and technique is a response to a felt need for an improved communication form to deal with problems of gestalt or holistic thought. Just as the folk models alluded to by Moore and Anderson emerged in a societal context as needed, games become a modern equivalent.

WHAT IS A FUTURE'S LANGUAGE

For the moment, let me identify seven basic requirements that must be met by any future's language:

(1) Ability to convey gestalt or holistic image.

(2) Permit the specification of detail at any appropriate level, in the context of the holistic image.

(3) Structured to permit the pulsing of specific, tangible inquiries or alternatives to permit correlation with the holistic image and any significant detail.

(4) The ability to display, make explicit, or permit the recording of explicit linkages between major segments of the holistic imagery; the creation of an awareness of feedback.

(5) A nonelitist, universal possibility for use; a basic catholicity of design.

(6) A future orientation (implying *any* time frame past or future *other* than the present).

(7) They must be basically transient in format to permit the restructuring or more careful articulation of the problem as viewed by those participating.

Clearly, gaming has not preempted as *the* future's language. However, if certain rules, concepts, or principles are employed consistently, the game product can certainly qualify in a wide variety of situations.

GAMES AND SOCIALIZATION

BESTY WATSON

Rutgers University

Games and play are cultural phenomena that have been with us throughout human history. The playing of games exists in all cultures, and indeed can be viewed as a universal feature of human societies. Huizinga (1950), in his important work *Homo Ludens*, puts forth the thesis that much of human culture is derived from play. He maintains that play creates in man a capacity for detachment, a respect for rules, a spirit of curiosity and inquiry, and the ability to develop and use strategies and tactics. It is the subtle interrelationships of these cultural "rules of the game" that form a stable foundation for society.

Game playing is an inherent part of man's world-building activity, for adults as well as children. Marshall McLuhan (1964: 235) has noted this sociocultural characteristic of games:

> Games are popular art, collective social reactions to the main drive or action of any culture. Games like institutions, are extentions of social man and of the body politic, as technologies are the extentions of the animal organism. Both games and technologies are counter-irritants or ways of adjusting to the stress of the specialized actions that occur in any social group. As extentions of the popular response to the workaday stress, games become faithful models of a culture. They incorporate both the action and the reaction of whole populations in a single dynamic image.

GAMES AND PLAY IN SOCIAL LIFE

Games and play are particular ways of structuring human interaction, which have several functions in social life. First, through his participation in games, the individual internalizes the rules, in themselves reflective of the norms, values, and skills of his social group. Second, game playing aids in the formation of one's self-image, which is derived through social interaction. Third, games serve as vehicles through which children and adults alike can practice communication skills, try out or learn new roles, and develop strategies for social interaction. Fourth, games allow people to explore potential or future reality. Fifth, games are fun; thus they provide a sanctuary from the reality of everyday life, and offer a respite from work and daily routine. Lastly, games serve as mechanisms for the social integration of individuals, because they provide companionship, friendship, and a chance for participation in recreational groups.

In most modern cultures, the chief occupation of children is play. As a child undergoes the first phases of socialization, he learns to control his behavior in light of another individual's attitudes, either toward that behavior or toward the environment. The child can be said to be learning "to take the role of the other," in that he responds to himself and to the world as he anticipates that the other would respond.

At first the young child's experience is only with his family or what Mead calls "significant others." At this stage, the games of small children are imitative of highly specific roles; they play "mommy" or "dress up" in adult clothing. As the child grows older, he gradually learns a less personalized and more complex form of role-taking as expressed in his developing ability to participate in organized games. In football, for example, the acting out of a highly specific individual role is not required; the player adjusts his behavior from moment to moment and does so in light of what a number of others are doing and of the rules and objectives of the game. In performing his role, he responds to a "generalized other": "the organized community or social group which gives to the individual his unity of self" (Mead, 1934: 154). According to Mead (1934: 159), the game is "an illustration of the situation out of which an organized personality arises."

Through game playing the child gradually internalizes the point of view of the community as a whole, and in turn utilizes these internalized attitudes as his own when he acts. These shared attitudes then take on a "taken-for-granted" nature as part of the child's personality. The world of the parents, of society at large, becomes the child's world, thus enabling him to play social roles by society's rules (Berger and Luckmann, 1967: 25). Not only does game playing function in the formation of a socially acceptable self-image, but in so doing the game teaches those characteristics most valued within one's culture. Play teaches loyalty, competiveness, and patience (Huizinga, 1950: 53), and in the process it allows the society to reaffirm over and over again its cultural values.

Still another function of games in social life is to serve as vehicles through which the individual can practice communication skills, and experiment with strategies and tactics of social interaction. In games the individual is able to learn about and try out new or potential social roles. Social life in general, but especially "face-to-face" relations, can profitably be viewed in the context of a gaming structure because, as McLuhan (1964: 237) puts it, "All games are media of interpersonal communication." In Erving Goffman's analysis of social relations, human conduct is seen as concerned with maintaining a specific conception of self before others. In Goffman's view, people are gamesmen "working the system" for the enhancement of self, and social relations are not cemented so much by norms and values as by strategies of "tact and prudent sociability"

(1969). Central to Goffman's notion that what is often characterized as communication in interpersonal contexts is, in actuality, "assessment of the other within the situation." In other words, one's response to the expressive behavior of another can be based on the assessment of what he believes the other is *really thinking* rather than what information is objectively transmitted (1969). In our "face-to-face" relations with others, we often find ourselves in situations where calculative gamelike behavior emerges. In Goffman's accounts of interaction in daily life, we see how, as in game playing, each person maneuvers to discover the value of information given and where each person's move is fully dependent on the move of his opponent.

Viewing human interaction in "face-to-face" situations points to the fact that games can help us both explain and structure present reality, and perhaps more importantly in the case of the child, games allow us to explore potential or future reality. Because games are telescoped acts, we are able to think out in advance the consequences of our actions. Players can learn from the experiences of others, conjecture as to the consequences of their behavior before they act; where the game simulates an actual situation, they can see which kinds of alternative actions are most likely to attain the desired end. Opie and Opie point out in their book on children's games:

> It appears to us that when a child plays a game he creates a situation which is under his control, and yet it is one of which he does not know the outcome. In the confines of a game there can be all the excitement and uncertainty of an adventure, yet the youngster can comprehend the whole, can recognize his place in the scheme, and, in contrast to the confusion of real life, he can tell what is right action. He can extend his environment, or feel that he is doing so, and gain knowledge of sensations beyond ordinary experience [1969: 3].

Most people distinguish playing games from serious work. Work deals with "reality," with earning a living, with production. Play, on the other hand, is not a task or moral duty. It is usually thought of as unproductive except for the self-satisfaction and entertainment that it provides. Play is considered as devoid of important repercussions upon the reality of everyday life. Play is pretending, and it represents a stepping outside the world of daily responsibility. Play is an interlude in the "normal" activities of the day. Play is, in an important sense, "disinterested," thus providing temporary satisfaction and relief from the habitual routines of life.

Game playing functions as a sanctuary from work because of man's social nature. Man is social, not only in the sense that physical survival and mental development depends upon his acceptance in an ongoing human group, but social in the larger sense of his need to participate actively with other people and his desire to be in unision with them. Whether he serves as a spectator or as an active participant, games and recreational groups provide the individual with an "in-group" within which he can feel at home. Man, especially modern man, has created many social and cultural forms which tend in the long run to inhibit and even penalize the expression of purely human qualities in social interaction. Indeed, some social analysts insist in the narrow view of men as inherently driven by self-interest and aggressive, competitive instincts. This view does not take into account man's desire to make sense of and to integrate his personal intersubjective life with his external environment and his fellow men. By allowing us to partake in collective recreation, games become situations which permit simultaneous participation of many people in some significant pattern of their own collective lives.

GAMES AS SOCIAL MODELS

There are games and ways of playing in every culture and era of man, but the nature of play activities differs greatly from culture to culture. The type of games played by a group of people, their manner of playing, and their collective reactions to these games are a reflection of the type of society within which they live.

Differences between cultures, subcultures, or even the sexes are mirrored in the differences in their manner of play. Herbert Gans, in *The Urban Villagers* (1962), notes that when middle-class and working class subcultures are compared there is a marked difference in how leisure time is spent. In the middle-class, the use of leisure time is often centered around the nuclear family. Husbands and wives participate in recreational activities together, with their family, or with other couples. In contrast, among the working class there is almost total age and sex segregation in recreational activities; men spend their leisure time together with other men of approximately the same age; the women and children do likewise. In the working-class subculture described by Gans, the family does not play together as does the middle-class family which he refers to as a "companionship unit." In this case, differences in play indicate something about family structure.

In an article comparing the play habits of French and American children, Martha Wolfenstein offers an excellent analysis of cultural differences as manifested in play. She notes that while playing with an airplane, a group of small French boys' activity centered on "talking" about the plane, rather than actually flying it. When an attempt to fly the plane was finally made it was found that it didn't fly, but in spite of this failure in practical use, the boys retained the same active interest in the conversation. A similar group of American boys, Wolfenstein (1955) says, would not spend the greater amount of their play time discussing the plane. Instead they would immediately try alternate ways of making it fly and would lose interest in it altogether if it proved to be a "failure in action." Here the differences in the play of children reflect the French theoretical approach versus that of American pragmatism.

CONCLUSION

Before Huizinga's work, scholars tended to either overlook play or to consider it a waste of time, a meaningless artifact (Caillois, 1958). But to Huizinga nothing could be further from the truth: "To the degree that he is influenced by play, man can check the monotony, determinism, and brutality of nature. He learns to construct order, conceive economy, and establish equity" (Huizinga, 1950: 57). Games constitute a form of communication, a way of making ourselves and our world understandable. Games and gamelike behavior help us to learn to manipulate our environment and each other. Games and play allow us to participate in collective recreation. Games are extentions of our social selves. Perhaps the most important aspect of games as communicative structures is found in their instructive value; not only do games help children to "take on the world of adults" and to learn to use their culture, but they allow adults, as well as children, to experiment and be innovative with specific aspects of their social lives.

REFERENCES

Berger, Peter and Thomas Luckmann
    1967 The Social Construction of Reality: A Treatise on the Sociology of Knowledge. Garden City, N.Y.: Doubleday-Anchor.
Caillois, Roger
    1958 Man, Play and Games. New York: Free Press.

Gans, Herbert
    1962 The Urban Villagers. New York: Free Press.

Goffman, Erving
    1969 Strategic Interaction. New York: Ballantine.

Huizinga, Johan
    1950 Homo Ludens: A Study of the Play Element in Culture. Boston: Beacon.

McLuhan, Marshall
    1964 Understanding Media: The Extensions of Man. New York: McGraw-Hill.

Mead, George H.
    1934 Mind, Self, and Society. Chicago: Univ. of Chicago Press.

Opie, Iona and Peter Opie
    1969 Children's Games in Street and Playground. New York: Oxford Univ. Press.

Wolfenstein, Martha
    1955 "French Parents Take Their Children to the Park." Pp. 99-117 in M. Mead and M. Wolfenstein (eds.) Childhood in Contemporary Cultures. Chicago: Univ. of Chicago Press.

SOME PRINCIPLES FOR THE DESIGN OF
CLARIFYING EDUCATIONAL ENVIRONMENTS

OMAR KHAYYAM MOORE and ALAN ROSS ANDERSON

University of Pittsburgh

When the ordinary typewriter was an exciting novelty, Mark Twain, who was an early typewriter buff, called it a "curiosity-breeding little joker"—and so it was, then. The *talking typewriter,* invented by Moore and Kobler (Moore and Kobler, 1963; Kobler and Moore, 1966), is a contemporary novelty which also elicits a good deal of curiosity. There have been numerous popular articles about it, and most of those who have played with it find it fascinating. Unfortunately, the interest generated by this machine does not carry over, necessarily, to the theoretical ideas which lie behind it. We say "unfortunately" for good reason. The machine itself is less important than the principles which guided its construction. The talking typewriter is merely one of a large number of possible inventions which can be made, we think, using this same theoretical context as a guide.

Our main purpose in this chapter is to explain and to illustrate a set of principles, four in number, for designing learning environments within which even very young children can acquire complex symbolic skills with relative ease. We intend to show, as we go along, that these principles for designing *clarifying* educational environments (where by a "clarifying educational environment" we mean an educational environment aimed to make the student [subject? victim?] *clear* about what he is doing, and more generally, what is going on) are systematically related to both a theoretical analysis of human culture and an interpretation of the socialization process, i.e., that process whereby a human infant, beginning life as a biological *individual*, becomes a *person*—and whose infantile *behavior* is gradually transformed into adult *conduct.*

In order to understand these "design principles," we must first explain the

Omar Khayyam Moore and Alan Ross Anderson, "Some Principles for the Design of Clarifying Educational Environments," in David A. Goslin (ed.) HANDBOOK OF SOCIALIZATION THEORY AND RESEARCH, ©1969 by Rand McNally and Company, Chicago, pp. 571-592. Reprinted by permission of Rand McNally College Publishing Company. (Pages 592-603 not included here.)

AUTHORS' NOTE: *Most of the theoretical work reported here was supported by the Office of Naval Research, Group Psychology Branch, Contract SAR/Nonr-609 (16). With respect* theoretical system out of which they emerged. We will then go on to a statement of the principles themselves, and finally we will consider, in some detail, an illustrative application of these principles to the problem of designing an actual learning environment. At this point, we will have come full circle—the talking typewriter will appear in a meaningful context as one part of a learning environment.

It will also be obvious that the talking typewriter is itself fundamentally a "social science invention." Because it *is* a social science invention, it is difficult to use it or similar devices intelligently without an appreciation of the social scientific ideas on which it is based.

Finally, before turning to our first task, that of sketching out the theoretical system upon which the set of principles is founded, we wish to acknowledge the contributions to our thinking of George Herbert Mead (1932, 1934, 1936, 1938), the father of the symbolic interactionist position in social psychology, and Georg Simmel (English translation, 1959), the originator of the school of "formal" sociology. We regard these men as central figures in the creation of the kind of sociology which can yield applications at the level of both mechanical inventions and social inventions. It may seem strange to some that we believe that Mead and Simmel have ideas which lend themselves to applications. Mead and his followers often have been criticized for spawning ideas which lacked testable consequences—to say nothing of applications; and to the best of our knowledge we are the only ones who have seriously entertained the thought that Simmel was within a light year of a practical application. We hope to show here that learning ideas taken from these two men can be reshaped into working principles for designing educational environments.

THEORETICAL BACKGROUND[2]

Folk Models

We think that it is a mistake to regard the ordinary human being as an atheoretical or a nontheoretical or even an antitheoretical creature. Some contemporary behavioral scientists seem to assume that the ordinary man, a citizen in good standing in whatever community he lives, is woefully lacking in intellectual resources to guide him in managing his affairs. He is credited only with some folk sayings and proverbs, some practical knowledge, some skill at rule-of-thumb reasoning, some tradition-based explanations—and that is about it.

In a contrary vein, we suggest a different view of "man." We think that, early in human history, probably at about the time men developed natural languages, they also created models of the most important features of their relations with the environment.[3] These were relatively abstract models which collectively

to applications, the major source of support has been the Carnegie Corporation of New York. The cost of developing the "talking typewriter" (Edison Responsive Environment) was borne exclusively by the McGraw-Edison Company of Elgin, Illinois.

covered relations holding between (1) man and nature—insofar as nature is not random; (2) man and the random or chancy elements in experience; (3) man in his interactional relations with others like himself; and (4) man and the normative aspects of group living. Cultural structures falling within these four classes of models were created in early history by many unsung Edisons and Einsteins (perhaps it would be more appropriate to say that they were created by the unsung Meads and Simmels of prehistory). Consequently, there does not exist a society, however "primitive," that does not have cultural objects falling within these four categories of models. It is convenient to have a name for all four classes of models, a name that suggests their origin early in human history. We call them "folk models."

Every society, as far back as we have any evidence, has *puzzles* which, we suggest, stand in an abstract way for nonaleatory man-nature relations. Every society has some *games of chance.* According to our view of the matter, games of chance are abstract models of the aleatory aspects of existence. Every society has *games of strategy* in the sense of von Neumann (1928; von Neumann and Morgenstern, 1947). These games capture some of the peculiar features of interactional relations among men, relations in which no party to an encounter controls all of the relevant variables upon which the final outcome depends, though each controls some of these variables and each participant must take some account of the potential actions of others involved in the situation if he is to behave intelligently. Every society has *aesthetic entities,* i.e., art forms which we claim give people the opportunity to make normative judgments about and evaluations of their experience. All societies make use of these folk models in the socialization of the young and for the recreation, or recreational enjoyment, of those who are older. Simple forms of these models are internalized in childhood and more complex versions of them sustain us in adulthood.

It should be pointed out that until mathematicians had made formal analyses of the structure of some of these folk models, their depth and subtlety were not appreciated fully. Of the four classes of folk models distinguished above, two have received adequate formal treatment, specifically, the various mathematical theories of probability have all games of chance as models, and the various mathematical theories of games of strategy have all games of strategy as models. The formal analysis of puzzles is not in as satisfactory a state as are games of chance and games of strategy; however, we have suggested that the methods of natural deduction may help clarify the structure of puzzles.[4] When it comes to aesthetic entities, everyone is at sea and it is not known whether mathematical analyses of aesthetic objects, should such analyses prove possible, would result in *only one* or *more than one* distinct class of models. It is well to remember that until the work of von Neumann (1928) no one was in a position to make a mathematically rigorous distinction even between games of chance and games of strategy—so we should be careful about making the assumption that aesthetic objects would yield to but one overall formal theory. Regardless of this, the mathematical research into the structure of folk models has made it perfectly obvious to us that early man was not a simple-minded clod. It required inventors of genius to create these intricate objects, but even a child can begin to play with most of them. Not much in the way of technical expertise was needed to fashion the equipment used in connection with folk models: bits of wood, or stone would do for the "pieces" used in most board games; a primitive technology was no bar to the creation of conceptually complex cultural entities.

Historically speaking, man not only invented and developed these fascinating folk models, he also devised suitable techniques for seeing to it that they were mastered by the ordinary citizen. If we think of these models as constituting the basic theoretical arm of a society's culture, then it is quite important that everyone, or virtually everyone, learns them. To put it another way, if folk models are abstract schemata which help orient us toward a wide variety of problems, then we should get them down pat. With respect to their inculcation, observe that, in general, they are learned, but not taught. What is taught are the "rules of the game," and once the rules are understood, each participant is largely on his own, except when the models are perverted by professionalism.

In every society there are social norms which distinguish between serious matters on the one hand, and fun and games on the other. Usually, specific times and places are set aside for the enjoyment of folk models. Also, the stakes for winning or losing are kept at some nominal value insofar as profit and loss enter. In addition, there are norms which regulate expressions of feeling and emotion with reference to using folk models. During the course of playing with a model, one is permitted to experience a fairly wide range of feelings and emotions, but extremes are excluded. These models serve, as it were, as a school for emotional expression—this is a kind of "school" in which boredom is unlikely and uncontrolled emotional frenzy is forbidden. All in all, the set of norms governing the use of folk models, and the models themselves, have proved so successful that people have to be prohibited from playing with them too much, despite the conceptual depth of the materials with which they deal. If we think of the models as teaching devices, then they are instructional with respect to relatively universal features of man's environment—they are abstract symbolic maps of human experience. We also can see that they "teach" in ways that satisfy the following conditions:

(1) They are "cut off," in some suitable sense, from the more serious sides of the society's activity, that is to say, they are cut off from immediate problems of welfare and survival. For example, if a child is learning the intricacies of social interaction, the activity in which he is experiencing or practicing the interaction must allow him to make many mistakes without endangering the lives or futures of those around him, to say nothing of his own safety. Similarly, such rewards as he receives from the activity must not be too expensive to those around him, or again, the activity would have just those serious consequences which these models, as teaching devices, must avoid.

(2) But in spite of the fact that the teaching devices must avoid serious consequences, some motivation must be built into them, or else the learner may lose interest. If we rely on the distinction between activities that are intrinsically rewarding, and

those that are rewarding only as a means, or extrinsically rewarding, we may say that the rewards in the learner's activities must be intrinsic or inherent in the activity itself. We call such activities "autotelic": they contain their own goals and sources of motivation.

(3) And finally, these teaching devices, if they are to be theoretically relevant to the problems which are likely to be encountered outside the context of an autotelic environment, indeed must be models of serious activities.

Thus far we envisage a situation like this: Every society makes up abstract symbolic models of its most serious recurrent problems. Despite the complex structure of these models, everywhere they are learned with pleasure by ordinary people. Every society has social norms governing the use of its folk models and these norms have the effect of making the models autotelic, so even though the models are models of serious matters, they must be treated playfully.

The notion that materials, in the sense of the contents of everyday life, are somehow abstracted from the stream of living and reappear as the play forms of sociability is a distinctively Simmelian idea—we borrowed it from him. It is he who argues that:

Actual forces, needs, impulses of life produce the forms of our behavior that are suitable for play. These forms, however, become independent contents and stimuli within play itself or, rather, as play. There are, for instance, the hunt; the gain by ruse; the proving of physical and intellectual stength; competition; and the dependence on chance and on the favor or powers that cannot be influenced [Simmel, 1959: 42].

And he goes on to say that:

To the person who really enjoys it [play], its attraction rather lies in the dynamics and hazards of the sociologically significant forms of activity themselves. The more profound, double sense of "social game" is that not only the game is played in a society (as its external medium) but that, with its help, people actually "play" "society" [Simmel, 1959: 50].

What Simmel did not do was to carry through a mathematical analysis of his cherished "play forms" of human association. He did see the need for such a formal analysis; in fact, he called for the creation of a kind of "social geometry" which would be up to characterizing the structure of play forms. He did not see that probability theory does the trick for games of chance, and, of course, the work of von Neumann (1928) on games of strategy did not come along until ten years after Simmel's death. And we all are still waiting for an intellectual giant the size of von Neumann to do a satisfactory mathematical analysis of aesthetic play forms. Nonetheless, the basic program for "formal" sociology, as envisaged by Simmel, is being carried out, and we like to think of ourselves as helping a little.

It should be remarked that the possibilities for developing an appropriate "social geometry" are not limited to analyzing folk models. For example, normative systems are of obvious importance in interpreting human interaction. Prior to the past decade, very little had been done toward developing a mathematical analysis of such systems. It was partly in response to our sense of the need for a program of formal sociology in this area that we undertook studies in what is now called "deontic logic." This topic has been treated by a vast number of investigators since von Wright's seminal essay of 1951, and the improvement in our own understanding of the basic ideas involved can be seen by comparing the analysis we offered in 1957 (A. R. Anderson and Moore, 1957) with a 1967 version (A. R. Anderson, 1967).[5]

Another problem area which has significance for "formal" sociology is the mathematical treatment of the notion of relevance. Human affairs are conducted within universes of discourse in which some standards of relevance are presupposed, but the study of "relevant implication" was a neglected area in mathematical logic, so efforts were made to create the required formal machinery.[6] Interestingly enough, the improvement in our understanding of problems in deontic logic came about mainly because of this apparently unrelated work on the logic of relevance. This was one of those "unexpected" bonuses which we come to "expect" from what seems, on the surface, to be merely remote abstract consideration.

Another area of investigation should be mentioned which also fits into the program of "formal" sociology. It turned out that there were almost no mathematical analyses of the logical structure of questions and answers. Yet surely here, if anywhere, is a distinctively human preoccupation, namely, the asking and answering of questions. The formal treatment of questions and answers is called "erotetic logic," a term introduced in 1955 by Mary and Arthur Prior (1955). It is being pursued by our colleague, Nuel D. Belnap, Jr. (among others), whose *An Analysis of Questions* (1963) provides a substantial treatment of this topic with bibliographical references.

Personality

Turning now from folk models qua cultural objects to some of their implications for personality, there is another possibility with reference to them that Simmel did not pursue. If our folk models and his play forms have the theoretical importance we attribute to them, then they should be of help in analyzing the structure of human personality. We have given some thought to this matter and our considerations have led us in the direction of the work of George Herbert Mead. As a heuristic gamble, we were willing to assume that the major functional components of human personality, and the organization of these components, reflect the structure of the folk models. By taking this view of human personality, we were led to ask whether each of the four kinds of folk models corresponds to a characteristic attitude or perspective that a person might take toward his world. It is our thesis that each class of models does so correspond, and that the models build upon one another in a particular order.

In capsule form, our position is:

(1) Puzzles emphasize a sense of agency. We call this the "agent perspective." This is the outlook that perhaps Cooley (1902) had in mind when he spoke of "the joy of being a cause."

(2) Games of chance emphasize a sense of patienthood, i.e., being the recipient of consequences over which we have virtually no control. We call this the "patient perspective."

(3) Games of strategy presuppose an agent-patient perspective, but emphasize what we call the "reciprocal perspective." In Meadian terminology this would seem to be the perspective of a "significant other." For example, in playing bridge there is room for meaningful acts of agency and we are sometimes patient to all manner of outrageous happenings, some due to chance, some due to our opponent, some due to our partner, and even a few of our own doing. But the heart of the game (as von Neumann [1928] showed with beautiful precision) lies in the possible interrelations between the two opposing teams, each of which must take the other into account. This means that a genuine game of strategy, such as bridge, does not reduce mathematically into either the form of a puzzle or the form of a game of chance. This means, also, that a person who is looking at the world from the standpoint of the reciprocal perspective does not see another human being as merely puzzling or unpredictable, but rather he see him as someone who is capable of looking at him as he looks at the other.

(4) Aesthetic entities emphasize a sense of assessing, evaluating or judging. This perspective presupposes *significant others* in interaction, i.e., it presupposes entities that behave in terms of the other three perspectives. We call this judgmental stance the "referee's perspective." The point of view of a judge in a bridge tournament (or any given player when he looks at his own play or the play of others as if he were the judge) is not that of any player qua player, nor is it some sort of average or consensus of the player's viewpoints. The referee's concern ranges over the whole game—his viewpoint presupposes that there are players with their reciprocal perspectives. With reference to Mead's analysis of personality, we think that the concept of the referee's perspective is a plausible explication of his concept of the "generalized other."

We made the point in connection with the reciprocal perspective that it did not collapse, logically speaking, into either of the two perspectives upon which it builds, namely, the agent and the patient perspectives. The reason we said this is that the mathematical structure of games of strategy is not reducible to either puzzles or games of chance. We want to make a similar argument now about the referee's perspective—it does not reduce to or collapse into any or all of the other three perspectives. Our reason for being confident about this is that the mathematics of the referee's perspective, insofar as it is deontic or normative, does not reduce to ordinary extensional logic, nor to the logic of possibility, nor to the logic of probability. The referee's perspective is a logically distinct realm. We realize that not everyone would find this line of reasoning convincing—and perhaps it should not be relied on too heavily—but we remember that not long ago there were those who thought it very unlikely that deontic logic could be set up on a solid footing, namely, on a basis which would not immediately collapse into the standard extensional systems.

A human being who has been socialized in the sense of Mead, i.e., an individual who has acquired a social self, should be able to take any of the four perspectives mentioned above. What is more, he should be able to handle them one by one, in pairs, in triples and in one superordinate quadruple—depending, of course, on the nature of the problem with which he is confronted. We say that his social self is constituted, in part, by the organization of these perspectives. A social self is neither something that anyone is born with, nor does it comes about automatically through the processes of physiological maturation; rather, it is an achievement in learning which we think is guided in part by autotelic folk models.

We agree with the general Meadian analysis about how the social self, as an organization of perspectives, emerges out of a matrix of social processes—and how it, in turn, may affect these same processes. We appreciate particularly the suggestions Mead makes concerning the process whereby the interplay among human beings begins with a "conversation of gestures" and leads on to symbolic interaction. This interaction takes place through the use of "significant symbols"—these symbols being defined in terms of a common universe of discourse. This universe of discourse, in turn, gains its relevance by virtue of its systematic relations to the set of social processes out of which the social interaction arises. For us, among the most significant of what Mead calls "significant symbols" are the symbolic complexes which we have dubbed "folk models." Of course, natural languages as systems of significant symbols are of prime importance, too. We agree completely with Mead about this.

Mead was well aware of the importance of play and of games as part of the process whereby a social self is acquired. In fact, he made a distinction between playing and taking part in a game to drive home his point that the development of the human personality takes place in a series of interrelated phases. A young child may play in the sense of taking the role of a series of significant others, but until he grasps the structure of the rules which make a game a game—that is, until he can govern his ongoing conduct in the light of what we call the referee's perspective, or in Mead's terminology, the "generalized other"—the child is only *playing* and not *gaming*. And, to the extent that he is not up to handling games, he is only partially socialized. Clearly, it is compatible with the Meadian position to assume that conduct which flows from a mature social self would involve the use of each of the perspectives in the solution of challenging problems. It is convenient to think of each perspective as a part of the social self. As has been made clear above, we think that there are at least four such parts of a socialized human being: agent, patient, reciprocator, and referee.

Information Processing

Any problem worthy of the full talents of an adult human being requires that he carry out a great deal of information processing with respect to it. Obviously, this information processing is subject to some kind of control. The question is "What kind?" In terms of an engineering analysis, there are two major kinds of control systems that we can consider: those of the *open-loop* variety and those of the *closed-loop* variety.

An open-loop control system is one which exercises its control in a way that is independent of the output of the system. Open-loop systems, generally speaking, are not bothered by problems of instability.

A closed-loop control system is one in which the control is somehow dependent upon the system's output (or some symbolic representation of that output).

Closed-loop systems tend to suffer from various forms of instability. As was remarked before, human beings must have some sort of a control system to govern their information processing. Judging from the tendency of human beings to become unstable, we guess that the human control system is of the closed-loop variety. As a matter of fact, we assume a great deal more than that—we posit that the four perspectives constitute a subsystem which functions as part of an overall control system governing information processing.

It may seem to some that our brief discussion above about control systems is very remote from a Meadian analysis. We do not think that this is the case at all. Mead attempted to formulate a number of ideas which, in retrospect, can be recognized as brilliant anticipations of concepts which later received explicit treatment along the lines that he suggested. The Meadian notion of an attitude or perspective is a case in point. Let us listen to Mead a bit as he tried to tell his students and colleagues what he meant by an attitude.

> Present results, however, suggest the organization of the act in terms of attitudes. There is an organization of the various parts of the nervous system that are going to be responsible for acts, an organization which represents not only that which is immediately taking place, but also the later stages that are to take place. If one approaches a distant object he approaches it with reference to what he is going to do when he arrives there. If one is approaching a hammer he is muscularly all ready to seize the handle of the hammer. The later stages of the act are present in the early stages—not simply in the sense that they are all ready to go off, but in the sense that they serve to control the process itself. They determine how we are going to approach the object, and the steps in our early manipulation of it. We can recognize, then, that the innervation of certain groups of cells in the central nervous system can already initiate in advance the later stages of the act. The act as a whole can be there determining the process [Mead, 1934: 11].

When Mead advanced this analysis of an attitude, his remarks were interpreted by many to be sheer teleological nonsense. Now, we understand these ideas much better—he is saying that ongoing human activity is subject to a closed-loop control system. No contemporary engineer would regard this as metaphysics (in the ba-a-ad sense).[7]

Information Processing for Human Beings

We posit a control system which governs information processing that is sufficiently reflexive to allow a human being to stand back from himself in order to view himself as an object. What is more, this control system must allow him to see himself from the standpoint of any of the perspectives while he is planning or executing actions. This hypothetical control system must make provision for the fact that we can and do soliloquize. It must both permit and control internal dialogues such as: "I would like to do X, but I am not sure that it would make me happy. My father approves, but he doesn't understand. My mother doesn't approve and she does understand. It's illegal, but my friends say it is good." We all go round and round like this, looking at the world in terms of what we can do, what might happen to us, what our friends and enemies think, and what the referee might say. Sometimes these considerations get out of hand and we become bogged down in repetitious and viciously circular chains of reasoning. Sometimes we fail to consider a problem from some important perspective.

The upshot is: we assume that a fully socialized human being, in a state of good emotional health, would have a control system which permits him to consider himself in a reflexive way from the standpoints which are represented abstractly in cultural terms by the four autotelic folk models.

Order of Mastering Perspectives

If the social self consists in part, at least, of an organization of perspectives, and if these perspectives are learned, then the question can be raised as to whether they are acquired in some particular order. Is the socialization process some sort of ordered sequence? We believe so and our analysis of human developmental phases follows our interpretation of Mead quite closely. We assume that the agent and patient perspectives are the first to develop—they are "twin-born," to use Mead's expression. The notion of an agent is linked to that of a patient—but it may take an infant some time to discover that this is so; there are indeed studies which indicate that it takes a while for new born infants to begin to understand the difference between their own bodies and their environment. But even if the agent-patient pair is twin-born, we still think of this pair as one term in a relation with the reciprocal perspective. Finally, the complete development of personality involves the pair-wise combination of the complex term agent-patient-reciprocal with the referee perspective. As was remarked earlier, the building up of this system into its most complex form does not mean that all parts need be involved in the solution of all problems; the system is sufficiently flexible, if its development goes well, so that the "parts" can be used one at a time, or in various combinations. As one can see, this is a fairly complex system even without further complications; there are some, on which we would now like to comment.

Complications Concerning Kind-Heartedness

Up to the present point, this summary of our position about human personality has made only passing reference to feeling and emotion. Are we to imagine that human beings are engaged mainly in the processing of information sans an involvement with affect? This would be an odd view of human nature, though the Meadian system tends to be odd in just this way. Mead had little to say about feeling and emotion. At the very least, we believe that it is essential to posit a *system* of feeling and emotion, and to make some assumptions about this system. This obviously is a complex topic. Here we will mention only a few of our assumptions—some which will help us later in the task of formulating principles for designing educational environments.

(1) Each perspective is directly connected to the system of feeling and emotion so that the control system gets a "reading" from the system of feeling and emotion about reactions to plans and the execution of plans, or, more generally, about ongoing activity. This means that we can have, but need not have, "mixed feelings and emotions." For example, a mountaineer, in thinking about rappelling down a cliff, may feel elation as an agent, anxiety as a patient, shame at the possibility of showing fear in front of his climbing companions (in a reciprocal sense), and guilt for rappelling at all since he is a family man and knows he should not take such risks—his own (kind-hearted) referee perspective says he is out of bounds.

(2) The system of feeling and emotion is so organized that, under some circumstances, at least, it is possible to change the scale of feeling and emotion without necessarily altering its relative proportions. For example, in playing a game of chess, we can run through a wide gamut of emotions in, as it were, "attenuated" form—we can experience token amounts of anxiety, fear, etc., without literally panicking. Of course, the system of feeling and emotion may get out of control as in some kinds of mental illness—the scale of intensity may be shifted in the direction of gross exaggeration, on the one hand, or flatness of affect, on the other.

(3) We not only have feelings and emotions, but each individual, in a reflexive way, can learn about his own reactions. The possibility of gaining some reasonable self-control with respect to affect, depends upon learning to recognize, differentiate and generalize about this vital aspect of ourselves; in other words, feeling and emotion can be schooled—and the use of autotelic folk models is normally part of this educational process.

Interlude

Someone who has followed the heuristic ideas presented above about human culture and the socialization process might be tempted to ask the following question: "If you regard these autotelic folk models so highly as guides to action, if they, indeed, represent abstractly so many salient features of human existence, and if they provide a basis for the structure of human personality, then do we need scientific models as opposed to folk models?" Our answer is that we do need scientific models. Folk models have served man well during most of his history, but there is something radically wrong with them with respect to their present theoretical relevance—something has happened which has rendered them worse than obsolete.

So long as the ordinary man lived out his life within the context of a static social framework, these models matched his world: the models themselves are essentially static entities. For instance, in any play of the game of chess, the rules—that is, the boundary conditions—remain constant. There may be plenty of lively action going on within this stable frame of reference and the participants may feel a wide range of emotions, but the rules are both fixed and inviolable in a normative sense. If you are working a puzzle, say, a jigsaw puzzle, the picture to be completed does not change as you work on the puzzle, and the pieces preserve constancy of size and shape. If you go to see a play two nights in a row, it remains the same play with trivial variations; the actors do not change their lines because you have seen it before, though you may appreciate it more thoroughly on seeing it the second time. The basic point we are attempting to make is that folk models mirror the static quality of unchanging or imperceptibly changing societies. The folk models in this respect are like the Newtonian conceptions of space and time—both presuppose a frame of reference which is invariant with respect to all that goes on within it.

Today we live in a new world, a world in acceleration, a dynamic, fluid world. In the 1940s, the major industrial societies underwent a massive acceleration in technological development. This increase in the rate of technological change was so large, as far as its social consequences are concerned, as to amount to a difference in kind rather than one of degree.[8] Because of this, we, along with many others, have come to divide technological history into two main periods— the primitive, from the dawn of human history to the 1940s, and the modern, from the 1940s on. In order to make this case we draw graphs of technological functions, plotting, on a time scale of 10,000 years, such things as the speed of travel, the force of explosives, the size of objects which can be manipulated with precision, the number of people who can be included simultaneously within one communication network. The curves for these and many other technological functions bend sharply upward at about this time and they are now heading off the graph. Of course, there are some who are unhappy with the notion on pinpointing this acceleration in the decade of the 1940s. They prefer to think in terms of a series of accelerations, each jolt larger than its predecessor. The time span for this series is taken to be the first half of the twentieth century. In any case, we agree with those who see a radical change.

Many aspects of this radical change in technological capability have become matters of grave concern. For example, most reasonably well-informed people understand that because the first fission device multiplied the explosive force of previous weapons a thousandfold, and a few years later a fusion device multiplied explosive force a million times, mankind now is in a position to do something it could never do before—to wit, it can commit suicide. All of this boggles the mind, but the aspect of the matter to which we want to draw attention here is the significance of the new technology for the socialization process.

We think that one important result of this technological leap is that we are in transition from what we have called a "performance" society to a "learning" society. In a performance society it is reasonable to assume that one will practice in adulthood skills which were acquired in youth. That, of course, has been the traditional educational patterns for human beings, and it is reflected in our linguistic conventions. We say that a medical student, for instance, learns medicine and the doctor *practices* it. There is also the practice of law, and in general, adults have been the practitioners of the skills which they learned as apprentices. In contrast, in a learning society, it is not reasonable to assume that one will practice in adulthood the skills which were acquired as a youth. Instead, we can expect to have several distinct careers within the course of one lifetime. Or, if we stay within one occupational field, it can be taken for granted that it will be fundamentally transformed several times. In a learning society, education is a continuing process—learning must go on and on and on. Anyone who either stops or is somehow prevented from further learning is reduced thereby to the status of an impotent bystander.

We assume that the shift from a *performance* to a *learning* society calls for a thoroughgoing transformation of our educational institutions—their administration, their curricula, and their methods of instruction. Education must give priority to the acquisition of a flexible set of highly abstract conceptual tools. An appropriate theoretical apparatus would range not only over the physical and biological sciences, but over the subject matter of the behavioral and social sciences as well. What is required is the inculcation of a deep, dynamic, conceptual grasp of fundamental matters—mere technical virtuosity within a fixed frame of reference is not only insufficient, but it can be a positive barrier to growth. Only symbolic skills of the highest abstractness, the greatest generality, are of utility in coping with radical change. This brings us back to the folk models which are inculcated in childhood. If they "teach" a conception of the world which is incompatible with a civilization in acceleration, then we have the challenge of creating new models appropriate for these changed and changing circumstances—we need models that are fundamentally dynamic.

In the next section of this chapter we will present some very general principles for designing educational environments. It will be apparent at once that we have tried to learn some lessons from the thousands of years of human experience with autotelic folk models, but as was indicated above, we do not think that is is wise to be bound to them in any exclusive sense. The usual kinds of autotelic folk models could get along very nicely with sticks and stones on their physical side; however, dynamic models for a learning society seem to require the imaginative use of a much more subtle technology.

PRINCIPLES FOR
DESIGNING CLARIFYING ENVIRONMENTS[9]

Our task now is to state and explain a set of four principles for designing educational environments. Any environment which satisfies all four of these principles will be said to be a "clarifying environment."

It will be seen that the first three principles to be treated are directly related to the notion of a folk model. The fourth principle seeks to make provision for the fact that we live in a world undergoing dynamic change.

(1) Perspectives Principle: One environment is more conducive to learning than another if it both permits and facilitates the taking of more perspectives toward whatever is to be learned.

(2) Autotelic Principle: One environment is more conducive to learning than another if the activities carried on within it are more autotelic.

(3) Productive Principle: One environment is more conducive to learning than another if what is to be learned within it is more productive.

(4) Personalization Principle: One environment is more conducive to learning than another if it: (1) is more responsive to the learner's activities, and (2) permits and facilitates the learner's taking a more reflexive view of himself as a learner.

The statement of the foregoing four [10] principles is sufficiently cryptic to make even a phrenologist happy. In spite of this fact, we believe they make some sense; and we forthwith proceed to try to explain the sense we think they make.

Perspectives Principle

The perspectives to which this principle refer are, of course, the four discussed in the previous section, namely, agent, patient, reciprocator and referee. This principle assumes, ceteris paribus, that learning is more rapid and deeper if the learner can approach whatever is to be learned:

(a) from all four of the perspectives rather than from just three, from three rather than from just two, and from two rather than from only one; and

(b) in all combinations of these perspectives—hence, an environment that permits and facilitates fewer combinations is weaker from a learning standpoint than one that makes provision for more combinations.

Another aspect of environmental flexibility with respect to the assumption of perspectives has to do with the attitude the learner brings to the environment each time he enters it. Imagine a learner who, one day, is filled with a sense of agency—he is in no mood, for instance, to be patient to anything or anybody. An environment will be more powerful from a learning standpoint if it lets him start off with whatever perspective he brings to it, and then allows him to shift at will.

As a parenthetical remark about shifting from one perspective to another, we think that young children do not have what is sometimes called a short "attention span," but they do have a relatively short and unstable "perspective span." This is one reason why there is little use in trying to deliver a lecture to a young child—he is not up to assuming the stance of a patient for very long at a time. But he can stay with the same topic or subject matter if he is permitted to run through a rather wide range of perspectives in whatever order he pleases.

When experts in education maintain that formal schooling is unsuitable for the very young child, the use of the word "formal" denotes the typical classroom situation in which most acts of agency are allocated to the teacher, the referee's role is also assigned primarily to the teacher, and the assumption of the reciprocal perspective in the form of interacting with peer-group members is forbidden through rules which are against note passing and which impose silence. About all that is left to the child is to be patient to the acts of agency of the teacher. This undoubtedly is an unsuitable learning situation for most young children—and the perspectives principle says that it is not as conducive to learning as a wide variety of alternative arrangements.

Another way to get the flavor of the perspectives principle is to pose the question as to why amusement parks amuse the young, but pall so rapidly. Think about the merry-go-round, the roller coaster, the fun-house with its surprises, etc.—it is apparent that what all of these "amusements" have in common is the rapid, involuntary shift in viewpoint within the context of one basic perspective; specifically, each exploits some facet of patienthood. Any environment which tends to confine people to one basic perspective is apt to become boring rather quickly. Of course, the symbolic level of amusement-park entertainment is relatively low, too, although it is high in its appeal to simple feelings and emotions. Consequently, a few trips to an amusement park go a long way.

Clearly, the theater is a more subtle form of entertainment; with it the shifts in perspectives are largely symbolic in character, rather than grossly physical. However, like the amusement park, the theater shares a weakness. Both force us to be spectators—patient to what goes on. An amusement-park ride hauls us through a predetermined course without any opportunity for changes due to our own acts of agency; similarly, plays run their predetermined courses. Though the patient perspective is salient at the amusement park and the theater, the referee's stance comes into the picture, too, as we assess and evaluate what goes on. There is also the vicarious opportunity to place ourselves in the roles of others, as for instance, when we witness the screams and squirmings of others on a roller-coaster ride.

It should be noted that these and related forms of amusement are changing. Recent innovations in motion pictures permit the audience to vote from time to time on how they want things to come out. This is a step, though a crude one, in allowing for the agent perspective in entertainment. Some new amusement-park rides give a few controls to the passengers. Also, turning our attention for the moment to "cultural" entertainment, the traditional museum seems to be on its way out—more and more displays are subject to some sort of control by the visitor. So, we see the amusement park, the theater (at least motion pictures), and the museum moving away from the boredom inherent in a confining perspective. They are coming closer to satisfying the perspectives principles, which (to repeat) says that "one environment is more conducive to learning than another if it both permits and facilitates the taking of more perspectives toward whatever is to be learned."

Autotelic Principle

For an environment to be autotelic it must protect its denizens against serious consequences so that the goings on within it can be enjoyed for their own sake. The most obvious form of protection is physical. There are sports which come perilously close to violating their own autotelic norms because of physical risks—mountaineering is one. When a mountaineer is asked why he climbs, the fact that this question arose indicates something is amiss. People do not go about asking bowlers, chess players, and tennis players, to take one mixed bag of players, for deep reasons to justify their activities. Mountaineers, like racing-car drivers, are always trying to prove that their sport only appears to be dangerous—they argue that it is not hazardous for those who are properly trained.

When it comes to designing educational environments, especially those concerned with the acquisition of intellectual skills, almost everyone is pretty well agreed to keep physical risk out. True, there are some advocates of corporal punishment; and we should remember that there is the occasional fanatic, such as a teacher we once knew who thought that the only way to do mathematics was in an ice cold room. He began each class by throwing open the windows, even on bitterly cold days, which gave a kind of chilly introduction to algebra.

It is relatively easy to keep physical risks out of educational environments though there may always be the school yard bully who punishes the scholar for his scholarship, and today, big-city schools increasingly require policemen to maintain order. Even so, it is more difficult to keep psychological and social risks out of an educational environment. If a student feels, while taking an exam, that he may disgrace himself and blight his future by failing to make a mark high enough to get into some special program, or if he feels that learning is simply a means of staying on a gravy train with stops only for prizes, honors, and scholarships on the way to success, then the whole learning environment is shot through with high psychological and social risks. For a learning environment to be autotelic, it must be cut off from just such risks.

Granted the nature of our present public and private school systems, and their relation to the broader society, it is doubtful that, at this time, more than a small fraction of the school day could be made autotelic. As a matter of fact, it is very difficult to arrange matters so that even a preschool child can have as much as thirty minutes a day that is really his, in the sense that none of the significant adults in his life is in a position to manipulate him, and where the things to be learned in the environment have a chance to speak persuasively to him in their own tongue.

Most contemporary education is nonautotelic; in fact, it prides itself on its nonautotelic status—school counselors carefully explain the financial and social rewards of further schooling. Through public service announcements, officials plead with dropouts to come back, and again, the basic argument given for returning to school is for rewards—financial and social. We never have heard a public service announcement which said something like, "Come back to school. Algebra is better than ever!"

The school day is so crammed full of activities that are planned to lead directly to the goal of at least one college degree, that a student seldom has the leisure to follow out the implications of an interesting problem, should he have the social misfortune of becoming intrigued by something truly puzzling. If a highly competent student works very hard, he may win a little extra time in which he can entertain a few ideas without having to cash them in at a science fair or some other parody of independent thinking.

Not only is the educational system largely nonautotelic in character, but the traditional folk models are in danger of being swept away. Little-league baseball replaces vacant-lot baseball. Amateur athletics in general seem to be turning into quasi-professional activities. On the more intellectual side our puzzles have been incorporated into the structure of tests—all current tests of ability are really a series of short puzzles.

Regardless of all of this, the autotelic principle states that the best way to learn really difficult things is to be placed in an environment in which you can try things out, make a fool of yourself, guess outrageously, or play it close to the vest—all without serious consequences. The autotelic principle does not say that once the difficult task of acquiring a complex symbolic skill is well under way, it is then not appropriate to test yourself in a wide variety of serious competitions. It is a common misunderstanding of the notion of an autotelic environment to assume that *all* activities should be made autotelic. Not so. The whole distinction requires a *difference* between a time for playfulness and a time for earnest efforts with real risks.

Productive Principle

Our statement of the productive principle is enigmatic at this point because we have not yet clarified the term "productive," though we have made implicit use of the concept of "productivity" in our prior discussion of folk models. So let us be explicit now.

We will say that one cultural object (a cultural object is something that is socially transmissible through learning) is more productive than another cultural object if it has properties which permits the learner either to *deduce* things about it, granted a partial presentation of it in the first instance, or *make probable inferences* (Peirce, 1955) about it, again assuming only a partial exposure to it.

Some examples may help. A perfect instance of a productive cultural object is a mathematical system. We can give the learner some axioms, some formation and transformation rules, and then he is at liberty to deduce theorems on his own. The logical structure of the system is what makes it productive. However, we are not always in a position to deal with such beautifully articulated structures. A case in point is the periodic table of elements. Its structure is productive on the basis of probable inference as opposed to deductive inference. Our evidence for productivity in this case is that empty cells in the table have been filled in with elements having the predicted properties. But compare the periodic table with an alphabetical arrangement of the same elements. The latter is less productive than the former by a country mile. In order to be more precise about all of this we would need a general theory of "probable inference" as well as a theory of "deducibility." We hope that a crude characterization of productiveness will be sufficient for our present purposes.

Turning back to the principle itself, it says, again ceteris paribus, that of two versions of something to be learned, we should choose the one which is more productive; this frees the learner to reason things out for himself and it also frees him from depending upon authority.[11]

Folk models, taken collectively, are good examples of productive cultural objects. To illustrate, once the simple rules for playing chess are mastered, it is not necessary to consult anyone in order to go on playing chess. It is true that one may be playing badly, but the structure of the rules for playing are sufficiently productive to guarantee that it is bad chess and not bad checkers that is being played.

Now that we have cleared things up a bit, someone might wonder why anyone would bother to state the productive principle as a principle for designing educational environments. Surely, people would not select the less productive of two versions of something to be learned. Yes, they would! The example above concerning the periodic table did not just pop into our heads. We observed, not long ago, a science teacher who had his students learn the atomic numbers and the atomic weights of the elements *in alphabetic order.* This is a tough task, and only a few of the children could manage it. Doubtless, you say, this is a rare aberration. Again, we beg to disagree. Let us take, as our case in point, the teaching of reading in the United States.

As everyone knows, or is supposed to know, there are two contrasting kinds of orthographic systems. On the one hand, there is the ideographic sort in which knowing some "words" give almost no clues as to how to handle the next written word. The Chinese system of writing is of this kind; it is barely productive at all.[12] On the other hand, there are many systems of writing which are alphabetic. Once the learner has cracked the code which relates the written and spoken versions of the language to each other, he can write anything he can say, or he can read anything that has been written. (The only things sacrificed by not referring to authority are the niceties of spelling and punctuation, but the phonetics carry the meaning.) Such alphabetic systems of writing are productive cultural objects, even, we should point out, in the case of a child of our acquaintance who spelled the word for eyes as "is." Given our usual spelling habits, this looks at best like an imaginative leap at an attempt to spell "eyes," but as Moore has pointed out elsewhere (1963), English orthography has more coherence than it is given credit for. However, many of the standard textbooks for teaching beginning reading used in our country today treat written English as if it were Chinese.

Personalization Principle

This principle, unlike the others, has two distinct parts: the idea is that the environment must be both (1) responsive to the learner's activities, and (2) helpful in letting him learn to take a reflexive view of himself. The explanation comes in two pieces.

(1) The responsive condition. The notion of a responsive environment is a complex one, but the intuitive idea is straightforward enough. It is the antithesis of an environment that answers a question that was never asked,[13] or, positively stated, it is an environment that encourages the learner first to find a question, then find an answer. The requirements imposed upon an environment in order to qualify it as "responsive" are:

(a) It permits the learner to explore freely, thus giving him a chance to discover a problem.

(b) It informs the learner immediately about the consequences of his actions. (How immediate is "immediate" will be discussed later.)

(c) It is self-pacing, i.e., events happen within the environment at a rate largely determined by the learner. (The notion that the rate is *largely* determined by the learner and not *wholly* determined by him is important. For example, some hyperactive children rush at their problems so much that the consequences of their actions are blurred—there must be provision for slowing down the learner under some circumstances; also, there are occasions when he should be speeded up. Nonetheless, it is basically self-pacing.)

(d) It permits the learner to make full use of his capacity for discovering relations of various kinds. (No one knows what anyone's full capacity for making discoveries is, but if we hand the learner a solution we certainly know we are not drawing upon his capacity.)

(e) It is so structured that the learner is likely to make a series of interconnected discoveries about the physical, cultural or social world. (What this amounts to depends, of course, upon what kinds of relations are being "taught" within the environment.)

The conditions for responsiveness taken together define a situation in which a premium is placed on the making of fresh deductions and inductions, as opposed to having things explained didactically. It encourages the learner to ask questions, and the environment will respond in relevent ways; but these ways may not always be simple or predictable. For a learner to make discoveries, there must be some gaps or discontinuities in his experience that he feels he must bridge. One way that such discontinuities can be built into a responsive environment is to make provision for changing the "rules of the game" without the learner knowing, at first, that they have been changed. However, it will not do to change the rules quixotically—the new set of rules should build upon the old, displacing them only in part. Such changes allow the learner to discover that something has gone wrong—old solutions will no longer do—he must change in order to cope with change. In other words, if you want a learner to make a series of interconnected discoveries, you will have to see to it that he encounters difficulties that are problematic for him. When he reaches a solution, at least part of that solution should be transferable to the solution of the next perplexity.

Finally, though a responsive environment does respond, its response has an integrity of its own. It is incorrect to think of a responsive environment as one which simply yields to whatever the learner wants to do—there are constraints. To take a trivial example, if the question is how to spell the word "cat," the environment permits the learner to attempt to spell it K-A-T—there is no rule against trying this, but he will not succeed that way, where by "succeeding," we mean both getting a satisfactory response from the environment, and learning the sort of thing the environment was devised to help him learn. Without the latter condition the environment would not be informative.

(2) The reflexive condition. One environment is more reflexive than another if it makes it easier for the learner to see himself as a social object. We previously made the point, the Meadian point, that the acquisition of the social self is an achievement in learning. Unfortunately, some of us are underachievers. One reason, we think, for our ineptitude in fashioning ourselves is that it is hard to see what we are doing—we lack an appropriate mirror. The reflexiveness which is characteristic of maturity is sometimes so late in coming that we are unable to make major alterations in ourselves. "The reflexive condition" is fairly heavy terminology; all we mean is that if an environment is so structured that the learner not only can learn whatever is to be learned, but also can learn about himself qua learner, he will be in a better position to undertake whatever task comes next. It facilitates future learning to see our own learning career both retrospectively and prospectively. It is a normal thing for human beings to make up hypotheses about themselves, and it is important that these hypotheses do not harden into dogma on the basis of grossly inadequate information.

We find it not at all surprising that athletic coaches have made more use of reflexive devices in instruction than have classroom teachers. This does not surprise us because of our confidence in play forms. It is in the realm of sports that motion pictures of learning and practice have come into wide usage. Coaches go over games with their players, spotting weaknesses, strengths, etc.— they do not forget their opponents, either. Of course, motion pictures used reflexively have limitations, but surely coaches have taken a step in the right direction.

The four principles presented above, perspectives, autotelic, productive, and personalization, are offered as heuristic guides for constructing educational environments. Undoubtedly, they are vague and ambiguous; the critical question is whether they are so deficient as to be useless. We do not think that they are totally without merit. In the next section we offer an application of the principles to show what can be made of them. . . .

NOTES

1. From time to time we will try to acknowledge our indebtedness to various authors, but a complete line of those authors to whom we feel indebted would be impossible for either of us to provide.

2. Since this section is mainly a summary of our own ideas, some worked out jointly, some separately, we have felt free to paraphrase our own papers without specific references. However, anyone who wishes to go more deeply into these ideas should read A. R. Anderson and Moore (1959, 1962, 1966); Moore (1957, 1958, 1961, 1964, 1965a, 1965b, 1968); Moore and A. R. Anderson (1960a, 1960b, 1960c, 1962a, 1962b, 1968).

3. In common with many contemporary philosophers, we acquire a certain sick feeling when hearing talk about "man," or even worse, "Man." When one reads translations of Aristotle, and finds that "Man is a rational animal," one has the idea that something deep is going on, but obvious parallels ("Whale is a large animal," "Mouse is a small animal") make the locution seem as ludicrous as it is.

We nevertheless defer to a tradition, with the understanding that when we use the term "man" we are referring to human beings, and that, in consequence, all the appropriate verbs should be in the plural.

4. A good simple treatment of natural deduction is contained in Fitch's text in symbolic logic (1952). Experimentation on human higher-order problem-solving which takes into consideration natural deduction is rare. However, some of our colleagues have attempted to take this into consideration, see Carpenter, Moore, Synder and Lisansky (1961). The preparation for an approach to natural deduction in terms of experimental techniques was worked out largely by Scarvia B. Anderson and Moore in a series of studies beginning in 1952 (S. B. Anderson, 1955, 1956, 1957; Moore and S. B. Anderson, 1954a, 1954b, 1954c).

5. Though the interest in deontic logic is our common concern, most of the relevant work under this project has been done by A. R. Anderson (1956a, 1956b, 1958a, 1958b, 1959a, 1962). For a reference to von Wright's essay, and a reasonably complete bibliography as of 1966, see the reprinted version of A. R. Anderson (1956a).

6. Again, the logic of relevant inference and entailment is of common concern. In this case the work has been done by A. R. Anderson (1957, 1959b, 1963); A. R. Anderson and Belnap (1958, 1959a, 1959b, 1959c, 1961a, 1961b, 1963); A. R. Anderson, Belnap and Wallace (1960); Belnap (1959a, 1959b, 1960a, 1960b, 1960c, 1960d, 1967); and Belnap and Wallace (1961), building their work on an important paper of Ackermann (1956).

7. We are well aware that "metaphysics" has an honorific sense, stemming from Aristotle's attempt to figure out how the universe ticks, and a pejorative sense, stemming from the logical empiricist rejection of theology and its sister-disciplines (e.g., Mariology). A good bit of what we are trying to convey in this chapter is probably metaphysics, in what we *hope* is a "good" sense.

8. We yield to none in insisting on our inability to make the difference between "kind" and "degree" clear, and this is not the place to try to get into the matter. But it does seem apparent that the difference between two chicken eggs of a slightly different size is a difference of degree, whereas the differences between either of the two eggs or a behemoth is of rather more startling proportions; the latter we think of as a difference in kind.

9. It should be made clear that though this chapter is in a certain sense a joint venture, the experimental and sociological part of the work belongs entirely to Moore. There is no particular point in trying to disentangle our contributions to what goes on here, beyond noting that the present chapter represents the results, some of which have been reported elsewhere, of about ten years of collaboration. Formulation and application of the principles to follow are the results of Moore's work.

10. C. S. Peirce observed somewhere that he had a "certain partiality for the number three in philosophy." From what follows, the reader will observe that *we* have a par-

tiality to its successor. On the other hand, even Peirce occasionally gave way to four (1868).

11. Though we do not know exactly how to characterize "productivity," we can give at least one clear example. The "natural deduction" methods of Gentzen, Jaskowski, Fitch (see Fitch, 1952, for references), and others are "productive" in that they help the students figure out what is going on. By contrast Nicod's single axiom for the propositional calculus prompted Irving Copi to quote Dr. Johnson's alleged remark about a woman preaching: it is "like a dog's walking on his hind legs. It is not done well; but you are surprised to find it done at all."

12. There is some slight productivity in the fact that such characters as those for tree, grove, and forest have a reasonable connection: a tree looks like 木 , a grove like 林 , and a forest like 森 . Similarly the character 口 means (among other things) mouth, entrance, opening, hole. But whoever would have guessed that 龜 meant turtle?"

13. We are all familiar with situations where we are given information we did not want to have. Our earlier discussion indicates our belief that many school children are in this situation (as we both were), and we suppose that the adult analogue is wading through all that dreary stuff about soap, while waiting for the eleven o'clock news on TV.

REFERENCES

ACKERMANN, W. Begrundung einer strengen Implikation. *Journal of Symbolic Logic,* 1956, 21, 113-128.

ANDERSON, A. R. *The formal analysis of normative systems.* Technical Report #2, Contract #SAR/Nonr-609(16). New Haven, Conn.: Office of Naval Research, Group Psychology Branch, 1956. (a) Also in N. Rescher (Ed.), *Logic of action and decision.* Pittsburgh: Univer. of Pittsburgh Press, 1967.

ANDERSON, A. R. Review of Prior and Feys. *Journal of Symbolic Logic,* 1956, 21, 379. (b)

ANDERSON, A. R. A reduction of deontic logic to alethic modal logic. *Mind,* 1958, 67 n.s., Journal of Symbolic Logic, *1957, 22, 327-328.*

ANDERSON, A. R. A reduction of deontic logic to alethic modal logic. Mind, 1958, 67 n.s., 100-103. (a).

ANDERSON, A. R. The logic of norms. *Logique et Analyse,* 1958, In.s., 84-91. (b)

ANDERSON, A. R. On the logic of "commitment." *Philosophical Studies,* 1959, 10, 23-27. (a)

ANDERSON, A. R. *Completeness theorems for the systems E of entailment and EQ of entailment with quantification.* Technical Report #6, Contract #SAR/Nonr-609 (16). New Haven, Conn.: Office of Naval Research, Group Psychology Branch, 1959. (b) Reprinted in *Zeitschrift fur mathematische Logik und Grundlagen der Mathematik,* 1960, 6, 201-216.

ANDERSON, A. R. Reply to Mr. Rescher. *Philosophical Studies,* 1962, 13, 6-8.

ANDERSON, A. R. Some open problems concerning the system E. of entailment. *Acta Philosophica Fennica* (Helsinki), 1963, fasc. 16.

ANDERSON, A. R. Some nasty problems in the formal logic of ethics. *Nous,* 1967, 6, 345-360.

ANDERSON, A. R., & BELNAP, N. D., Jr. A modification of Ackermann's "rigorous implication." (Abstract) *Journal of Symbolic Logic,* 1958, 23, 457-458.

ANDERSON, A. R., & BELNAP, N.D., Jr. A simple proof of Gödel's completeness theorem. (Abstract) *Journal of Symbolic Logic,* 1959, 24, 320-321. (a)

ANDERSON, A. R., & BELNAP, N. D., Jr. Modalities in Ackermann's "rigorous implica- tion." *Journal of Symbolic Logic,* 1959, 24, 107-111. (b)

ANDERSON, A. R., & BELNAP, N. D., Jr. A simple treatment of truth functions. *Journal of Symbolic Logic,* 1959, 24, 301-302. (c)

ANDERSON, A. R., & BELNAP, N. D., Jr. Enthymemes. *Journal of Philosophy*, 1961, 58, 713-723. (a)

ANDERSON, A. R., & BELNAP, N. D., Jr. The pure calculus of entailment. *Journal of Symbolic Logic*, 1961, 27, 19-52. (b)

ANDERSON, A. R., & BELNAP, N. D., Jr. Tautological entailments. *Philosophical Studies*, 1961, 13, 9-24. (c)

ANDERSON, A. R., & Belnap, N. D., Jr. *First degree entailments.* Technical Report #10, Contract #SAR/Nonr-609(16). New Haven, Conn.: Office of Naval Research, Group Psychology Branch, 1963. Also in *Mathematische Annalen*, 1963, 149, 302-319.

ANDERSON, A. R., BELNAP, N. D., Jr., & WALLACE, J. R. Independent axiom schemata for the pure theory of entailment. *Zeitschrift fur mathematische Logik und Grundlagen der Mathematik*, 1960, 6, 93-95.

ANDERSON, A. R., & MOORE, O. K. The formal analysis of normative concepts. *American Sociological Review* 1957, 22, 1-17. Also in B. J. Biddle & E. Thomas (eds.), *Social role: Readings in theory and applications.* New York: Wiley, 1966.

ANDERSON, A. R., & MOORE, O. K. *Autotelic folk models.* Technical Report #8, Contract #SAR/Nonr-609(16). New Haven, Conn.: Office of Naval Research, Group Psychology Branch, 1959. Also in *Sociological Quarterly*, 1960, 1, 203-216.

ANDERSON, A. R., & MOORE, O. K. Toward a formal analysis of cultural objects. *Synthese*, 1962, 14, 144-170. Also in M. W. Wartofsky (ed.), *Boston studies in the philosophy of science, 1961/1962.* Dordrecht, Holland: D. Reidel, 1963.

ANDERSON, A. R., & MOORE, O. K. Models and explanations in the behavioral sciences. In G. J. DiRenzo (Ed.), *Concepts, theory, and explanation in the behavioral sciences.* New York: Random House, 1966.

ANDERSON, S. B. Shift in problem solving. *Naval Research Memorandum, Report #458.* Washington, D. C., 1955.

ANDERSON, S. B. Analysis of responses in a task drawn from the calculus of propositions. *Naval Research Laboratory Memorandum, Report #608.* Washington, D. C. 1956.

ANDERSON, S. B. Problem solving in multiple-goal situations. *Journal of Experimental Psychology*, 1957, 54, 297-303.

BELNAP, N. D., Jr. Pure rigorous implication as a *"Sequenzenkalkül."* (Abstract) *Journal of Symbolic Logic*, 1959, 24, 316. (b)

BELNAP, N. D., Jr. Tautological entailments. (Abstract) *Journal of Symbolic Logic*, 1959, 24, 316. (b)

BELNAP, N. D., Jr. *A formal analysis of entailment.* Technical Report #7, Contract #SAR/Nonr-609(16). New Haven, Conn.: Office of Naval Research, Group Psychology Branch, 1960. (a)

BELNAP, N. D., Jr. Entailment and relevance. *Journal of Symbolic Logic*, 1960, 25, 144-146. (b)

BELNAP, N. D., Jr. First degree formulas. (Abstract) *Journal of Symbolic Logic*, 1960, 25, 388-389. (c)

BELNAP, N. D., Jr. EQ and the first order functional calculus. *Zeitschrift fur mathematische Logik und Grundlagen der Mathematik*, 1960, 6, 217-218. (d)

BELNAP, N. D., Jr. *An analysis of questions: Preliminary report.* Santa Monica, Calif.: System Development Corporation, 1963.

BELNAP, N. D., Jr. Intensional models for first degree formulas. *Journal of Symbolic Logic*, 1967, 32, 1-22.

BELNAP, N. D. Jr., & WALLACE, J. R. *A decision procedure for the system E_i of entailment with negation.* Technical Report #11, Contract # SAR/Nonr-609(16). New Haven, Conn.: Office of Naval Research, Group Psychology Branch, 1961.

CARPENTER, J. A., MOORE, O. K., SNYDER, C. R., & LISANSKY, E. S. Alcohol and higher-order problem solving. *Quarterly Journal of Studies on Alcohol*, 1961, 22, 183-222.

COOLEY, C. H. *Human nature and the social order.* New York: Scribners, 1902 (1922, 1930). P. 217.

FITCH, F. B. *Symbolic logic.* New York: Ronald, 1952.

KOBLER, R., & MOORE, O. K. *Educational system and apparatus.* U. S. Patent #3,281,959. 27 figures, 51 claims granted, 12 references. Also granted in many foreign countries, 1966.

MEAD, G. H. *The philosophy of the present.* LaSalle, Ill.: The Open Court, 1932.

MEAD, G. H. *Mind, self and society.* Chicago: Univer. of Chicago Press, 1934.

MEAD, G. H. *Movements of thought in the nineteenth century.* Chicago: Univer. of Chicago Press, 1936.

MEAD, G. H. *The philosophy of the act.* Chicago: Univer. of Chicago Press, 1938.

MOORE, O. K. Divination—a new perspective. *American Anthropologist,* 1957, 59, 69-74. Also in W. A. Lessa & E. Z. Vogt (Eds.), *Reader in comparative religion: An anthropological approach.* (2nd Ed.) New York: Harper & Row, 1965.

MOORE, O. K. Problem solving and the perception of persons. In R. Tagiuri & L. Petrullo (Eds.), *Person perception and interpersonal behavior.* Palo Alto, Calif.: Stanford Univer. Press, 1958. Pp. 131-150.

MOORE, O. K. Orthographic symbols and the preschool child—a new approach. In E. P. Torrence (Ed.), *Creativity: 1960 proceedings of the third conference on gifted children.* Minneapolis: Univer. of Minnesota, Center for Continuation Study, 1961. Pp. 91-101.

MOORE, O, K. *Autotelic responsive environments and exceptional children.* Report issued by The Responsive Environments Foundation, Inc., Hamden, Conn., 1963. Also in J. Hellmuth (Ed.), *The special child in century 21.* Seattle: Special Child Publications of the Sequin School, Inc., 1964; and O. J. Harvey (Ed.), *Experience, structure and adaptability.* New York: Springer, 1966.

MOORE, O. K. Technology and behavior. In *Proceedings of the 1964 invitational conference on testing problems.* Princeton: Educational Testing Service, 1964. Pp. 58-68.

MOORE, O. K. From tools to interactional machines. *New approaches to individualizing instruction.* Report of a conference to mark the dedication of Ben D. Wood Hall, May 11, 1965. Princeton, N. J.: Educational Testing Service, 1965. (a) Also in J. W. Childs (Ed.), *Instructional technology: Readings.* New York: Holt, Rinehart & Winston, 1968 (in press).

MOORE, O. K. Autotelic responsive environments and the deaf. *American Annals for the Deaf,* 1965, 110, 604-614. (b)

MOORE, O. K. On responsive environments. *New directions in individualizing instruction.* Proceedings of the Abington Conference, 1967. Abington, Penna.: The Abington Conference, 1968.

MOORE, O. K. & ANDERSON, A. R. *Early reading and writing, part 1: Skills.* 16 mm. color and sound motion picture. Pittsburgh: Basic Education, Inc., 1960. (a)

MOORE, O. K. & ANDERSON, A. R. *Early readings and writing, part 2: Teaching methods.* 16 mm. color and sound motion picture. Pittsburgh: Basic Education, Inc., 1960. (b)

MOORE, O. K. & ANDERSON, A. R. *Early reading and writing, part 3: Development.* 16 mm. color and sound motion picture. Pittsburgh: Basic Education, Inc., 1960. (c)

MOORE, O. K., & ANDERSON, A. R. Some puzzling aspects of social interaction. *Review of Metaphysics,* 1962, 15, 409-433. (a) Also in J. H. Criswell, H. Solomon & P. Suppes (Eds.), *Mathematical methods in small group processes.* Stanford, Calif.: Stanford Univer. Press, 1962. Pp. 232-249.

MOORE, O. K. & ANDERSON, A. R. The structure of personality. *Review of Metaphysics,* 1962, 16, 212-236. (b) Also in O. J. Harvey (Ed.), *Motivation and social interaction.* New York: Ronald Press, 1963.

MOORE, O. K., & ANDERSON, A. R. The responsive environments project. In R. D. Hess & R. M. Baer (Eds.), *Early education.* Chicago: Aldine, 1968.

MOORE, O. K., & ANDERSON, S. B. Modern logic and tasks for experiments on problem solving behavior. *Journal of Psychology,* 1954, 38, 151-160. (a)

MOORE, O. K., & ANDERSON, S. B. Search behavior in individual and group problem solving. *American Sociological Review,* 1954, 19, 702-714. (b)

MOORE, O. K., & ANDERSON, S. B. Experimental study of problem solving. *Report of Naval Research Laboratory Progress,* 1954, August, 15-22. (c)

MOORE, O. K., & KOBLER, R. *Educational apparatus for children.* U.S. Patent #3,112,569. 6 figures, 13 claims granted, 6 references. Also granted in many foreign countries, 1963.

PEIRCE, C. S. Some consequences of four incapacities. *Journal of Speculative Philosophy,* 1868.

PEIRCE, C. S. *Philosophical writings of Peirce.* New York: Dover, 1955.

PITMAN, J. Man—the communicating animal, par (verbal) excellence. In A. C. Eurick (Ed.), *New approaches to individualizing instruction.* Princeton: Educational Testing Service, 1965. Pp. 49-60.

PRIOR, MARY, & PRIOR, A. Erotetic logic. *Philosophical Review,* 1955, 64, 43-59.

SCOTT, D., A short recursively unsolvable problem. *Journal of Symbolic Logic,* 1956, 21, 111-112.

SIMMEL, G. *Georg Simmel, 1858-1918: A collection of essays.* K. H. Wolff (Ed.) Columbus: The Ohio State Univer. Press, 1959.

VON NEUMANN, J. Zur Theorie der Gesellschaftsspiele. *Mathematische Annalen,* 1928, 100, 295-320. Reprinted in A. H. Taub (Ed.), *Collected works.* Vol. 6. New York: Pergamon, 1961.

VON NEUMANN, J., & MORGENSTERN, O. *Theory of games and economic behavior.* Princeton: Princeton Univer. Press, 1947.

INTRODUCTION:
In Defense of Games

JAMES S. COLEMAN

Johns Hopkins University

T he fascination of games is a curious matter. It must arise in part from the arbitrary setting aside of the vaguely defined but complex and deadly serious rules which govern everyday life, and substitution of a set of explicit and simple rules whose consequences vanish when the game is over. But if I were to attempt to explain this fascination in gerneral terms I should never succeed. I will instead introspect, and ask what it is that fascinates me, not as a player in a game but rather as a sociologist in constructing them. For in this there is certainly something to be explained. Why should self-respecting sociologists, who could be working in research directions that would gain far more recognition from colleagues, instead toy with games in a field—educational sociology—that has long languished in the cellar of the discipline?

Let us even take as given, for the sake of argument, that games have remarkable potentials for learning—that they could transform the techniques by which children learn in schools and thus transform children that schools presently leave untouched or mildly "educated." Important as such effects are, they are no reason for the sociologist to excite himself about games, no reason to leave his other work for the fascination of constructing and testing games. The sociologist is not, after all, educator; his task instead is to study, and hope to better understand, society and social organization.

What is it then that fascinates me and the other students of social organization? What makes us abandon the proper behavior of sociologists and fix instead on games that may induce learning in children?

To come to an answer requires first a closer look at the very notion of games and the peculiar relation they bear to social life. A game—nearly any game, not merely those termed "simulation games"—constitutes a kind of caricature of social life. It is a magnification of some aspect of social interaction, excluding all else, tearing this aspect of social interaction from its social context and giving it a special context of its own. Even those games that are farthest from those

Reprinted from the **American Behavioral Scientist**, Volume 10, October 1966. ©1966 by Sage Publications, Inc.

described in this issue exemplify this. A boxing or wrestling match abstracts from its context the direct physical violence that resides in social life and re-creates this violence under a set of explicit rules. When I was a boy in the Midwest, cornhusking contests abstracted one activity from the life of farmers, established a set of rules, and gave this activity a temporary but central position for the participants.

This unique relation of games to life can be seen even better in other ways. The informal games of young children appear to be crucial means for learning about and experimenting with life. One of the most perceptive students of the social and intellectual development of young children, Jean Piaget, has observed this development in the simple games children play, such as the games of marbles. It appears that for children, games are more than a caricature of life; they are an introduction to life—an introduction to the idea of rules, which are imposed on all alike, an introduction to the idea of playing under different sets of rules—that is, the idea of different roles, an introduction to the idea of aiding another person and of knowing that one can expect aid from another, an introduction to the idea of working toward a collective goal and investing oneself in a collectivity larger than himself. It appears that games serve, for the young child, all these functions as an introduction to life.

Still another aspect of this special linkage between games and life is provided by a recent development in psychiatry—a turn from the emphasis on traumatic events in the patient's life, on oedipal complexes and mysterious fixations, toward an emphasis on the often destructive behavior strategies that adults use toward others. I refer principally to the book *Games People Play*, by Eric Berne. While Berne's use of the term "games" to describe these strategies extends the meaning beyond that used in these pages, the very use of the term indicates the close liaison between explicit games and the behavior people engage in as part of everyday life.

All these illustrations are intended to convey the intimate connections between games and social life. If there is such an intimate relation, and a few sociologists are wagering their professional lives that there is, then games themselves in all their forms become of great potential interest for the sociologist. It may be that he, just as the young child, can gain insight into the functioning of social life through the construction and use of games.

But beyond this there are certain special characteristics to the games described in this issue, and to the games that sociologists find of particular interest. Some games involve the interaction of a player with his physical environment—for example, a maze or a jigsaw puzzle, or block puzzles, or a cornhusking contest, or a pole vault. These games abstract from life either certain physical skills or certain intellectual skills of inference from physical evidence. Other games, such as number puzzles or crossword puzzles, involve interaction with a symbolic environment, in these two instances an environment of numbers and an environment of language.

Such abstractions of activities from life hold some interest for the sociologist, but much less interest than another class of games which abstract from life some elements of social relations or social organization. Many games incorporate some aspects of such relations, but a few games incorporate enough such relations that a special term has been used to describe them: social simulation games. Such games pluck out of social life generally (including economic, political, and business life) a circumscribed arena, and attempt to reconstruct the principal rules by which behavior in this arena is governed and the principal rewards that it holds for the participants. Such a game both in its construction and in its playing then becomes of extreme interest to the student of social organization. For from it he may learn about those problems of social relations that are his central concern. The game may provide for him that degree of abstraction from life and simplification of life that allows him to understand better certain fundamentals of social organization.

It is this, then, that makes the sociologist fascinated with a certain kind of game—the possibility of learning from this caricature of social relations about those social relations themselves.

But the question immediately arises, how could a professional sociologist and a sixth- or twelfth-grade child possibly learn about social life from the same game? My answer is that it is precisely appropriate that this be so. For children have too long been taught things that are "known," and have too seldom been allowed to discover for themselves the principles governing a situation. It may well be in the physical sciences that the young student and the professional scientist cannot learn from the same environment. (Yet Einstein, in a paper he wrote to explain the meandering of streams, mentioned that although the mechanical principles involved were well known and simple in application to the problem, he had met very few physicists who understood these principles well enough to use them in the simple application. I suspect that a perceptive high school student would suffer no serious disadvantage relative to a professional physicist.) But it is certainly the case that the professional sociologist and the young student can learn from a single game.

It is also true that what is meant by learning is quite different for the sixth grader and the professional sociologist. The sixth grader is learning to incorporate this experience into his own life, learning to recognize the dominant aspects of this social environment so that he can respond appropriately to them when he meets such an environment in his own life. The professional sociologist is learning how to describe in general terms the functioning of this system of relations, learning to fit the system of relations to an abstract conceptual scheme.

This, then, is the fascination of the sociologist with social simulation games— the opportunity to learn about social organization by forming a caricature of such organization and then observing this caricature. This is supplemented by his fascination in seeing young children learn as much about such environments as he himself can know how to transmit through the construction of the game.

NOTES ON DEFINING "SIMULATION"

R. GARRY SHIRTS

Simile II

There is a simple schema which I have found useful for defining "gaming and simulation" activities in my own thinking and when talking with others. No claim of ultimate truth is made for this particular classification scheme, nor do I even suggest that the scheme has tied down most of the loose ends, or created logic-tight definitions. Even so, it has been extremely helpful to me in the past and I hope may be helpful to others.

The basis of this classification system is three kinds of activities: simulations, contests, and games. These three basic activities are represented thus·

Reprinted from the "Occasional Newsletter," Spring 1973, with the permission of SIMILE II.

If we overlap the boxes, we create four new categories:

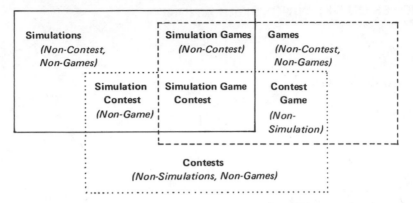

Definitions and examples of the resulting seven groups follow.

Simulations (Non-Contest, Non-Game)

The term "simulation" has taken on a special meaning among gamers. While everyone seems to be able to say what simulation is not, few people can say what it is. The problem partially develops from our insistence on differentiating between the noun *simulation* and the infinitive *to simulate*. Traditionally a simulation is something more than that which simulates; the term "simulation" has been reserved for the modeling or simulation of *systems* which can be represented in part by mathematical or quasimathematical formulas.

In the classification system proposed here, however, that distinction is disregarded. A simulation, rather, is anything which simulates or models reality. Listing representative examples of "simulations" from the very abstract to the concrete, we arrive at a surprisingly varied array of activities:

> **Simulations**
>> *Mathematical Formulas*
>> *Models of*
>>> *Physical Systems*
>>> *Military—Industrial Systems*
>>> *Social Systems*
>> *Role-Playing*
>> *Film*
>> *Literature*
>> *Painting*
>> *Sculpture*

The first four items unquestionably qualify as simulations in the traditional sense. Role-playing, however, is generally regarded as a lower-class cousin, and film, art, and sculpture as members of unrelated though honorable families. They are included here as "simulations," because:

(1) The sculptor, artist, film-maker, and writer, are, in fact, simulating reality much of the time.

(2) Recognition of the similarity of the artist-as-simulator and the engineer/social scientist/educator-as-simulator may serve as a means of bridging the communication gap between C. P. Snow's two cultures, the technocrats and the artists.

(3) It is possible that we can learn something about modeling from the artists. After all, they have been wrestling with such questions as validity, the relationship of the model to reality (abstract vs. representational art), the role of criticism, the effect of the medium on the message, and a myriad of mutual problems for several centuries.

Contest (Non-Simulation, Non-Game)

The essence of this kind of activity is competition. The competition may be between man and man, man and himself, man and nature, or nature and nature. Sometimes there are formal rules and sometimes the only rules are those determined by the act of trying to win.

The contest and the game are frequently confused. Many contests are games but not all; neither are all games contests. The difference between the contest and the contest-game is discussed below under the Game-Contest category. Examples of contests are: elections, competition between businesses, man's struggle against his environment, etc.

Contests

(Non-Simulation, Non-Games)

Man vs. Man
Man vs. Himself
Man vs. Nature
Nature vs. Nature
Examples: Business competition, political contest

Games (Non-Contest, Non-Simulation)

Bernard Suits, in the *American Philosophy of Science,* (XXXIV, 1967, 148-156) has, in my opinion, done an excellent job of defining a game. He says several things in the article but the most important to me is the notion that a game is an activity in which people agree to abide by a set of conditions (not necessarily rules) in order to create a desired state or end. The conditions that the participants agree to abide by may well involve *inefficient* ways of accomplishing the desired state or ends. For example, rather than getting a golf ball in the cup by the most efficient method, which is probably walking over and placing it in the cup, we agree to get it into the cup by hitting it with a metal stick with a small flattened surface on the end. We agree to run around a track to cross a line instead of running directly toward the line. We knock down bowling pins by rolling a heavy ball down a narrow wooden lane instead of just walking over and knocking them down. The notion of inefficiency is extremely important in the definition of a game. Not that every game has to use inefficient means to accomplish its end but that inefficiency is a meaningful possibility.

Another important idea is that the condition the game is created to produce may be something other than winning, such as body movement, laughter, creativity, embarrassment, etc. My children play a game called "Truth or Dare": you ask another person a question and he must either tell the truth or do some daring or ridiculous stunt; either way it produces laughter and fun but no competition is involved. The point is that many people want to limit games to the notion of competition, but games have been created for many other reasons. Opie, in *Children's Games in Street and Playground* (1969), points out that when children are confined to schoolgrounds, they tend to play competitive games with winners and losers (what we are calling here "contest games"). However, when they are not confined by the school boundaries and can roam at will on the streets and in the fields, they tend to play non-competitive games which create laughter, physical exercise, and body contact.

Examples of non-competition, non-simulation games include many of the encounter games where the purpose is to create an open climate of trust; many of the theatre games and, finally, much of what we generally classify as "play" are non-competitive, non-simulation games.

Games

   Non-Contest, Non-Simulation

      Encounter Games
      Many of the Non-Simulation Theatre Games
      Much of What Is Generally Described as "Play"

Contest Game (Non-Simulation)

Many people limit their definition of games to this category. But this is only one type of game—a game in which the conditions one agrees to abide by are designed to create competition and winning. Educational game publishers receive many games to be considered for publication. Most are straight role-playing situations which generally are not published because almost any teacher can build them with very little effort. The second most popular type of game received is contests, which are frequently called simulations incorrectly by their authors. For example, there is a game which purports to be a simulation of the electoral college. What it consists of is a series of questions which the participants answer and if they answer correctly, they are given so many electoral votes. The process of the game is in no way simulating the process of the electoral college. It is a game-contest pure and simple.

Sports, gambling, mathematical games, and word games are all examples of non-simulation contest games. The difference between a pure contest and a contest game is the relative importance of the conditions under which the contest is conducted. In a pure contest, inefficiency in the rules is not a possible alternative. In other words, in the pure contest whenever possible the rules or conditions of the contest must be related to what is won as efficiently as possible. For example, a businessman would not consider setting up a business next to his competitor in order to make the competition keener and more enjoyable. He might move next to his competitor for other reasons, but not to create more competition.

Another way to differentiate between the game contest and the pure contest is to realize that in the pure contest the participants will always be seeking ways to reduce the competitive aspects of the situation and at the same time increase their chances of winning, regardless of whether it makes the competition fairer.

Still a third way is to realize that in the pure contest the participant may want to impose rules or conditions which would give him an unfair advantage over his opponent, whereas in a game contest a participant would not seek to establish such rules since that would destroy the prupose of the game.

Contest Game

(Non-Simulation)

   Examples:
   Sports
   Gambling
   Mathematics Games
   Word Games

Non-Contest Simulation Games

In this category are those activities which are games designed to simulate reality but are not contests. For example, that harmless activity known as "Ring Around the Rosies" is really a non-contest simulation game having to do with the Black Plague.[1] The ring around the rosies is a pustule; the pocketful of posies is the pus; "ashes, ashes" means that the pustule goes black; and "All fall down" means that everyone dies. The child's games of "Store" and "Assembly-line" are other examples.

Simulation Game

(Non-Contest)

Examples:
Store
Assemblyline
Ring Around the Rosies

Non-Game Simulation Contest

In this category are activites which are contests and simulations but not games. For instance, suppose an industrial engineer were interested in determining which of two methods for warehousing a product was most efficient. He might simulate a contest between the two methods to determine which one to adopt.

Simulation-Contest

(Non-Game)

Example:
Competition Between Two
Different Systems for Warehousing
Goods

Simulation Game Contest

This is the category in which most of the experiences which are generally called "educational simulations" and "simulations games" belong—for example, SIMSOC, STARPOWER, BALDICER, INS, etc. These experiences are contests because they are concerned with the allocation of scarce resources such as money, influence, time, space, etc. They are simulations because they model reality and games because (1) the participants agree to abide by a set of conditions in order to create an experience, and (2) inefficient means such as communicating by written message rather than through talking, are frequently incorporated into the rules.

Simulation Game Contest

Examples:
Crisis, SimSoc, Starpower, Dangerous
Parallel, Democracy

SUMMARY

In reading this over there are three ideas which are especially important to me. One is the relationship between art and our type of simulation activity and how we might profit from seeing these two 'disparate' activities as different approaches to the same problem. The second is Suits' definition of a game which I heartily recommend as required reading for anyone interested in the topic. And third is the recognition that games are frequently designed to create such conditions as laughter, physical contact, and trust as well as competition.

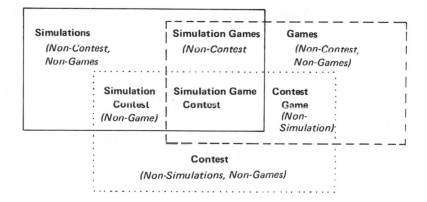

NOTES

1. Thanks to Ted Rodgers of the Hawaii Department of Education for this piece of information.

INTRODUCTION TO GAMING-SIMULATION TECHNIQUES

R. H. R. ARMSTRONG and MARGARET HOBSON

University of Birmingham

Historical Development

Gaming-simulation techniques are of long standing, such games as *chess, go* and *shogi* having been developed from war games used in the Indian subcontinent, China and Japan some thousands of years ago. Modern war gaming dates from the latter part of the eighteenth century, when the Prussian military establishment became conscious of the need to revise the training of combat officers.[1] Whilst in the last thirty years the training element has remained important, gaming techniques have been employed in such areas as strategic planning (the Japanese attack on Pearl Harbor), tactical operations planning ('hunt and kill' patterns for submarine search), weapon use and development (analysis of air and ground combat), and the attempt to define the characteristics and implications of new weapons systems (the pattern of deployment of nuclear weapons by NATO in Europe).

Although primarily a post-war development 'crisis' games were developed during the inter-war period. In such games role playing is linked to a scenario of an actual or hypothetical crisis situation within the area of international relations. The 'crisis' game owes much in terms of its form to the 'free-play' kriegsspiel developed from the Prussian war-game exercises.

It was not until 1956 that the American Management Association, in cooperation with IBM, explored the idea of 'business war games.' Since then the growth of business games, mainly for training purposes, has been rapid. The approach has been applied also to the examinations of operational problems (stock control), though such developments have not been as prolific as the military equivalents.

Applications to problems of public administration and land use planning date from 1960 with the demonstration game POGE.[2] Two better-known examples, METROPOLIS[3] (Duke) and CLUG[4] (Feldt) date from 1964 to 1965 respectively, and have served to stimulate developments in the public service sector.

Reprinted from ALEA Manual with the permission of the copyright holders, Margaret Hobson and Robert H. R. Armstrong.

Elements of Gaming-Simulation

Thus paradoxically the use of games for training and operational planning is both ancient and novel, but in all the applications of gaming-simulation certain common elements can be found. These are:

(a) People playing *roles* not necessarily corresponding to those they assume in the real-life situation;

(b) a *scenario* defining a problem area or a given 'state of the system';

(c) an *accounting system* designed to record such decisions and events together with their consequences, as are taken or occur during play;

(d) some algorithm(s) (implicit or explicit) which indicate(s) *operating procedures* for playing and controlling the exercise.

Different games place the emphasis on different elements. Thus some games are almost entirely role-playing exercises whilst at the other extreme certain computer simulations emphasize the accounting system.

Computer Simulations

At this point it may be useful to distinguish a gaming-simulation exercise from a computer or machine simulation. Gaming-simulations will always employ all four elements—roles, scenarios, accounting systems, and operating procedures—and at least the major roles will be represented by human players. In the computer simulation whilst the four elements are present they are represented in symbolic or analogue form within a model. Thus the relatively 'free' decisions taken by role players in a gaming-simulation exercise are replaced in the computer or machine simulation by programmed responses to a series of alternatives.

Gaming-simulations which use a computer employ it as part of the accounting system. Even the most 'sophisticated' model is employed as a basis for processing information and responses generated by the human players; this is still true where the programmed responses are deliberately designed to be 'random.' Hence the emphasis is not upon the logic or inner consistency of the model but upon either or both the relationships:

(a) between the roles represented by human players;

(b) between the players and the scenario.

Hence, gaming-simulation cannot be a 'predictive device. The presence of human players affords opportunities for the absurd and the irrational to dominate. Conditions of play will vary from exercise to exercise, not the least important variable being the personalities of the players.

Apparent differences in outcome between gaming-simulation and computer-simulation approaches arise as a consequence of three sets of factors:

(a) differences in expectations as to the nature of the outcomes and the manner in which they are interpreted;

(b) differences in the purposes for which the two approaches are used, leading to

(c) differences in the types of situation or problem areas to which the two approaches are applied

Items (b) and (c) are discussed further below.

Gaming-Simulation Techniques

The elements of gaming-simulation—roles, scenarios, accountancy systems and operating procedures—encompass a range of alternatives for use in both construction and presentation of exercises. This section outlines some of the possibilities.

(1) Roles and role playing. There are two aspects to be considered in relation to roles:

(a) role definition:

(b) role allocation.

Roles may be defined to correspond to their real-life counterparts or may be an amalgam of certain interests or interest groups which have selected characteristics in common. Either approach may be used in a gaming-simulation exercise, or the two may be combined. The objective, whichever approach or combination is adopted, is to introduce into the exercise what are seen as the 'key' decision-making groups. The approach to definition may be to provide a minimum framework with points of reference serving as the basis for the development of the role during play. Alternatively, the approach may be one designed to ensure 'consistent' application of a more prescriptive definition throughout an exercise. The definition itself may include statements of objects, values or attitudes inherent in the role, or more detailed activities which are mandatory, appropriate, unacceptable or prohibited. Statistical, financial and other information descriptive of the position for that role may also be included.

Given the definitions, the allocation of roles to players can be undertaken in a number of ways. A player may take a role that matches or is closely akin to his own real-life role. An alternative is the 'role-reversal' approach, where a player is required to play a role other than his real-life one. Where roles correspond to individual decision makers in real-life situations, some attempt has been made to match players' personality characteristics to those of the person whose position they will be playing. In the attempts to simulate the outbreak of World War I, separate runs were undertaken with 'matched' and 'unmatched' personalities,[5] Other attempts to simulate more recent crises by one of the great powers have suffered from the lack of well-informed Chinese communist leaders!

(2) Scenarios. The scenario in a gaming-simulation exercise defines the situation presented to the players at the start of the exercise. The scenario may be presented in two parts—the first providing a framework for the exercise as a whole, and the second detailing points of reference for the individual roles.

The scenario provides information. This may be in the form of written reports, diagrams, maps, physical models, statistical information and financial statements. Normally several of these methods are used in combination, and the information may be provided in manual form, displayed in the area where the exercise is to be held, or made available to roles on specific request.

The scenario may relate to a past, present or future situation; thus the attempt to simulate the outbreak of World War I is situated in the past,[6] the North-East Corridor transport study in the present,[7] and many weapons systems simulations some twenty years in the future.[8] In addition a scenario set in the present may contain some information relating to the past, and also forecasts of the future.

Given the initial scenario, the dynamic element introduced by role playing (whatever the interpretation adopted by the players) may lead to changes in the form and content of the scenario. It is the function of the accounting system to monitor and process the activities of the roles and update the scenario.

(3) The accounting system. The accounting system may present:

(a)   a series of cumulative totals for the exercise as a whole;

(b)   a series of cumulative totals for the individual roles;

(c)   an autonomous model which processes the individual items of information or cumulative totals. This is the only 'model'[9] using the term in its strictest sense— which is employed in the exercise, in that it contains in-built assumptions relating to behaviour and response. In both manually operated accounting systems and computer- or machine-based accounting systems, the assumptions may be open to challenge and discussion, either during or at the close of play. Any changes proposed by the players (and normally a degree of consensus is required) can be relatively easily substituted in the manually operated system. In the computer- or machine-based system, the proposed changes may be more complex, possibly requiring some rewriting of parts of the programme.

(4) Operating procedures. Operating procedures for a gaming-simulation exercise may be discussed under three heads: (a) Procedures for the conduct of the exercise; (b) procedures for roles/role-players; (c) procedures for operation of the accounting system.

(a)   Procedures for the conduct of the exercise. Such procedures are normally concerned with the simulation of time and sequence of play. The basic real-time unit selected (e.g., 1 day, 1 month, 1 or more years) may be treated in one of three ways—time allocated in the exercise may be greater than, equivalent to, or significantly less than the real-time unit. Each simulation time unit is normally described as a 'round' or 'cycle.' Four major types of activity may occur in any such round—planning, decision making, accounting and assessment. These activities may run in parallel or, at the other extreme, may be separated to occupy specified time periods within the round. A common approach links planning and decision making in one time interval, and accounting and review in the remaining time of the round, though many variations are employed.

(b) Procedures for roles/role players. Procedures under this head are designed to direct the attention of players as to when or how specific activities may or are required to be performed and recorded during the sequence of play. Such procedures may relate to economic, social or political relationships between players, and between the role(s) and the changing game environment. The procedures may specify that an activity is undertaken at a particular time and/or recorded on an appropriate form or other display. Whatever the specific nature of the activity, the general aim is to provide links between roles, the scenario and the accounting system in order to disseminate, generate new, and revise the base information contained in the exercise.

(c) Procedures for the operation of the accounting system. Whatever type of accounting system is used (computer-based or manual) the relevance of the new and revised information produced is dependent upon the systematic use of the accounting system. Hence, the procedures must define what information is required for accounting purposes, and the sequence in which (or time at which) it is to be processed. The first requirement can be met by careful design of the forms passed to the controlling team in each round, on which role players record details of decisions or transactions. An alternative approach requires that players undertake some specific action (manipulate some piece of equipment) in respect of each type of decision taken.

Sequence and timing in relation to information transmitted to the accounting system are best defined in terms of role for the controlling team. The main objectives underlying such rules for the controlling team in relation to sequence and timing will be to regulate the inflow and processing of information and to ensure that the feedback to players is undertaken systematically, and in a manner that matches the time scale being used in the exercise.

In practice, the four elements of roles, scenarios, accounting systems and operating procedures are more closely inter-related than is suggested in the above account. (These are outlined schematically in Figure 1.) It is in fact possible to combine certain or even all aspects in one 'presentation.' The greater the degree of combination, the more abstract the exercise becomes. Thus in many recreational games, representation becomes symbolic: in gaming-simulations an attempt is made to move away from abstraction whilst still employing fundamentally similar elements.

Uses and Applications of Gaming-Simulation Techniques

Validation of the effectiveness of gaming-simulation can be discussed only in relation to the purposes for which it is used. Suffice it to say here that the subject is a controversial one, though no more so than that of validating the effectiveness of many other educational and decision making aids.

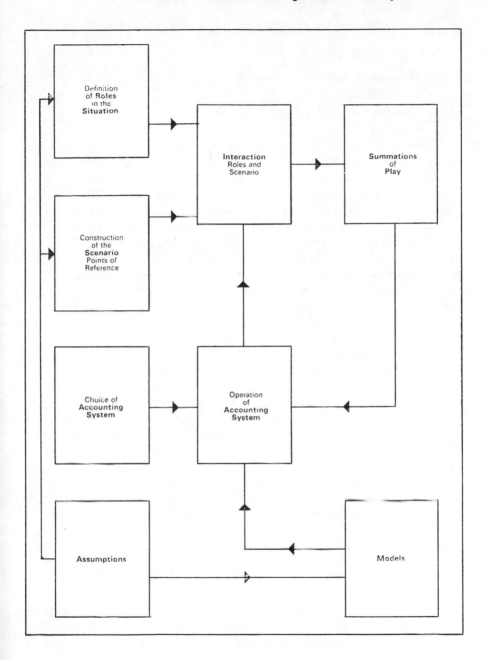

FIGURE 1. Components of a Gaming-Simulation Exercise

There are four main areas in which gaming-simulation is commonly used:

(a) Education and training.

(b) Decision making and policy formulation.

(c) Research.

(d) Operational investigations.

The considerations determining its use in the first of these four areas are different from those applying to the other three. In an educational setting the object is to create an environment within which students may learn about the 'total' situation through the medium of their own activities. With young inexperienced students a substitute for 'experience' is being provided, and as the aim may be to teach particular lessons the exercises used may be highly structured. By contrast where 'experienced' professionals are taking part the aim is not so much to teach specific lessons as to provide an opportunity for exploration of and experiment with situations with which they are familiar. In these cases the exercises are less structured, players having more freedom to direct the course of an exercise.

In considering the remaining three applications, it is necessary to assess the relevance of this approach in relation to other approaches and techniques. Figure 2 provides a schematic classification of techniques, related to the two dimensions of calibration (ability to measure) and rationality (consistency of behaviour in relation to stated or implicit objectives). Given this classification gaming-simulation techniques are seen as making their contribution in the sector bounded by little calibration and much irrationality. In this case the 'results' of gaming-simulation activities are neither likely nor intended to have the quantitative precision of those associated with the sector bounded by calibration and rationality. Exploration of an area by the use of gaming-simulation techniques may lead to the clearer definition of the key elements in a given situation, and ultimately the employment of the quantitative techniques. Such clearer definition may stem from 'results' obtained in one or more of the following areas:

(a) Identification and understanding of the interactions between two or more roles, e.g., initiation of contacts, their timing and purpose, leading to an evaluation of the resulting opportunities for cooperation or likelihood of conflict.

(b) Identification of information requirements and the use of information by the roles.

(c) Identification of the problem opportunity areas created by complexes of decisions which may not always be the direct concern of the roles represented in the exercise.

(d) Exposure of assumptions underlying the decision-making behaviour of key groups in a situation.

These types of result may be obtained from 'insights' gained collectively or individually during the course of an exercise or they may arise as the consequence of post-exercise evaluations, not necessarily immediately following play.

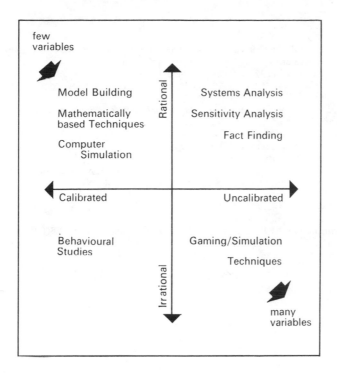

FIGURE 2.   A Schematic Representation of Some Techniques Available in the Stiuations Described by the Two Dimensions of Rationality (consistency of behaviour) and Calibration (measurability).

In some cases the results may be obtained by using an exercise on a number of occasions.

It should be emphasized that results may not always have the same significance for all players. Different players may take part in successive runs of the exercise or the same players may participate in the same or different roles.

There may be occasions on which there appear to be few positive outcomes, either during or immediately following play. However, individuals often 'recognize' in real life similarities with the game situation over a longer period of time. In this connection it is important that the circumstances arising during play are treated as a yardstick and the attempt to transpose game experience directly into real-life situations is resisted.

In this introduction the emphasis has been upon the construction and use of gaming-simulation exercises. There are many situations in which the techniques employed can be abstracted from an exercise and restructured for use in providing agendas for discussion, displaying information and presenting new concepts. Their application in these latter areas is an even newer, but rapidly growing extension of techniques derived from ancient war-gaming.

NOTES

1. Extensions of the Prussian method were developed later by Russia, Britain and the United States of America.

2. Hendricks, F. H. (1960), *Planning Operational Gaming Experiment,* a paper presented to the North California Chapter of the A.I.P. Meeting on 'New Ideas in Planning,' 19 November 1960.

3. Duke, R. D. (1964), *Gaming Simulation in Urban Research,* Institute for Community Development, Michigan State University, East Lansing, Michigan.

4. Feldt, A. G. (1965), *The Community Land Use Game,* Ithaca, New York, Miscellaneous Papers No. 3, Division of Urban Studies, Center for Housing and Environmental Studies, Cornell University.

5. Hermann, C. F. and Hermann, M. G. (1967), 'An attempt to simulate the outbreak of World War I,' in *American Political Science Review,* Vol. LXI, No. 2'

6. Ibid.

7. Abt Associates Inc. (1967), *Northeast Corridor Transportation Games, Game Administrator Handbook,* Cambridge, Mass.

8. Wilson, A. (1968) *The Bomb and the Computer,* Barrie & Rockliff, The Cresset Press.

9. The term 'controlling model(s)' is often used to cover all three aspects of the accounting system. Here the word 'model' is being used loosely.

Part II

ELEMENTS OF DESIGN AND CONSTRUCTION

GAMING-SIMULATION AND SOCIAL SCIENCE:
Rewards to the Designer

CATHY S. GREENBLAT

"**P**laying games" has, in recent years, become a popularized expression pointing to actions designed to avoid a host of things, such as reality, 'meaning,' and each other. Our ideas about games are quite to the contrary; far from thinking of games as ways of hindering communications, we believe that games and gaming-simulations are important learning-communications aids. Some of the argument was presented in the previous section. Here we add the notion that games can be viewed as communication devices in that they are *languages of social theory* to use in "talking" to students, colleagues, and oneself. Where we have previously focused attention upon the more holistic view the *player* can gain from participation, here we urge that the social scientist interested in theory construction may find the translation of his ideas into a gaming model to be intellectually exciting and productive. Design of a game—the systematic translation of understandings into an operating model—and subsequent examination of the model through observation of play of it can lead to a refining of theoretical formulations and consequently to a higher level of social scientific understanding.

REASONS AND REWARDS

Raser (1969: 15-19) has suggested four reasons for simulating:

(1) Economy: "It's cheaper to study and experiment with a simulation than with the real thing, and costly mistakes can be avoided by 'running it through in advance.'

(2) Visibility: "Simulations aid visibility by making certain kinds of phenomena more accessible for observation and measurement, and by introducing clarity into what is otherwise complex, chaotic, or confused."

(3) Reproducibility: "Simulations are valuable because they allow phenomena to be reproduced, and thus enable the experimenter to derive statistical probabilities when the outcome is uncertain, and/or enable him to vary numerous aspects of the system in ways that yield profitable insights into how the system operates. In other words, simulations allow controlled experiments to be made that would otherwise be impossible."

(4) Safety: "Simulations are used for safety purposes, both to protect human beings while they are being trained or studied, and to produce laboratory analogues of dangerous phenomena that we need to study."

Simulation brings with it several kinds of rewards. All focus upon the idea that for a number of kinds of problems, a gaming-simulation may be a more productive way of conceptualizing elements and relationships, whether one's purpose is teaching or refinement of theory. Many people who have worked in several media have found that gaming models represent happy middlegrounds between the looseness of verbal tools and the rigidity of mathematical formulae. Thus Feldt (1972: 1) argues that

> the advantages of gaming models over verbal and mathematical models are that a game provides relatively specific designation of concepts and at the same time avoids the semantic problems inherent in most verbal models. Furthermore, the degree of precision required in defining a concept and its relationship to other components of the model is much less than that required in most mathematical models. Thus the game model builder may deal with relationships and components which are not easily quantifiable.

The rewards of building a gaming-simulation model rest upon several crucial features of the process:

(1) The process of design forces greater clarity in thinking about critical elements, and pushes one to think at various levels of abstraction.

(2) The design process forces a search for concreteness; it demands an explication and articulation of theory and of conditions under which relationships hold beyond what is usually demanded of the social scientist.

(3) To make the game work, the social scientist must develop an overall or systemic understanding of his topic; thus he is pushed to synthesize.

(4) The model, once developed, can be experimented with readily, and thus serves as a fruitful tool for exploration of theory.

Let's treat these ideas one by one.

CRITICAL ELEMENTS AND LEVELS OF ABSTRACTION

Like a book, a simulation is an explicit statement of what the designer believes about some aspect of reality. The "referent system" simulated is not "reality" itself, but, rather, a set of theoretical propositions about reality. The design process thus entails delineation of theory, construction of a conceptual model, and translation of this into an operating model or simulation. If the translation is good, the gaming-simulation should operate in much the same way that the real world system operates.

There are two critical points here: first, it is not the whole system that is modelled, but rather selected elements. And second, the model doesn't need to *look like* the referent system, but should *behave* like it. The decision to simulate thus forces the designer to decide what the essential features of the system are. He must abstract from reality, making explicit what has often been implicit. He has likewise to decide upon the appropriate level of abstraction for his purposes. His job is to transform and substitute—to find substitute mechanisms and surrogate functions. The solutions can take quite different forms: some substitutions represent reductions in scale, while others entail the introduction of properties that are different in form but that produce the same *consequences.*

The process of thinking about *what it is* that really is important and how to represent it so the system will "work" may foster new insights, for one must think at different levels of abstraction.

Players in CLUG (the COMMUNITY LAND USE GAME), for example, are asked to bid for land, decide upon the location and extent of public utilities, set tax rates, construct buildings, decide upon renovation of deteriorated buildings, etc. Of course, they are only "playing" at these things; they don't bring in bulldozers or moving men to displace tenants in urban renewal areas; but the game asks them to make the same decisions and to go through the same steps the real-life developer or planner goes through: it is a *reduction in scale.*

Likewise, in some international relations games, such as INS, DANGEROUS PARALLEL, or CRISIS, "diplomats" confer with their ministers, examine indicators of social, political, and economic states of their countries, meet formally and informally with delegates of other "countries," etc. Again the parallels between the real-world and simulation behaviors are quite close; the latter are abstracted from the large number of things done by politicians and reduced in scale.

In other games, on the other hand, it is the *function* rather than the *structure* that is considered important. It may be the feelings or attitudes experienced by real-world role players that one wants to simulate in the model, rather than the specific mechanisms that create them. In STARPOWER, for example, players accumulate points through trading and bargaining with little colored chips. Surely this is not a familiar or regular activity for most players, yet the mechanism of trading chips generates the desired outcome variable: differential success in the "marketplace."

In HORATIO ALGER, a simulation of problems of social welfare, some players are allowed to work each round while others are only periodically permitted work roles. To generate the frustration of nonworkers necessary to make the system operate as the real-world poverty-welfare cycle works, the designers have included a "work" element: those who are employed are given tinker toys and told to construct something. It's amazing how alienated players who have no tinker toys to use become—and I'm speaking of college students, faculty, and other adults! Obviously, the verisimilitude of playing with tinker toys vis-à-vis going to work is relatively weak; yet the device makes the simulation operate much like the simulandum. I worked recently with some students designing a game which includes differential work rewards for those with differing amounts of education. They are planning to offer different types of "jobs," and they are borrowing the tinker toy work idea and expanding upon it. Some players each round will be told exactly what to do—their task will entail building units of small parts of a larger whole. Others who acquire more education in the game will be given much more independence and creativity in what they can do with their tinker toy sets. In such a way, we hope to generate some of the different work-related feelings engendered in real-life work situations, and to try to trace their consequences.

In BALDICER, a simulation concerned with population explosion and limited capacities to deal with it, an important variable is differential ability to generate enough produce to feed the nation's people. Each player has responsibility for feeding the population of a country of a specified population. At one point in the round, he or she is given a sheet on which to write "PUSH, PULL, DIG, SWEAT" as many times as he can in thirty seconds. The number of completed phrases equals the number of thousands he can feed that year without obtaining outside assistance. Obviously, this is a far cry from planting, fertilizing, harvesting, etc., but it symbolically represents differential capabilities, generates scarcity for some, and propels players to examine interdependencies.

Finally, in a gerontology game, designer Fred Goodman wanted to simulate the decreasing physical-geographic mobility that comes with increasing age. He rejected mobility "points" to be forfeited and decided to use ropes. Players are tied to their chair legs, with differential lengths of rope representing the amount of mobility they have. Some people then can get around more than others, and, as players age, their ropes are shortened. Homologue, rather than analogue, is employed to communicate the critical elements of restricted mobility.

THE SEARCH FOR CONCRETENESS

The second point made above was that game design demands that one think quite concretely about the system. Constant questions about relationships are generated as the designer works to establish linkages between game elements.

Again, an example will perhaps make this clearer. Last year some of my students decided to build a "college game" to show the options available to students in deciding how to spend their time, and the short- and long-term consequences of the various alternatives. It was something of a variant of the LIFE CAREER GAME and GHETTO. In the initial process of design, they allowed players to 'invest' time in studying, social life, etc., and gave them rewards of different types for these activities. But then the question arose of long-term consequences. Was it just the degree, for example, that was important in attaining a job? Or did differential academic success (an A average as opposed to a C average) have differential consequences for obtaining a job as well as for graduate school admission? And what *was* the relationship between grade point average and probability of admission to graduate school? The same kinds of questions arose concerning social life. Soon they started looking in books and journals for data on the relationship between college grades and jobs attained, salaries, etc., and for the relationship between extent of dating and age of marriage for college graduates. The demand for concreteness stemmed from the demands of game design.

In another such exercise, a group of psychiatric nursing students began design of TERMINEX, a game to sensitize doctors to some of the problems involved in dealing with patients with terminal illnesses. Their thesis was that different "types" of patients should be told of their state in different ways: some should be told outright, others more slowly, etc. Soon after beginning to construct the conceptual model and the gaming-simulation elements, however, they found it necessary to specify quite directly just what type of strategy was appropriate with what "type" of patient. The general proposition had to be translated into quite specific statements in order to build payoff matrices for strategies employed for each patient type. Again a literature search for data was undertaken.

In the process of design, then, unanswered questions are often unearthed. Again and again one must ask "what happens in this case?"

THE SYSTEMIC UNDERSTANDING

Closely related to this demand for making critical elements explicit and making conditions and relationships concrete is the third point: in order to make the game work, the designer must develop an *overall* or *systemic* understanding. He has to ask not only about definitions and specific linkages, but also about overall interconnections among roles, goals, resources, and rules. How do they fit together? Simple examples of this cannot be offered, but the reading selections on the development of SIMSOC and THE MARRIAGE GAME illustrate the process of going from theory to operating model, trying to integrate the various elements. The nature of the challenge is conveyed by Gamson (1971: 306-307) in the following quote from the selection which appears in the readings section:

> I have found the process of continual modification of SIMSOC a peculiarly absorbing and rewarding intellectual experience. It has a concreteness and immediacy which is lacking in much of the intellectual work I do. If I am trying to understand why, for example, a particular social movement developed at one time and not another, I will usually struggle through a sequence of questions until I achieve some vague sense of closure. This suffices until enough questions have been raised about my explanation to give me again a vague sense of uneasiness.

> The process is somewhat different with the development of a simulation. It is as if I have a complicated Rube Goldberg device in front of me that will produce certain processes and outcomes. I want it to operate differently in one way or another but it is difficult to know where and how to intervene to achieve this purpose because the apparatus is delicate and highly interconnected. So I walk around it, eyeing it from different angles, and imagine adding a nut here or a bolt there or shutting it off and replacing some more complicated parts.

Each of these interventions must take the extremely specific form of a rule. To intervene, one must play a mental game in which the introduction of every specific change must be weighed in terms of how the whole contraption will operate. Such mental games force one to develop a clear picture of what the apparatus looks like and why it operates as it does. Each hypothetical alternative must be put in place and imagined in operation. Finally, an explicit choice has to be made and the game actually run under the altered rules, and one has a chance to discover whether he really understands the contraption or not. When a rule-change has the effect it is supposed to have, the experience can be very exciting—as exciting as predicting a nonobvious outcome in any social situation and having it turn out correctly.

EXPLORATION AND EXPERIMENTATION

Creation of the model, subsequent play with it, and systematic manipulation of parameters may unearth new questions and provide a means of exploration and experimentation. These topics will be explored in greater detail in the chapter on uses of gaming-simulation for research.

The possibilities for experimentation via a simulation are increased if one considers that it is possible to simulate a *proposed* or *hypothetical* system as well as an existing "real" one. A colleague, for example, recently showed me a "Utopian, futuristic" paper he wrote. The two questions it had raised from readers were "But how would we ever get there from here?" and "It sounds good, but would it work?" By incorporating the paper's ideas into a gaming-simulation and manipulating it, he might be able to find some tentative answers to these questions. Gaming-simulations, then, may permit us to hypothesize more realistically about "unreal" but hoped-for states such as nonviolence, absence of war, recognized interdependence, etc.

FURTHER ELABORATION AND EXAMPLES

Some of the selections offered in the reading section describe different elements of the values and rewards for the social scientist who designs a gaming-simulation. The Greenblat and Gamson papers already referred to offer examples of the process of translation of theory to gaming-simulation. In "Social Processes and Social Simulation Games," James Coleman outlines general dimensions of the simulation process. And in "Sociological Theory and the 'Multiple Reality Game,'" some alternative design approaches are suggested.

REFERENCES

Feldt, Allen
    1972   "Operational Games as Educational Devices." In CLUG: Community Land Use Game, Player's Manual. New York: Free Press.

Gamson, William A.
    1971 "SIMSOC: Establishing Social Order in a Simulated Society." Simulation and Games 2 (September).

Raser, John
    1969 Simulation and Society. Boston: Allyn & Bacon.

THE GAME DESIGN PROCESS

RICHARD D. DUKE

INTRODUCTION

The game design process has perhaps been reinvented as frequently as the wheel; combining the two *may* be unique. In any event, the process as presented in Figure 1 illustrates at least the major considerations that the game designer must confront.

The game design "wheel" is derived from the perspective of gaming as a communications device. Hence, the emphasis on thoughtful review of alternative modes of presenting the material (should you really use gaming in this instance?); the concern for clear specification of *what* is to be conveyed; and the emphasis on the evaluation of the completed product (whether the players benefit as intended).

The process is iterative in actual practice, although it usually entails some "muddling through" by the designers. Nevertheless, it is important to visualize the process in its entirety, to anticipate each phase from the outset of the process.

There are four distinct phases to game design:

(1) During the *initiation* phase, the designer analyses the communication problem at hand. This requires taking into consideration the nature of the client, intended use of the product, the audience, the subject of the message and exactly what communication purpose is to be served (e.g., questionnaire, interdisciplinary dialogue, training). Pragmatic constraints of cost and time must also be considered. On the basis of this analysis, an appropriate medium for communication must be chosen.

(2) If the choice *is* gaming-simulation, the designer continues around the wheel to the *design* stage. A conceptual map must be developed (i.e., a general statement of the ideas to be conveyed through the game). A review of the state-of-the-art of gaming—that is, of the existing repertoire of gaming techniques—helps determine the nature of the game being designed. The basic

AUTHOR'S NOTE: *This paper has been elaborated in far greater detail in* Gaming: The Future's Language. *Beverly Hills, California: Sage Publications, 1974, especially Section III.*

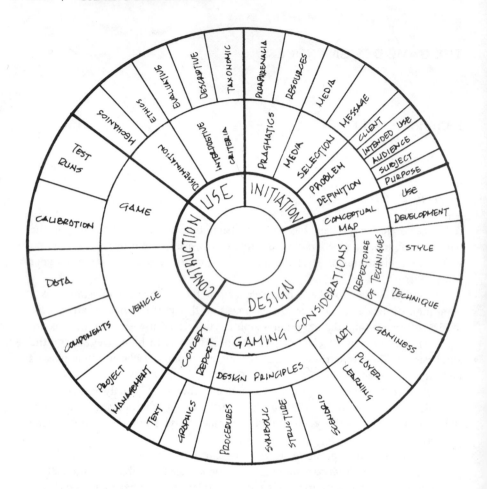

FIGURE 1. The Game Design Process

components (scenario, symbolic structure, and procedures) are developed in a concept report which outlines the form the game will take and the work that must be done to bring the game to completion.

(3) During *construction,* the gaming vehicle is created and appropriate data loaded, calibrated, and tested. Finally, the game is reproduced, operators are trained, and the game is put into operation.

(4) The phase of *use* of the game entails a set of responsibilities of designer to operator, operator to player, and user to designer. Interpretive criteria for the classification of the game should be provided by the designer. These four phases are now presented in greater detail.

FOUR PHASES OF GAME DESIGN

Initiation

At the present stage of development in the field of gaming, most if not all games represent "happenings" rather than the products of a deliberate design process. One prominent designer insists that his games convey no "message"; rather, he contends, they are free and ethereal situations which the participant can direct from 'inside' the game. The function of a book may well be to establish creative or innovative thought by the reader, thought which may go well beyond the specific content of the book. Nonetheless, the author of the book must intend some message or purpose (e.g., to inspire creative thought) before a coherent book can be assembled. Similarly, the author of a game must have an express and coherent purpose or "message" to guide the construction of a game. Only the clear articulation of this purpose permits the rational selection of gaming as the appropriate communication mode.

Equally important to the clear expression of message or purpose is the careful definition of the intended audience. Surely Dr. Seuss's formal written transmission to other adults would not be recognizable to the children who love his popular books. The game is more occasion-specific than any other form of communication; it is imperative for a game designer to have in mind his intended audience, their motivations for participation, and the typical conditions of use.

The more precisely the designer has articulated the message or purpose of the game, audience characteristics and motivations, and the pragmatic considerations controlling the use of the product, the more effectively various design considerations can be employed to ensure a successful game. This will also give some assurance that gaming is an appropriate media choice.

Design

Game design is a combination of (a) mimicry of existing game formats and styles; (b) an elusive but real 'art'; and (c) design principles.

Mimicry of existing games styles. There is a marked tendency for prominent gamers to have an identifiable style. This is so pronounced that it is common parlance among the gaming set to identify and/or quickly explain a game by indicating that it is a typical "Professor Jones game," inserting one of perhaps a dozen well-known names. Unfortunately, the neophyte is likely to ape the "Professor Jones" of his acquaintance but somehow fail to reflect the sensitivity of the master in its new application. The beginner would do well to have mastered as broad a repertoire of games as possible emphasizing different authors and different subject areas, if the full benefit of the past experience of the profession is to be enjoyed.

The "art" of gaming. The "art" of gaming, like any art, is probably best learned from experience with the medium; nonetheless, one clue can be offered as to its nature. All games are basically iterative in their structure, reinforcing the hunch that this somehow improves learning potential; this is probably achieved by both successively defining the totality of the problem in increasing detail and by positive reinforcement through repetition.

Design principles. Principles of design are categorized here into three general areas:

(1) Use of symbolic structure: A game is really a "language" which entails the integrated use of (an) existing language(s) as well as a 'game-specific' language designed for the particular game. (A language is defined as a shared symbol set subject to conventions of use.) These definitions suggest that two levels of skill are required of the game designer; first, the clear articulation of the game-specific language to ensure rapid and effective player use; and, second, careful integration of this new and unique language with each of the other modes of communication employed in this particular game. Any symbolism which is unique to the game represents a hurdle to the players until it has been assimilated. New symbols should only be introduced for specific purposes and players must be trained in their meaning.

(2) Use of scenario: The use of scenario seems to parallel its use in a novel or in the performing arts (legitimate theater, movies, TV). In each case, it becomes an integral part of the technique for conveying the "story" or plot. In gaming, the extent and character of the scenario employed is governed by the considerations indicated above under "initiation." Probably no single failure in game design is more common than an inappropriate use of scenario (too complex or simple, inappropriate to the audience, timing unsuccessful), and this can inevitably be traced to either an initial lack of clarity by the designer relative to purpose, audience, and condition of use or to the insensitive or inappropriate use by the game operator.

(3) Game procedures: There has been an undue emphasis on *rules* in gaming-simulation, perhaps as a result of the strong heritage of game-theoretic applications. A much more productive concept is "procedures," intended as a more flexible term to cover all mechanics of play, including any essential rule structure. Because the game may be viewed as an "environment for learning," it is essential that players be able to interact with the game, often in ways not initially perceived by the designer. In so doing, they may feel it necessary to change the structure of the game. Because of game design considerations, certain conditions may be inviolable (e.g., the requirement of cycles as iterative experiences or the calculations inherent to a particular model). These may well be called "rules" implying a necessary finality or rigidity. On the contrary, if the players are permitted or encouraged to alter, amend, or enrich procedures within the basic gaming structure (e.g., moving from a non-existent definition of acceptable player behavior in the game through rudimentary and finally to an advanced articulation in successive cycles of play), we can maximize learning without the labored and unnecessary specification of an elaborate rule structure.

Conceptual mapping and the concept report. Gaming is best understood as a communication form; each game is very specific to some precise communications purpose. The relatively high cost of the technique argues for precision in the design and parsimony in the construction and use of the game. Two powerful tools are available to meet these objectives—conceptual mapping, and its interpreted expression through the gaming technique, the concept report.

Games are most frequently employed for conveying complex systems. It is readily acknowledged that the system may not be understood at the time of game design, and that the basic purpose of the game may be to either extract concepts from a knowledgeable audience, or to assist some research team in the articulation of the system. In each case, the basic objective of the game should be expressed as a conceptual map—an internalized, organized, "gestalt" comprehension of the complex reality being conveyed. Failure to achieve this will most likely lead to an ultimate lack of precision in the game product.

The articulation of the conceptual map permits the systematic review of gaming considerations (repertoire of techniques, art, theory, and principles of design) to achieve the most effective game design for a particular communication purpose. Having achieved this end, it is highly desirable to commit this to writing (text, diagrams) in the form of a concept report before beginning construction. This serves two purposes: (1) it becomes a blueprint to guide the construction process; and (2) more important, it becomes the basis for evaluating the final product. If no documentation of purpose or objective exists, and no coherent review of gaming considerations is presented, the final game will not be subject to intelligent evaluation.

Construction

Many professional gamers like to bypass the first two sectors of game design and jump headlong into construction. As a consequence, there may be a failure to achieve precision of design, the careful engineering of the construct to meet some precise communication need, and almost certainly a loss of parsimony in the construction activities themselves. Nonetheless, there are two reasonable explanations of this tendency to begin in the middle. First, the concept of games as a problem-specific language yielding to orderly rules of design has not been generally recognized; and second, it is hard work to answer all the questions that must be raised to establish detailed game specifications, and the difficulty is compounded by the element of risk inherent in making a written commitment which might later serve as an indictment of the author. The process of constructing a game inevitably forces the designer to confront these questions anyway, although in a less systematic way. Part of the art of good game construction lies in the ability to make a simultaneous solution of many variables. And so, as recognized earlier, we can expect the designer to benefit from an orderly and sequential concept of game design, even though in practice many liberties may be taken.

If a gaming project is of any magnitude, various project management mechanisms may be required. Note that if proper care has been taken in developing the concept report, construction can be a highly organized and efficient process employing standard management practice. If, however, construction is coterminous with an effort at determining objectives and the methodology to be employed, at least marginal chaos is to be expected.

Construction, in either case, entails identifiable components (boards, paraphernalia, models, etc.), the collection of data required for loading (if necessary) and their joint assembly into an initial (usually rudimentary) game. This must be carefully subjected to calibration and evaluation before release for professional use. Countless hours of participant time have been squandered with immature game products. These often result in the unnecessary aggravation of a captive audience and sometimes bring about the permanent alienation of some players from the gaming technique. This phase, even more than those preceding it, requires that attention be deliberately redirected to the basic game objectives in order to maximize the final fit of the game.

Use

Earlier, many efforts at game design were described as "happenings." If there is a lack of precision in the design of games, it is unfortunately doubly true of many game runs, even though the game itself is cleverly conceived and carefully executed and tested. What is the underlying cause?

If we substitute "book" for "game," we have a clue to this failure. Very sophisticated systems exist and are in routine use which enable "book" to be identified and secured (readily finding the 50 that might be relevant to your problem out of a library of perhaps 1,000,000). Common practices allow reasonably precise evaluation on the grounds of appropriateness, validity of content, level of presentation, etc., simply by examination of the physical object "book" (this examination may be superficial at first and become more intense as the search narrows to a few books that might be appropriate. At this stage, several passages, even chapters, might be read to complete the evaluation).

Returning to "game," we find no useful parallels—whoever understood the *game* of Monopoly by reading the dreary and endless rules presented in fine print? In short, we have not developed, as a profession, interpretive criteria that are in common usage. The gamer must trust to luck, hoping to improve his batting average through experience and personal contacts.

Three interpretive criteria must be established as routine convention among gamers: (1) a taxonomic system must be employed for filing purposes. This could simply be done by endorsing some existing system currently in use for books, such as the Dewey Decimal System; (2) a brief written abstract in standardized format should be used to reveal design specifications as well as the author's purpose, subject, and intended context of use. Remember, the game is a highly specific communication device—if the author didn't know where *he* was aiming, how do *you* know what you will hit? (3) Finally, standardized procedures for evaluation must be formulated, to assist reviewers, whether for institutional endorsement as in a standard class use of the product, individual selection for one time use, or for game review purposes in journals.

During both the design and construction phases, mechanics and ethics of dissemination must be considered. Mechanics (packaging, training operators, etc.) are straightforward, but important. Final effective use of the game will, in many cases, be constrained by mechanics and associated costs. These should be thoughtfully anticipated since problems often yield to sophistication of technique.

Of increasing urgency is a sense of ethical responsibility for the product, both in its design and use. Academic tradition and legal precedent are probably adequate for any commercial versions that may be involved (e.g., is the product in the public domain, or private property? Has adequate courtesy been given to those whose games were aped?). These are the responsibility of the designer and yield readily to standard convention.

New ethical problems emerge in the *use* of a game, and these are always the responsibility of the operator, although in many cases this responsibility must be shared by the designer. If the designer's conceptual map is a speculation, an alternative to be explored, should it be presented as reality? What are the obligations to the player to avoid injury that occurs beyond the game environment? Ethical questions of gaming are only now beginning to emerge—each designer shares the responsibility for careful thought.

CONCLUSION

Once the game design process is understood as a "gestalt," it can be employed in discontinuous fashion and at various levels of detail. The four phases discussed above are of one fabric, and the art of game design requires their simultaneous solution.

The readings included in this section have been selected to elaborate upon and exemplify the points made above. The selection "Specifications for Game Design" offers a concrete set of questions the game designer should address, and Feldt and Goodman's "Observations on Game Design" offers some suggestions about ways to answer some of the critical questions. Two examples are offered by game designers who describe the process of design as they worked it through: William Gamson in "SIMSOC: Establishing Social Order in a Simulated Society" and Cathy Greenblat in "From Theory to Model to Gaming-Simulation: A Case Study and Validity Test." Greenblat's other paper describes a means of making games correspond more closely to reality by including a "multiple realities" component where appropriate. Finally, Coleman addresses the general design considerations relevant for games that simulate social processes.

FROM THEORY TO MODEL TO GAMING-SIMULATION:
A Case Study and Validity Test

CATHY S. GREENBLAT

T hose involved in the design, use, and evaluation of gaming-simulations agree on neither terminology nor taxonomy. I propose to stay out of the fray by offering a working definition with no intention that it serve as a seminal statement. For present purposes, then, a simulation is an operating model of central features of a system; that is, it shows *functional* as well as *structural* relations. Some simulations are operated by a computer; Monte Carlo or other techniques are employed to generate decisions or actions, and the consequences are yielded by the simulation model. Other simulations are operated by human players, thus saving the designer from the need to build in psychological assumptions. The actions of participants in these latter simulations are those of game players—that is, they perform a set of activities in an attempt to achieve goals in a limiting context consisting of constraints and of definitions of contingencies. The hyphenated term "gaming-simulation," then, seems to best represent this hybrid form in which game activities are performed in simulated contexts (see Raser, 1969; Greenblat, 1971, for elaboration).

Like a book, a gaming-simulation is an explicit statement of *what the designer believes* about some reality he is attempting to simulate. The referent system, you will note, is not "reality" itself, but rather a set of theoretical propositions about reality. The process of design thus entails (1) delineation of the theoretical base, (2) construction of a conceptual model of the system, and (3) translation of this model into an operating model or simulation (for elaboration of these points, see Inbar and Stoll, 1972; Duke, 1974).

Several types of validity of such a model can be discussed, depending upon the use to which it is to be put. For heuristic purposes, it may suffice that the model has high "face validity"—that is, it "feels" or "seems" to players to be like the referent system, although there may in fact be serious distortions. Solar system models in planetariums typically are accurate simulations of the relative *size* and *velocity* of the planets, but not of their differential weights, since the latter is usually not taught via the models.

For research purposes, theoretical and empirical validity are probably more crucial; that is, more concern must be expressed with the criteria of inclusion and exclusion in going from theory to model to operating model; and more concern must be expressed with the way in which the behavior of players actually simulates the real-world behavior of their counterparts. For planning purposes, too, empirical validity is important.

Sarane Boocock (1972: 30-31) has described the types of validity most relevant to game models:

(1) *common sense or face validity:* the extent to which the game structure or outcomes "seem" to reproduce the simuland. This corresponds to face validity, logical validity, or the logical approach as described in the general research literature, and to verisimilitude in the gaming literature;

(2) *empirical validity:* the closeness of fit of game structure or outcome to other measures serving as "criteria" of the phenomena under study. This is similar to or includes what the literature calls predictive validity, event validity, variable parameter validity, the empirical approach, and comparison of gross outcomes;

(3) *theoretical validity:* the degree to which the game structure or outcomes conform to some theoretical or logical principles. This is close to, though not synonymous with, construct and hypothesis validity.

Most gaming-simulations have been examined only in terms of their face validity, although a few studies report extensive attempts to ascertain empirical and/or theoretical validity (cf. Russell, 1972; Coplin, 1966; Guetzkow, 1968).

The purpose of this paper is to describe the process of going from theory to conceptual model to a gaming-simulation of marital decision-making: THE MARRIAGE GAME: UNDERSTANDING MARITAL DECISION-MAKING (Greenblat, Stein, and Washburne, 1973). Through this case study, the theoretical validity of the model can be examined.

THEORETICAL ROOTS

THE MARRIAGE GAME emerged from the belief that marital decision-making has to be understood in terms of two sets of elements: those in the society at large, and those in the individual making the decisions. It posits that marital interactions take place between individuals who live in a world of external social facts, many of which have been internalized, affecting conceptions and values.

The elements and relationships of the theory are summarized in Figure 1. The left side indicates that the opportunities available to system members derive from two elements: (1) general system states (reflected in unemployment rates, health conditions, levels of technological achievements, etc.) and (2) generally accepted definitions of categorical identities of individuals (such as age, sex, marital status, etc.) Thus, job opportunities vary with the overall state of the economy, and the opportunities for males and females, young and old, Black and white, often differ. These opportunities carry with them objective costs and rewards; occupational status-roles, for example, bring resources of property, power, and prestige given to all who occupy them.

The right side of the diagram indicates that individuals living within the society possess (1) categorical identities (age, sex, race, etc.) defined by the society as good and bad, better and worse; and (2) sets of values resulting from prior experiences or "biographies." The individual's choices from the opportunities provided by the society reflect both the identities and the personal value-frame. A woman choosing a job chooses from a limited set available to her *because* she is a woman, and chooses in terms of values she has developed through her life's experiences. The subjective costs and rewards experienced from this choice result from both the objective costs and rewards attached to the opportunity and the extent to which these types of costs and rewards are valued by the individual. Thus, the occupational prestige received by two people *objectively* may be the same, but be experienced by one as much less because of the lower value he attaches to prestige. The subjective costs and rewards received from new experiences may also modify the value-frame used to make future decisions.

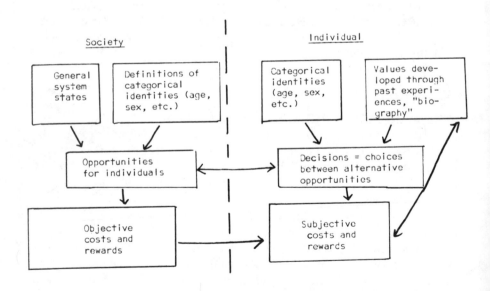

FIGURE 1. Theoretical Basis of the MARRIAGE GAME

BUILDING THE CONCEPTUAL MODEL

In ideal terms, the design process of a gaming-simulation begins with the move from theory to delineation of a conceptual model; when this has been completed, the designer is ready to move to considerations of how to translate this model into a gaming format. In actual history, however, much of the process of conceptual modelling is an ongoing one, interwoven with the construction process. Elements are added to the model as the analogues, homologues, and iconic elements built into the vehicle force the designer to acknowledge the incompleteness of his understanding or formulation. At a later stage, test runs with the game point out additional missing elements or elements imperfectly linked with others (cf. Gamson, 1971).

The discussion to follow thus somewhat bends the truth; the game design was a heuristic process for us and taught us much about the marital system and its operations that we did not realize when we began. The model to be presented, then, was mostly a precursor to the game; to some extent, however, it was emergent from the design process.

Our concern was to focus upon the decision-making elements of marriage, their consequences, and their interconnections. The model we developed was an exchange theory model which, roughly stated, holds that each decision made by an individual will accord him certain costs and certain rewards. Three basic elements of the model therefore had to be defined and the linkages specified: (1) major decision points; (2) the units of cost and reward involved in each; and (3) the factors accounting for similarities and differences in the amount of cost and reward experienced.

Factors Affecting Costs and Rewards

The third decision was critical and decided first; the relevant categorical identities we enumerated included age, sex, marital status, parental status, social class, race, subculture, and region. It was clear at an early stage that considerable limitation of these sources of diversity of opportunity and rewards would be required if the game were to be playable. Some reduction of diversity was accomplished through assignment of characteristics to players, but this role-casting solution was kept to a minimum in the belief that, if extensive, it would decrease the pedagogic value of play. Age and marital and parental status sources of diversity are initially limited through role-assignment. Diversity in marital and parental status derives from subsequent decisions. All players are told that they are single and childless as the game commences, and that they are at an age at which their culture brings pressures to marry and rewards and punishes in accordance with conformity to this pressure. During the course of play, they must decide whether to marry, stay single, or live together without marrying. Once married, periodic decisions must be made concerning continuation or termination of the relationship. Varying costs and rewards are assigned to those of different sex, marital, and parental statuses at relevant points in the game.

Diversities arising from class, race, and regional variations have been handled in a quite different way. The model specifies the nature of the decision areas and the units of consequence, but is data-free. The *basic* game has been constructed to represent the situation we believe pertains for middle-class college graduates in the metropolitan areas of the United States, and variability within this context is even further minimized, for this population is defined fairly homogeneously for game purposes. In a later chapter, substitute matrices, tables, and rules are provided, along with directions for use for those who wish to alter the parameters and create variations. Thus one can represent the marital system of the lower class or the working class. Regional and racial variations that diverge from these class variations can be added by the interested teacher or student with minimal effort. Other parameters, such as the sexual freedom present in different subcultural groups, can likewise be varied through alteration of matrices, creating different constraint and reward systems. These can be sequential alterations from one run to another, or participants can play from different versions at the same time to simulate a more heterogeneous society.

Finally, variations stemming from differences in personal value-frames are put into the model and brought into the game through the device of requiring players to weight each of the units of cost and reward, yielding a final weighted total score for each. Hence there is no automatic route to success in the game. Higher total scores reflect greater success in maximizing those subscores seen by the player as most important.

Decisions, Costs and Rewards

We turn then to the first two considerations: decision points and units of cost and reward. We have reduced the many units of cost and reward present in marital systems to roughly quantifiable ones. There are seven kinds of points a player can gain as a result of the decisions he or she makes in the game: security points, respect and esteem points, freedom points, enjoyment points, sex gratification points, marital and parental status points, and ego support points. As noted earlier, players must assign value weights to each of these at the beginning of each round. We believe these reflect the differential importance people attach to the specific units. The differential importance attached to the whole marital system can be brought in only by the individual player confronted by the need to evaluate and interpret his final score. This aspect of the model is conveyed to the players by telling them:

THE MARRIAGE GAME has no winners or losers because it is a simulation of real life, and in real life the relative contribution of marital rewards to overall satisfaction with life is different for different people. Those whose rewards come primarily from their marriages require higher marital "scores" or greater marital satisfaction than do those whose marriages constitute only one of several sources of satisfaction, and thus may be willing to sacrifice some marital rewards for rewards in other areas. For example, they may pursue occupational goals at the expense of their family lives. Therefore it is difficult to make direct comparisons between marriages. As you play THE MARRIAGE GAME your goal should be to maximize those rewards you most value rather than to 'beat' other players [Greenblat et al., 1973].

The decision points in the model represent those we considered most critical over the course of a year of marital interaction: these include decisions concerning jobs, basic budget expenditures (housing, maintenance, transportation), allocation of free time, sexual relations, use of contraception, luxury purchases, interaction with others, giving of ego support, deciding upon marital status for the next year. In real time, of course, these are not made sequentially but from day to day. The simulation design posed constraints in the form of the necessity to calculate consequences, leading to the need for a time-frame which would permit such calculations. One round of play of approximately fifty minutes represents one year of simulated marital interaction. In approximately six and a half hours, then, a player can experience a seven-year sequence of courtship, early marriage, and early parenthood. The time compression is thus of a magnitude of approximately ten thousand to one. This creates some distortions, of course, but it also is the source of the vehicle's pedagogic promise—it allows the experiencing of the general nature of the marital decision-making process in a much foreshortened time.

Figure 2 graphically represents the conceptual model for THE MARRIAGE GAME. Factors leading to variation in opportunities and/or costs and rewards are indicated on the left. In the center are the decision areas, presented in the order they must be dealt with in the game. On the right the nature of the cost and reward consequences of such decisions are indicated.

OPERATIONALIZING THE MODEL

This model has been operationalized in the participants' manual in the form of instructions and matrices indicating the choices available and the consequences for players with different identities. These cannot all be discussed here: one example should indicate their general character. Figure 3 shows the job options charts for men and women in the basic (i.e., middle-class) version of the game. Notice that costs and rewards are affected not only by the player's sex, but by his or her marital status, parental status, and prior savings. (These factors were stated earlier in Figure 2.)

The male role player selects a full-time job, a full-time and part-time job, a part-time job, or no job at all from the first chart and records the consequences. The female role player makes her selection from the second chart. Charts and/or instructions are available for other decisions.

Factors affecting opportunities and/or costs and rewards: Sex · Marital status · Parental status · Other

Decision areas

1. JOB CHOICE
2. GENERAL EXTERNAL EVENTS (e.g. sickness)
3. FIXED COSTS PAYMENT (e.g. repayment of debt, taxes, alimony)
4. BASIC BUDGET DECISIONS
   A. Housing
   B. Maintenance
   C. Transportation
5. INTERACTING WITH OTHERS
   A. Allocating free time
   B. Sexual relations
      1. Partner:
         a. spouse
         b. non-spouse → chance of VD
      2. Amount
      3. Contraception → chance of conception
         a. Yes
         b. No → birth → abortion
   C. Luxury purchases
   D. Choosing primary partner
   E. Comparing with others
6. ASCERTAINING QUALITY OF SEX RELATIONS
7. RATING PRIMARY PARTNER AND RELATIONSHIP
8. CALCULATION OF SCORES: OBJECTIVE AND SUBJECTIVE
9. DECISION: MARITAL STATUS FOR NEXT ROUND

Other possible consequences / Cost-Reward consequences: Security · Esteem · Freedom · Sex gratification · Enjoyment · Marital-parental status · Ego-support

Other factors noted: Savings; Occupational status; Overall quality of relationship

FIGURE 2. The Conceptual Model for the MARRIAGE GAME

Chart 1: Job Options for Men

Option	Income	Freedom Points	Esteem Points	Other Effects
Full-time job	$10,000	100	100	—
Full-time and part-time jobs	16,500	33	90	Forfeit 100 sex gratification cards.
Part-time job only; less than $3500 savings	6500	133	65	—
Part-time job only; $3500 or more savings	6500	133	135	—
No job; less than $10,000 savings	0	200	50	—
No job; savings of $10,000 or more	0	200	150	—

Chart 2: Job Options for Women

Option	Income	Freedom Points	Esteem Points	Other Effects
Full-time job:				
If no children	$6500	100	65	—
If children	6500	100	55	—
Full-time and part-time jobs:				
If single	9500	50	65	Forfeit 100 sex gratification cards.
If married, with no children	9500	50	55	
If married, with children	9500	50	45	
Part-time job only:				
If single	3000	150	60	—
If married	3000	150	65	—
No job:				
If single	0	200	55	—
If married	0	200	70	—

FIGURE 3. Job Options Charts

CONCLUSION

In summary, we believe THE MARRIAGE GAME meets the test of theoretical validity fairly solidly. Many elements are omitted but that, after all, is the *nature* of the simulation procedure. We do believe that the elements included and the relationships between them represent the major elements of a theory of marriage.

REFERENCES

Boocock, Sarane S.
    1972 "Validity-Testing of an Intergenerational Relations Game." Simulation and Games 3 (March): 29-41.

Coplin, William
    1966 "Inter-Nation Simulation and Contemporary Theories of International Relations." American Political Science Review 60 (September): 562-578.

Duke, Richard D.
    1974 Gaming: The Future's Language. Beverly Hills: Sage Publications.

Gamson, William A.
    1971 "SIMSOC: Establishing Social Order in a Simulated Society." Simulation and Games 2 (September): 287-308.

Greenblat, Cathy S.
    1971 "Simulations, Games, and the Sociologist." American Sociologist 6 (May): 161-164.

Greenblat, Cathy S., Peter J. Stein, and Norman F. Washburne
    1973 THE MARRIAGE GAME: UNDERSTANDING MARITAL DECISION-MAKING. New York: Random House.

Guetzkow, Harold
    1968 "Some Correspondences between Simulations and 'Realities.' " In Morton Kaplan (ed.) New Approaches to International Relations. New York: St. Martin's Press.

Inbar, Michael and Clarice Stoil
    1972 Simulation and Gaming in Social Science, New York: Free Press.

Raser, John
    1969 Simulation and Society. New York: Allyn & Bacon.

Russell, Constance J.
    1972 "Simulating the Adolescent Society: A Validity Study" Simulation and Games 3 (June): 165-188.

SIMSOC:

Establishing Social Order in a Simulated Society

WILLIAM A. GAMSON

University of Michigan

Developmental processes frequently seem more orderly and logical in retro-spect than at the time. It is easy to forget, after the fact, that many decisions were made for poorly articulated and ephemeral reasons. Good sense can be made of many choices which are better attributed to good fortune. I mention this because I am aware that the account I am about to give of the development of a simulated society (SIMSOC) will inevitably exaggerate the self-conscious planning involved in the evolution of this game.

This paper will be divided into three sections. First, I wish to describe the initial stimulus to the development of SIMSOC and my objectives at the beginning. Then I will describe my troubles and triumphs in actually running the game in various versions over a period of about four years. Finally, I will make some general observations about developing game simulations, based on my own experience with SIMSOC.

GETTING STARTED

Two general predisposing factors combined with two more immediate events in directing my energy toward developing a game simulation. The first pre-disposing factor was simply an attraction for games. I enjoy playing games, and perhaps more important, I have always manipulated rules of existing games in efforts to remove defects or to make the games more exciting. Second, I have been concerned for some time about making the learning process less passive. This concern is reflected in an interest in cooperative work-study programs such as the one I experienced as an undergraduate at Antioch College and in courses which have a workshop atmosphere—for example, those in which the class collectively attempts to carry out some research project.

Reprinted with permission of Macmillan Publishing Company, Inc., from SIMULATION AND GAMING IN SOCIAL SCIENCE by Michael Inbar and Clarice Stoll. ©1972 by the Free Press, a division of Macmillan Publishing Co., Inc.

More immediate events played on these predispositions. During the spring semester of 1964, I was scheduled to teach a large introductory social psychology course. It has previously been taught in the conventional manner of such courses at large state universities—with a large lecture which would involve a less passive experience. My thoughts focused on substituting a lab for the lecture, but I was uncertain what form such a lab should take.

At about the same time, I had occasion to review Harold Guetzkow et al. (1963) and to discuss the INTER-NATION SIMULATION with some observers and students who had participated in it. I was impressed with the sense of involvement it produced and the greater salience certain aspects of international relations seemed to have for these students. Of course, I had no way of knowing if the game produced these animated concerns or if the students had had them from the beginning. But the fact remained that they *believed* they had learned something new, and, if this belief was an illusion, it was at least the sort of illusion a student ought to get at college.

If important aspects of the international system could be simulated, surely there were many social psychological concerns that would lend themselves to such treatment. Here was a possibility for an unusual and exciting social psychology lab, if I could think of a game simulation appropriate to the course material. My point of departure contained two premises which have remained constant through several revisions of the basic game.

(1) The demands of a game simulation as a learning device and as a useful research device are, to some degree, in conflict. If one is to interpret the outcomes of a game simulation, it is necessary to put some fairly severe restrictions on the interaction of the participants. If everyone is allowed to communicate freely and simultaneously with everyone else, it will be difficult to know and control enough of the conditions of the simulation to be able to attribute the outcomes to any particular parameters. While it may not be necessary to control interaction completely, the demands of research push toward systematic record-keeping and well-channeled and observable transactions among participants.

In contrast, the effectiveness of a game simulation as a learning device is due in large part to its motivational impact on participants. Its ability to involve them in the learning process is enhanced by spontaneity and the opportunity for improvisation. Heavy demands on recording and controlling interaction, it seemed to me, would interfere with such spontaneity and would slow down the pace of the game for reasons which contributed little to the learning process.

It is possible to overdraw the conflict and one can, I suppose, attempt to strike some balance in restricting interaction. But the balance is likely to produce a different and less appropriate game than one would have developed for either single purpose (learning or research). My own task, I believe, was greatly simplified by a clear commitment to a single purpose. It enabled me to introduce complications and innovations quite readily and to eliminate a great deal of paper work and recording which served no educational purpose. Where the main action of the INTER-NATION SIMULATION centered around the filling out of some rather elaborate forms, the main action in SIMSOC increasingly centered around a series of complicated verbal transactions, and the forms became simpler as the game developed.

(2) The second premise concerned the central purpose of the simulation I intended to develop. I wished to center attention on a central problem of collective decision, a problem epitomized by the "mixed-motive" or bargaining game (compare Schelling, 1960). Such games are characterized by a mixture of mutual dependence and conflict, of partnership and competition among the players.

I wished students to confront this problem collectively and as an organizational problem. Given the advantages to each individual of improving his relative position, how could he manage collectively to deal with his common interest? The dilemma is well illustrated by the "free rider" problem (compare Coleman, 1966). It is to everyone's advantage to see to it that certain collective goods and services are produced. But it is even more to the advantage of any individual *not* to participate in the cost if all others do, since he will get the benefits anyway. In confronting this fundamental problem of collective decision, I expected students to confront a number of issues of deviance and social control, of legitimacy and the development of social norms, of leadership and role differentiation. These issues were central to the content of the social psychology course for which SIMSOC was to serve as a lab.

PROBLEMS OF DEVELOPMENT

It should be obvious that the points of departure described above offered only the mildest sort of constraint on the form and content of SIMSOC. In developing a playable game from such considerations, a great many features had to be specified, and many of these decisions were made casually or on the basis of practical considerations. Some of these vaguely considered details turned out to have important consequences for the effectiveness of the game. It seems useful to me to describe the evolution of the game, identifying features which were omitted in later versions and the reasons for eliminating them. I will also describe features which were added and the effects of these additions. But before proceeding, some basic description of SIMSOC seems in order.

The game attempts to provide a rich environment for sustained exploration and learning, rather than to illustrate the impact of environmental constraints on the participants. As in the INTER-NATION SIMULATION (INS), the goals of the players are not sharply defined; rather, they are given certain resources, opportunities, constraints, and general objectives and are then put on their own (compare Raser's description of the INS, 1969).

The game is designed to be played for a period of from five to ten sessions, allowing from one and a half to two hours per session. Groups ranging in size from twenty to sixty can play it in any setting in which there are four or more rooms in reasonably close proximity. Players are dependent on others for survival and for achievement of their goals. Some of their goals conflict with those of other players, and all players are competing over the allocation of rather scarce resources which they need to continue functioning.

The society has certain basic groups—economic, political, mass media, and others—but it has no government. The game starts with a collection of diverse interests, and it is up to the players to create institutions (especially governmental ones) and a normative order as the need arises. To do this, they meet with other individuals of like interest, bargain with individuals who have different interests, travel about to influence players in other regions, threaten others with arrest and confiscation of their resources, and generally "wheel and deal." Inevitably some stay strictly in their own region and do not initiate much interaction. At the end of each session, there is an occasion for investing resources in a manner which affects the "National Indicators" of food and energy supply, standard of living, social cohesion, and public commitment. Everyone is affected by what happens to these indicators, which are altered not only by direct investment, but by various other actions of the members of the society (including absenteeism, arrests, unemployment, and death).

The action of the game typically centers around the ability of collectively oriented leaders to win legitimacy and the compliance and cooperation of others. Any group or individual with sufficient resources can create a police force, and a major axis of conflict has frequently been between relatively popular leaders without a force and a small group with little legitimacy which possesses a force and is challenging the leaders. There are many variations on the basic theme, and some SIMSOCs have borne little resemblance to it. Peaceful SIMSOCs have generally been dull, and, as the discussion below will indicate, much of the tampering with the rules in different versions has aimed at strengthening private and subgroup interests and weakening the emphasis on the common interest. Fortunately, only a small number of deviants is necessary to keep things interesting, and there are usually a few individuals who—perhaps out of mischief or boredom—try to make trouble for any established leadership group.

Early Versions

The brief description above is intended to convey something of the type and flavor of the game. The full rules are too detailed to review here and run over twenty single-spaced pages in length. They have changed dramatically over the years, and it is an instructive exercise to examine some of the differences between the current version and various early versions of the game.

The original version of SIMSOC was designed to be played one hour per week over a six-to-eight week period among the 25 students in an honors section of an introductory social psychology course. It was played in one ordinary classroom, in which players were allowed to meet in buzz groups but could, by petition, call a meeting of the whole class. Each player had a private interest group to which he belonged. His points were determined by the value of his private interest group and by the state of the National Indicators mentioned above. Grades were attached to these points. The state of the National Indicators determined the average grade for the class as a whole, in a range from C— to B+; an individual's private group score determined whether he would get above or below this average. (In addition, of course, the lab grade was based on written work analyzing SIMSOC and relating it to other course material.)

The private interest groups (six in all) had a series of complicated effects on each other and on the National Indicators. An increase in the score of any given group raised the scores of some other groups and lowered the scores of still others. Increases in score for most private groups tended to lower the National Indicators, with the exception of one or two private groups whose interests were complementary with the National Indicators but competitive with interests of most of the other private groups. Players were given a complicated matrix describing these effects. The mechanism for raising the private group scores or National Indicators was resource units, which players were given in fixed amounts each session to invest at the end of the session. Players could also increase their "annual" income of resources by personally investing them, rather than putting them in their private interests or National Indicators.

This first version of SIMSOC had an interesting history. First of all, the problem of social order was apparently solved quite rapidly and effectively. After about the second session, the group met entirely as a whole and strongly discouraged any private meetings or communications. A set of three leaders developed, covering a diverse and conflicting set of private interests. They proposed and won acceptance of a plan in which all members of the society would turn over all resources to them, and they would invest all of them in the National Indicators with the exception of a small amount needed to keep the private groups all at an equal level. This plan had some technical difficulties which kept the leaders busy a good part of the time.

However, two events occurred which prevented the full achievement of the collectively oriented, smooth society envisioned. First, one individual and later a few others refused to turn their resources over to the leadership. They were subjected to strong social pressure and vilification from other class members, although the leaders treated them rather gently. However, most of these hold-outs *actually invested their resources according to the leadership's plan*. In short, they insisted on the right to do themselves what the developing norms dictated that they do, an interesting and subtle form of control with a context of apparent deviance. Furthermore, the exclusion of their resources from the general pool enabled the leadership to excuse a continued imbalance among the private groups, stemming from the technical difficulties of keeping them equal. They were able to blame the inequalities on the deviant societal members and thus divert criticism directed against their leadership.

A second event came near the end and took a dramatic form. One of the three leaders charged with carrying out the investment of the pool of resources departed from the general plan and invested the entire pool in his own private group. The hostility directed toward him became so intense that he stopped coming to the lab sessions. I personally was rather delighted, because this action reopened for the class the dilemma of social order that they had apparently so successfully solved, and in the remaining couple of weeks they began to wrestle again with the problem that had seemed quite simple to solve earlier.

I considered this initial version of SIMSOC a qualified success, but there were many problems that emerged. Now, six versions later, with approximately thirty independent trials, I cannot claim that all problems have been eliminated. There are two classes of problems: (a) those that were relatively superficial and even mechnical and proved not too difficult to eliminate, and (b) those that are related to the fundamental design of SIMSOC and have proved exceedingly stubborn. There are two problems in this latter category, and they remain only partially solved: (1) unequal participation, with some students extremely busy and others doing almost nothing, and (2) a need for meaningful private goals for the players to compete with the raising of the National Indicators as a goal. Before discussing the only partly successful attempts to deal with the stubborn problems, I will describe the removal of several minor bugs that infested early versions of SIMSOC.

Minor and Soluble Problems

(1) Consensus formation. If the members of a real society could gather in a single room, communicate about their problems, and put pressure on deviants, the difficulty of achieving consensus would be greatly simplified. If SIMSOC were to be a simulated society rather than a small group, its citizens should face the problem of achieving consensus *without the possibility of face-to-face communication among all members.* Consequently, in all subsequent versions of SIMSOC, including the present one, members are divided into four regions with provisions for travel between regions but with the restriction that there can never be more than fifty percent of the members of the society in any given region at any time.

(2) Focus on technical aspects. An inordinate amount of time in early SIMSOCs was spent working out the implications of the highly complicated matrix of interdependencies for investment policy. The major question for the class concerned "How do we best achieve our agreed-upon collective goal?" rather than "How do we achieve collective goals in the face of substantial conflict of interest?" The focus on technical aspects helped to transform the game from a mixed-motive situation to a pure coordination game. Furthermore, the focus on technical issues tended to be dominated by a few experts, leaving most societal members with very little to do. Through several revisions, the

devices for building in significant elements of conflict and private interest have become less and less technical.

Nevertheless, even in the current version, some subjects report that SIMSOC has "an overemphasis upon the economic aspects of society." Since there is no serious effort to simulate the economy, and economic complexity is now minimal, this reaction requires some interpretation. It arises, I believe, from the fact that competition over scarce resources inevitably becomes the focus of much of the action. The remark that economic aspects are overemphasized can be interpreted to mean that the game produces a society that is "too materialistic" for the taste of many players. This, it should be recognized, is a comment about the "quality of life" in their SIMSOC and not a comment about its value as a learning experience.

Another reaction, in a similar vein, suggests that SIMSOC imposes an individualistic and capitalistic order. There is a partial truth in this since, initially, resources are controlled privately. However, there is nothing to prevent or discourage collective control of one sort or another and this is, in fact, a frequent outcome. One SIMSOC user in a business school reported with some surprise that his students created a "full-fledged socialist state." Private control, then, is simply an initial condition that may or may not be maintained in the course of play as participants erect their own institutional order around the bare bones given them in the Participant's Manual.

(3) Instructor as participant. I had intended to limit my role to observer once the initial rules had been explained and the society was in process. This proved to be impossible in the early versions, because the rules were ambiguous on a number of points. I was continually being called upon to interpret rules and to adjudicate conflicts about interpretation. Since such adjudication invariably favored some individuals at the expense of others, it became more and more important to members of the society to control this source of power. Allowing myself to become involved as an important and influential member of the society seemed to be subversive of the basic purpose of the game.

It took me several revisions and trial runs to solve this problem, but it has now been solved by a simple and effective device. The interpretation of the rules was turned over to a group within the society. The rules grant them the right to resolve whatever ambiguities arise. This has the double advantage of removing the instructor from the society and presenting the members with an important additional problem of controlling a powerful group in the society with great potential for arbitrary and self-interested employment of its authority.

(4) Intervention in ongoing processes. In the early trial runs of SIMSOC, I attempted several times to add new rules while the game was in progress, each time with very poor results. These interventions tended to change the focus of the game from dealing with conflict among participants to dealing with conflict between class and instructor. They were received as unwanted intrusions whose purpose it was to disrupt the social order and as a kind of "cheating" by the

instructor to overcome the fact that the players had "beaten" the game. They also produced a kind of anomie in which players felt paralyzed to act, for fear that the ground rules for action would suddenly be changed. I resolved, finally, to refrain from changing any rules once the society was in process and to reserve changes for future iterations of the game.

The latest version, however, contains a provision for the introduction of certain "events." These events take the form of crises with which members of the society must deal (e.g., epidemics and natural disasters). Players are warned in advance that such events will occur as a normal part of the game although they are not told of their nature or when they will happen. The forewarning on these events and their inclusion in the written rules serves to remove them from the category of the instructor's arbitrary intervention and allows them to be perceived and accepted as a normal part of the game.

(5) Lack of structure and purpose. I wished to avoid a highly structured, role-imitation situation in favor of one which created a rich context for continued exploration and learning. This lack of specificiation about what was expected from the players was a source of anxiety for many of them and, because of their pressures for greater direction, a source of anxiety for me as well. I responded to this problem in three ways. First, I began to recognize that the uncertainty and confusion tends to disappear by the third session. The knowledge that this is true gives me more courage in being able to reject requests for specification. Second, I have made a point of calling attention to the lack of structure and to the confusion at the beginning about what is expected as a normal and desirable feature of the game. I hope it makes people feel less anxious to be told that it is normal and proper for them to feel anxious at this particular point in time. Third, I have made a number of concessions to structure, perhaps too many. These changes and concessions are only partly directed at relieving anxiety in the early stages; they have an additional purpose of relieving boredom and alienation which may develop in the late stages.

The major effect of the changes on the early sessions is to provide a fairly clear-cut task for the opening session. This seems desirable on two accounts: (1) It helps students over the initial shock which occurs when the explanation of the rules is completed and they are told, "All right, the society is in session." In the early versions, a considerable period of staring at each other and rereading rules took place at the outset. (2) Perhaps more important, it helps to prevent premature closure and the rapid acceptance of any suggestion for action which helps to resolve the ambiguity about what constitutes appropriate behavior. Under the current rules, some basis of subgroup orientation has an opportunity to occur before an elite can mobilize a mass of unintegrated individuals for collective action. The preliminary organization which takes place in the opening session makes the citizenry less accessible to rapid mobilization and reinforces some sense of private or subgroup interest below the level of the society as a whole (compare Kornhauser, 1959). This serves to keep the central problem of collective decision a more difficult one to solve.

(6) Ambiguity about obedience to rules. Once the interpretation of the rules was vested in a group of participants in SIMSOC, a new problem arose. The instructors made it clear that it was up to the society to enforce the rules. The players were told that this was an important problem with which they would have to deal. However, some players violated the rules with impunity, even doing such things as simply removing resources from other players' folders. Young ladies were known to seek refuge in the ladies' room to protect their resources—a modern sanctuary equivalent to the churches of olden day.

Much of the problem was created by confusion on the part of the instructors, including myself, about the difference between two kinds of rules. The difference may be thought of as that between natural constraints and man-made laws. We failed to distinguish the basic constraints which made the game possible from those which the players imposed to run their society successfully.

The current version of SIMSOC makes this distinction very sharply. The players are asked to accept the rules which the manual supplies as equivalent to natural constraints. The rule, for example, that one may not travel while under arrest should be taken as the equivalent of the physical prison which makes this impossible. For the simulation to have any meaning, it is necessary to accept such constraints.

In contrast, the members of the society may decide, through an elected legislative body, for example, to introduce certain regulation on travel. Here we emphasize the rules of the game *do not require* obedience to man-made regulations. It is up to the society to enforce such laws or norms or to tolerate deviance as its members see fit. This is not, we emphasize, a distinction between a constitution and the laws of a particular administration. Both of these are man-made laws and can be changed by common agreement, but the rules of the manual are not subject to change by the members of the society any more than we can change the law of gravity by common consent. Nevertheless, interpretation may be required about whether some action or agreement is possible or impossible under the meaning of some given constraint, and a group within the society is charged with this decision. This leaves some remaining tension about what is deviance within SIMSOC and what is simply cheating (or not playing the game), but the problem has occurred with much less frequency since the distinction has been made explicit in the rules.

(7) Mechanical difficulties. In the earliest version of the game, time pressure was omnipresent. It operated as a distorting factor because discussions were continually being cut short and acquiescence obtained on the grounds that time was running out and some action had to be taken. This fault was corrected in a shift to two-hour sessions, and experiments with one and a half hour sessions also appeared reasonably free of undue time pressure, especially when some flexibility about running over the limit existed.

Reinvolvement in the world of SIMSOC after a one-week layoff has generally presented some problem. Much of the involvement which was present at the end of the previous session is now dissipated and is not always easy to recapture.

Situations in which I have run SIMSOC continually over a two-day period have generally been more effective. This procedure has involved three one and a half to two-hour sessions per day, with half-hour breaks between.

Players are not informed which session is the final one, in order to avoid end-game effects. Typically, the game has ended amidst groans and protests when the players have assembled for a session. There is generally some frustration over termination at a point when many people feel they are about to execute some long-planned program of action or coup d'etat. However, ending the society while action is at its peak has certain educational advantages, although it takes longer for the players to adjust from their roles in SIMSOC to their roles as social analysts in the course. In this latter role, they frequently generate an excellent discussion about what would have happened in the next few sessions and why.

(8) Feckless rules. A number of provisions were included in the original version, some of them for poorly articulated reasons, which proved quite worthless although they did little harm. For example, the first version allowed death and reentry into the society as a new person. I believe I wished to include certain aspects of socialization, but it is difficult to recall at this late date the specific reason for the rule. In any event, the attempt was a double failure. Not only did it fail to capture any significant aspects of the process of socialization, but it produced a series of self-interested "suicides" which had absolutely nothing to do with actual suicide. Players who were unhappy with their position found this a convenient way to liquidate and begin afresh. Furthermore, those whose private interests conflicted with the National Indicators were advised by their fellow citizens to drop out in the interest of the society. Many players were happy to oblige with such altruistic "suicide." This aspect of the simulation hardly seemed integral to the central purposes and was simply abandoned in later versions.

A number of other features of early versions were dropped in subsequent revisions. There is no point in describing these in detail, for they contain no general lesson other than underlining the following observation: Mistakes with little or no effect are not as disturbing as those which produce unintended side effects, since they do not interfere with the processes being simulated or create new problems.

Stubborn Problems and Attempted Solutions

(1) Alienation and apathy. A major problem in the early versions of SIMSOC was the tendency for the society to become dominated by a collectively oriented elite which served the society well. This, of course, was no problem from the standpoint of SIMSOC members. They had engaged in a classic interchange of support for effective leadership. In a real society, they could then turn to private pursuits such as love and leisure. However, in SIMSOC, the educational ob-

jectives required that they participate in some active way in a societal role; neither they nor the instructors could see the point of their coming to class for two hours in the evening simply to write letters to friends or to study for examinations in another course.

I sought a solution to this problem which would provide enough structure so that every individual had to hold a job to live but, at the same time, would leave room for a great deal of innovation and ingenuity in achieving goals. The problem still remains only partly solved. Involvement remains unequal, but the inequality now seems more a product of choice and lack of interpersonal skill than of something intrinsic to the rules.

(2) Tying grades to performance. My intention in attaching grades to performance in the game was to make the conflict and the stakes more real and vivid. Far from increasing conflict, the introduction of this stake had the effect of all but eliminating conflict among the class members. Powerful norms developed against pursuing private interests, and as in the case of the instructor's intervention, the meta-game was between class and instructor. The class believed that the instructors (myself, or, later, several teaching fellows) wished to create conflict among them, and their problem centered on how to prevent this from happening. The fear of untrammeled conflict was frequently expressed in Hobbesian terms by the participants: "Unless we stick together this whole thing is going to degenerate into a situation in which everyone is going to be cutting everyone else's throat." Those who deviated did so, in my judgment, for reasons which had little to do with grades. Motivations such as curiosity and rebelliousness against the authority of societal leaders were much more often dominant.

Now, it might be argued that the condition described above is realistic and should be welcomed in the simulation. The fear of a short, nasty, and brutish existence if the fragile web of order is destroyed may be a genuine and important presence in society. The problem came not from a lack of correspondence between the simulation and the world but in the breakdown of the boundaries between the two and the transformation of the simulation. The "real" game became the familiar one the students were accustomed to playing— the college game. Furthermore, they had available to them a powerful set of controls over their fellows that is not usually present. Each student could very clearly see who would be affected adversely by any self-serving actions he might take, and those who were affected would know and be able to hold him accountable. Far from making the dilemma of social order more intense, the introduction of grades created such powerful forces for coordination and order that the dilemma was largely removed. Several societies very quickly become smoothly functioning welfare states with few or no challenges and little or no occupation for the mass of participants.

I became convinced that it was desirable to separate the student role from the role of citizen of SIMSOC if the educational purposes were not to be subverted. This involved basing all grades on the written work in connection with SIMSOC, a task which emphasized skill as an observer and analyst of events, rather than skill as a participant. Rewards for skill in the society became purely symbolic, but this change did not lead to any lessening of involvement. On the contrary, the students appeared able to play the game more freely and creatively and with enjoyment as conflict became less threatening. This change produced a more realistic simulation of the *balance* between the forces for order and the forces for conflict. It was this balance, and the dilemmas arising out of it, which I wished to simulate and would lose by making either set of forces predominate. Achieving this balance has been the most difficult problem of SIMSOC, with imbalance invariably coming from having too little conflict. Ironically, the removal of grades contingent on performance greatly increased self-interested behavior and thus made the basic dilemma of the game more realistic.

(3) Lack of force. An important omission in the early versions of SIMSOC was the absence of any functional equivalent of force. This exclusion turned out to be important for unexpected reasons. There has never been much need in SIMSOC for force as a facility to be used on behalf of the society as a whole. Social pressures are so strong when backed by a leadership with high legitimacy that few individuals have been able to muster the will or the resources to resist. The importance of force is that it offers some possibility for competing against legitimate power. The most important innovation added to SIMSOC was the provision for a police force or army that could be created by any individual or group of individuals with sufficient resources to do so. This police force, once created, can restrict the travel and confiscate the resources of any member of the society on any grounds it deems appropriate. One can, of course, also be created on behalf of a central government or leadership group, but the essential point is that access is not limited to such a group.

The use of the police force option has varied greatly in different runs of SIMSOC. Some societies had no police forces; some had only a highly legitimate one; some had a private or guerrilla force; and some had two or more forces engaging in what amounted to a civil war. But even in those societies in which no police forces were created, the possibility of resort to such a force by a regime or its opponents remained an important and fundamental constraint.

(4) Scarcity. A major difficulty in creating meaningful private interests has been the lack of anything to do with one's money (Simbucks) other than using it to raise the National Indicators. In early versions, players learned after a few sessions that they really did not need Simbucks and could not do much with them, especially when the leaders were providing for all members those few things on which money could be spent. They could not buy cigarettes, vacations, or grades with such Simbucks, so why seek them or try to hold onto them?

I have attempted to deal with this problem by creating a combination of scarcity and personal need. In the current version, travel tickets are severely limited in number so that they have real value as a social object on which to spend Simbucks. Furthermore, all members must provide for their subsistence through the purchase of subsistence tickets, and the failure to do so on successive weeks eliminates them permanently from the society. These subsistence tickets are also in scarce supply. In addition, certain events will occur which will require Simbucks for personal protection. For example, when an epidemic occurs, members will have to purchase immunity or run the risk of being eliminated from SIMSOC through contracting the disease. Other minor privileges may also be purchased. Hopefully, these provisions will make the acquisition and holding of Simbucks a more meaningful action than it has been in the past and will complicate the problem of providing sufficient money for keeping the National Indicators high.

(5) Inequalities. Early versions of SIMSOC had a strong bias toward equal opportunity. This bias was introduced partly for "fairness" when grades were attached to performance and partly without thought because such initial equality is so typically a part of games. The players bring a strong egalitarian norm into the game as well.

The most recent version of the game seriously departs from the equal opportunity bias. Heads of seven basic groups are designated at the beginning, and these seven individuals have complete control over the Simbucks assigned to those groups. The others have no money and must acquire it by getting one or more jobs with the basic groups. There are opportunities to depose the head of a group, but no way of guaranteeing control over the new head.

In addition, a limited number of people in the society own Travel Agencies and Subsistence Agencies. These agencies allow the owner to dispense a limited number of travel tickets or subsistence tickets each session. The instructor has the option of making the inequality cumulative by assigning these agencies to the seven group heads or diffusing power somewhat by assigning them to other individuals. In either event, the current SIMSOC will contain "have" and "have-not" members from the beginning, as well as conflicts and differences among the basic groups. It remains to be seen how successful these built-in inequalities will be against the strong egalitarian norms which players bring to the game.

(6) Collectively oriented groups. I made the mistake of introducing into SIMSOC a procedure for electing a council. Having such a procedure as an explicit part of the rules moves a long way toward solving at the outset the problems which the players are being asked to solve. The council has overwhelming legitimacy immediately and is difficult for any group to challenge except by the use of force. An important problem for the players is to go through the process of establishing governmental institutions which have the allegiance of members of the society. But the emergence of a state loses much of its interest and difficulty when it is virtually established by the rules of the game from the beginning. The current rules of the game eliminate any government, leaving the players free to create whatever institutions seem appropriate to them as the need arises.

Certain of the basic groups in earlier versions had essentially collective goals—for example, to advise the society on proper investment policy and to reward various kinds of socially useful contributions. These groups served to reinforce the already strong collective orientation of the members and have been eliminated in the current version in favor of groups with private goals which conflict with or at least are independent of collective goals.

Thus, my private war against e pluribus unum continues.

SOME GENERAL OBSERVATIONS

I have found the process of continual modification of SIMSOC a peculiarly absorbing and rewarding intellectual experience. It has a concreteness and immediacy lacking in much of the intellectual work I do. If I am trying to understand why, for example, a particular social movement developed at one time and not another, I will usually struggle through a sequence of questions until I achieve some vague sense of closure. This suffices until enough questions have been raised about my explanation to give me again a vague sense of uneasiness.

The process is somewhat different with the development of a simulation. It is as if I have a complicated Rube Goldberg device in front of me that will produce certain processes and outcomes. I want it to operate differently in one way or another, but it is difficult to know where and how to intervene to achieve this purpose because the apparatus is delicate and highly interconnected. So I walk around it, eyeing it from different angles, and imagine adding a nut here or a bolt there or shutting if off and replacing some more complicated parts.

Each of these interventions must take the extremely specific form of a rule. To intervene, one must play a mental game in which the introduction of every specific change must be weighed in terms of how the whole contraption will operate. Such mental games force one to develop a clear picture of what the apparatus looks like and why it operates as it does. Each hypothetical alternative must be put in place and imagined in operation. Finally, an explicit choice has to be made and the game actually run under the altered rules, and one has a chance to discover whether he really understands the contraption or not. When a rule change has the effect it is supposed to have, the experience can be very exciting—as exciting as predicting a nonobvious outcome in any social situation and having it turn out correctly.

Playing a game may be a more active experience than listening to a lecture, but developing a game is more active still. In fact, one begins to wonder if students learn anything from actually playing a game like SIMSOC. Involvement is, after all, a mixed blessing in the learning process. Some students become so involved in the game as an end in itself that they have difficulty analyzing its events in a detached fashion. In the discussions which follow the playing of SIMSOC, it is sometimes difficult to get players past self-justifying descriptions of their actions and motivations. I have seen students who have *continued* to maintain a deception which they had practiced in a completed game to a point where it obscures their own and others' understanding of what happened.

Other students, at least in an introductory course with a partly captive audience, just don't like games. Games require an active, instrumental, and controlling posture toward the environment. Those who are shy or passive, or who simply like to let things happen and take their enjoyment from whatever occurs, may not enjoy the whole experience.

However, it is not necessary to enjoy a game or to do well in the playing of it to learn from it. I would hypothesize that whatever learning takes place happens not at the time of playing, but afterward, and only if some conscious effort is made to convert the experience into some knowledge or insight. In SIMSOC, if students learn anything, it is because their participation produces a rich array of stimulus material centered around questions of social control and social order. Then, having played the game, they analyze and think about what happened in a manageably complex setting in which they were intimately involved. Students begin to construct their own picture of the Rube Goldberg device and to understand why obtaining compliance was so easy in one instance and so difficult in another. Once they have reached that point, it is not difficult to get them to consider the same questions about their own society or others which they have observed or read about.

One learns, then, not only from playing but also from discussing the experience and thinking about it afterward. The use of games for teaching seems to me to be most successful when the students are asked to make the jump from player to simulator. As players, they accept certain constraints which are given in the rules and use resources to achieve certain objectives specified or suggested by the rules. When the students become simulators, the *rules themselves* become the resources, and they can manipulate these resources—hypothetically or actually—to see whether the resulting process will take the form they believe it will. Developing simulations is too enjoyable and valuable a learning experience to be hoarded by professors.

REFERENCES

COLEMAN, J. S. (1966) "Foundation for a theory of collective decisions." Amer. J. of Sociology 71 (May): 615-627.

GAMSON, W. A. (1969a) SIMSOC: A Participant's Manual and Related Readings. New York: Free Press.

——— (1969b) SIMSOC: An Instructor's Manual. New York: Free Press.

GUETZKOW, H., C.F. ALGER, R. A. BRODY, R. C. NOEL, and R. C. SNYDER (1963) Simulation in International Relations: Developments for Research and Teaching. Englewood Cliffs, N.J.: Prentice-Hall.

KORNHAUSER, W. (1959) The Politics of Mass Society. New York: Free Press.

RASER, J.R. (1969) Simulation and Society. Boston: Allyn & Bacon.

SCHELLING, T. C. (1960) The Strategy of Conflict. Cambridge, Mass.: Harvard Univ. Press.

SOCIAL PROCESSES AND SOCIAL SIMULATION GAMES

JAMES S. COLEMAN

Games are of interest to a social psychologist or sociologist for at least two reasons. First, because a game is a kind of play upon life in general, it induces, in a restricted and well-defined context, the same kinds of motivations and behavior that occur in the broader contexts of life where we play for keeps. Indeed, it is hard to say whether games are a kind of play upon life or life is an amalgamation and extension of the games we learn to play as children. The book by Eric Berne, *Games People Play,* describing some socially destructive behaviors as games, gives persuasive argument that in fact the latter might be the case. And the perceptive observations by Jean Piaget of the importance of simple games like marbles for young children as early forms of social order with its rules and norms strengthen this view.

The second course of interest is the peculiar properties games have as contexts for learning. There are apparently certain aspects of games that especially facilitate learning, such as their ability to focus attention, their requirement for action rather than merely passive observation, their abstraction of simple elements from the complex confusion of reality, and the intrinsic rewards they hold for mastery. By the combination of these properties that games provide, they show remarkable consequences as devices for learning.

Both these topics are dealt with in other chapters in this volume. I want here to examine how a particular kind of game, a "social simulation game," can provide still another source of interest to the social scientist. A social simulation game, as I shall use the term here, is a game in which certain social processes are explicitly mirrored in the structure and functioning of the game. The game is a kind of abstraction of these social processes, making explicit certain of them that are ordinarily implicit in our everyday behavior. These games raise several questions: What is the way a simulation game characteristically mirrors social processes? What are the kinds of social processes most easily simulated in a game? What is the relation of construction and use of a game to, on the one hand observation, and on the other hand social theory? These are the questions I want to address in this chapter. I will use specific examples from games developed by the Hopkins groups as illustrations of the most important points.[1]

Reprinted from **Simulation Games in Learning,** edited by Sarane S. Boocock and Erling Schild. ©1968 by Sage Publications, Inc.

THE ROLE OF THE SOCIAL ENVIRONMENT

A social simulation game always consists of a player or players acting in a social environment. By its very definition, it is concerned principally with that part of an individual's environment that consists of other people, groups, and organizations. How does it incorporate that environment into its structure?

There are ordinarily two solutions, either or both of which are used in any specific game. One is to let each player in the game act as a portion of the social environment of each other player. The rules of the game establish the obligations upon each role, and the players, each acting within the rules governing his role, interact with one another. The resulting configuration constitues a social subsystem, and each player's environment consists of that subsystem, excluding himself.

Examples of this solution occur in most of the Hopkins games. In the PARENT-CHILD GAME, there are two subsystems: one is the parent and child, and the other is the community of parents and children. Each parent's principal interaction is with his child, but he has interaction also with the other players in the roles of parents in the game. And each child's principal interaction is with his parent, but he interacts as well with other players in the roles of children. The LEGISLATURE portion of DEMOCRACY consists of players in a single subsystem. Each player is a legislator, and interactions are with the other players in their roles as legislators.

A second way in which the social environment is embodied in a social simulation game is in the rules themselves. The rules may contain contingent responses of the environment, representing the actions of persons who are not players, but nevertheless relevant to the individual's action. A game using this solution can in fact be a one-player game, in which the whole of this player's environment, represented by the game, is incorporated in the rules.

An example of this solution is the LIFE CAREER GAME. In this game, the sole player begins as a young person, making decisions about his everyday activities and implicitly his future. The responses to these decisions occur through the environmental response rules, which represent the responses of: teachers in school, school admissions officers, potential employers, and potential marriage partners. But none of these roles is represented by a player in the game. The probable responses of persons in such roles to various actions of a player are embodied in the environmental response rules, and the actual responses are determined by these rules in conjunction with a chance mechanism. The player in the game plays for a score, and the only relation to other players is through a comparison of scores.

Most games use a combination of these two solutions. A portion of the environment is represented by other players, and a portion by the environmental response rules. An example is the game of LEGISLATURE. Players receive cards representing the interests of their constituents, and their score in the game consists of votes given by the hypothetical constituents, according to the environmental response rules which make these votes contingent upon the legislator's furthering of the constituents' interests. In some games also, alternative versions have a part of the environment as players in one version and as environmental response rules in another. For example, the complete game of DEMOCRACY includes a citizen's action meeting in which the players are constituents who determine collectively the votes they will give to their legislator contingent upon his action. This moves the behavior of the constituents from the rules into the area of play.

The embodiment of the social environment in the rules requires more empirical knowledge of the responses of the organizations of individuals than does the solution that represents them by players. For the players respond on the basis of their own goals and role constraints, and the game constructor need not know what these responses will be. In contrast, if the responses are part of the rules, the game constructor must know in advance the responses contingent upon each possible action of the players.

The representation of environment by players, on the other hand, requires greater theoretical acumen. For if the players' responses and thus the system of behavior, are to mirror the phenomenon in question, each player's goals and role constraints must be accurately embodied in the rules. For example, in the LIFE CAREER GAME, if the role of college admissions officer were to be represented by an actual player, the goals of the officer, together with his role constraints, must be approximated correctly in the rules, if his selection of candidates is to correspond reasonably well to reality.

The decision to represent a given portion of the environment in either of these ways depends in part upon the mechanics of the game. In some cases, there will be too many players if a given portion of the environment is represented by players; thus it must be represented by the rules. In other cases, such as the LIFE CAREER GAME, the player moves from one environment to another, so that each environment is only temporarily a part of the game.

Finally, it should be noted that every game selects only certain portions of the social environment to be included in either way. Some portions are left out, often because they introduce social processes other than the ones being simulated. For example, in the game of LEGISLATURE, interest groups acting as political pressure groups are explicitly excluded, because of the additional processes this would introduce, obscuring the one being simulated.

TYPES OF RULES

In the discussion above one type of rule was repeatedly mentioned, and described as "environmental response rules." This is only one of several types of rules that are necessary in social simulation games, and it is useful to indicate briefly the several types. This will give some better idea of the elements of which a social simulation game is composed. It is often stated that the rules of a game are like the "rules of the game" in real life, that is, the normative and legal constraints upon behavior. This, however, corresponds only to one type of rule necessary in any game.

The most pervasive type of rule in every game is the *procedural rule.* Procedural rules describe how the game is put into play, and the general order in which play proceeds. In a social simulation game, the procedural rules must follow roughly the order of activities in the phenomenon being studied. Sometimes, the procedural rules explicitly incorporate assumptions about the social processes involved. In the PARENT-CHILD GAME, for example, each round of play between parent and child consists of a sequence of four activities: first, discussion between parent and child in an attempt to reach agreement about the child's behavior; second, orders given by the parents in those areas where no agreement was reached; third, behavior decisions on the part of the child; fourth, decisions of the parent whether to supervise the child's behavior and possibly punish for disobedience. This sequence of activities explicitly embodies assumptions about family functioning. In some cultures, a different set of procedural rules would be necessary, for example, eliminating the first step. Or a more theoretically sophisticated version of the game would leave the sequence of activities undetermined, to be selected by the behavior of the players.

A subtype of procedural rule, found in all games, may be called the *mediation rule.* This is the set of rules specifying how an impasse in play is resolved, or a conflict of paths resolved. In basketball, when there is an impasse when players from opposing sides are wrestling over the ball, the referee calls a jump-ball. In social simulation games, mediation rules are necessary whenever two or more players conflict, and neither has the formal authority or the power to get his way. Mediation in the COMMUNITY RESPONSE GAME is necessary when two players attempt, each in ignorance of the other's action, to operate the same agency. A more important type of mediation is necessary in the ECONOMIC SYSTEM GAME, when workers and employer cannot agree on a wage. This impasse, if allowed to continue, would disrupt the game, just as similar impasses would disrupt the real economy if not subjected to mediation or arbitration.

A second type of rule, closely related to the first, is the *behavior constraint.* These rules correspond to the role obligations found in real life, and specify what the player must do and what he cannot do. They are often stated along with the procedural rules, but they are analytically distinct, for they represent the role specifications for each type of player. For example, in the COMMUNITY RESPONSE GAME, each player in a community role is constrained to use only ten units of "energy" in each time period; and if he decides to operate a community agency, he must devote a specified number of energy units to this activity.

A third type of rule is the rule specifying the *goal* and means of goal achievement of each type of player. In every game, all players have goals, and the rules specify both what the goals are and how the goals are reached. In a social simulation game, the goals must correspond roughly to the goals that individuals in the given role have in real life. Often, the correct specification of a goal is an important aspect of the theory embodied in the game. For example, in the COMMUNITY RESPONSE GAME, each player's goal is to "reduce his anxiety" as quickly as possible. This, together with the specification in the rules of the amount of anxiety he receives from uncertainty about family, from non-performance of community role, etc., constitutes a theory about behavior under conditions of disaster. Or in the CONSUMER GAME, each consumer's goal is to gain the maximum amount of satisfaction. This, together with the schedule of satisfaction received from each type of goods purchased, is based upon the economists' theory about consumer behavior. Insofar as the theory underlying goal specification is correct and the behavior constraints are correct, the behavior of the player should correspond to the behavior observed by persons in that role in real life. If the behavior of the player deviates greatly, it is very likely because the theory about goals of persons in that role is defective.

A fourth type of rule, referred to in the earlier section, is the *environmental response rule.* These rules specify how the environment would behave if it were present as part of the game. In the game of baseball, some fields with a portion of the outfield blocked off have a "ground rule double," for balls hit into that area. This rule is based on the probable outcome of play if the interference with play had not existed.

In all simulation games, the environmental response rules are more important. Since a simulation game is an abstraction from reality, the environmental response rules give the probable response of that part of the environment that is not incorporated in the actions of the players. In a social simulation game, most of the environmental response rules give the probable response of persons, groups, or organizations not represented by players. Examples are those in the LIFE CAREER and other games, as discussed in the preceding section.

There is finally one type of rule in all games as well as real life, which may be called *police rules,* giving the consequences to a player of breaking one of the game's rules. These rules sometimes specify merely a reversion to a previous state (corresponding to "restitutive law" in society), sometimes specify a punishment to the player who has broken the rules (corresponding to "repressive law" in societies). The principal function of the referee in games (besides applying mediation rules) is to note when rules are broken and apply the designated corrective action.

Ordinarily, the breaking of procedural rules leads merely to restitutive action, while the breaking of behavior constraint rules leads more often to repressive action, or punishment of the offending player. In many social simulation games, as in many parlor games, the breaking of a rule is corrected by the moral force of the other players, and their power to stop the game by refusing to play. In a larger game with more players and more differentiated areas of action, police rules are more necessary, as is a referee or policeman to note the delinquency.

THE ROLE OF BEHAVIOR THEORY

A game used as a social simulation is based upon certain assumptions, explicit or implict, about behavior. The similarity of the assumptions from one game to another suggests, as further analysis confirms, that social simulation games have a special kinship to a certain type of behavior theory. In addition, each game designed as a social simulation implies a quite specific theory about behavior in the area of life being simulated. These specific theoretical elements are principally manifested in the goal-specification rules, but also may form part of the behavior constraints, procedure rules, and environmental response rules. Examples of this relation between rules and theory are evident in the preceding discussion; more examples and a closer examination will be given below. First, however, it is useful to examine the general affinities of games to one type of behavior theory.

In every game, each competing unit has a goal specified by the rules, and means by which he achieves this goal. If the competing unit is a team, then all players on this team share the same goal, and have individual goals only insofar as they contribute to the team goal. If it is a player, then he has an individual goal. Even in the former case, individuals as persons (not as players) may have individual goals besides the team goal given by the rules—for example, to excel within one's own team. These goals, however, are not part of the explicit structure of the game, but arise because of the rewards they bring outside the game itself.

Whether the competing unit is an individual or a team, the game functions because each individual pursues his own goal. Thus, a social simulation game must necessarily begin with a set of individuals carrying out purposive behavior toward a goal. It is hardly conceivable, then, that the theoretical framework implied by a social simulation game be anything other than a purposive behavior theory. This means a definite theoretical stance on several issues: On the issue in social theory of expressing the assumptions of the theory at the level of the individual or at the level of the collectivity or social system, the use of games implies taking the former, individualist position. On the issue of purposive theory versus positivist theory (where behavior is described as a lawful response to an environmental stimulus), the use of games implies the purposive orientation. On the issue of purposive, goal-oriented behavior versus expressive theory (where the individual act is an expression of some inner tension without regard to a goal), the use of games again implies the purposive orientation. On the issue of behavior determined by personality or other historical causes not currently present versus behavior determined by the constraints and demands of the present (and possibly expected future) situation, the use of games implies the latter, the theory of present- and future-governed behavior. On the issue of purposive, goal-oriented behavior versus behavior governed wholly by normative expectations and obligations (as, for example, occurs in some organization theory, where the individual's interests play no role, and he is predicted to behave simply in accord with organizational rules), the use of games implies the former, goal-oriented position. The use of games takes as its starting-point the self-interested individual (except in the case where the competing unit is a team), and requires that any non-self-interested behavior (e.g., altruistic behavior, or collectivity-orientation) emerge from pursuit of his goals, as means to those individual ends. For this reason, social simulation games that use a collectivity, such as a family, as a team to form a competing unit, are not as theoretically complete as are those games in which the individual player is the competing unit. To specify a collectivity as a competing unit prevents simulation of those processes that induce the individual to realize his goals through investing his efforts in a collectivity's action, It may well be, of course, that for a given social simulation, one wishes to take those processes as given, in order to simulate others. For example, in the CONSUMER GAME, the goals of the finance officer of the department store are given as the profitability goals of the department store itself. Similarly, players acting as the consumers are given satisfaction points for purchases corresponding to the satisfaction of both husband and wife together, not corresponding to the satisfaction of one alone. For the purpose of this simulation, the question of how the department store manager induces the finance officer to act in the store's interest, or how the other family members induce the consumer to act in the family's interest, are not taken as problematic.

To state the theoretical position implied by the use of social simulation games does not answer all the theoretical questions that arise. Any given game makes certain specific assumptions about goals. For example, in the LIFE CAREER GAME, the question arises whether the satisfaction points that constitute the game's goal should be given at each time period, so that the player's score is his cumulative satisfaction over the period of play representing a number of years, or whether points should depend only on his final position at the end of the game (say at age twenty-five). This question becomes almost a philosophical one; but it must be resolved to appropriately motivate the players. Again, in the LIFE CAREER GAME, it is assumed that the individual represented by the player can derive satisfaction from several different areas of life, and that his behavior will depend in part upon the relative importance he attaches to these areas (e.g., family, self-development, financial success, etc.). Consequently, one decision in the game is a weighting of these areas by the player, in essence determining his own goals.

In the first level of the game of Legislature, each legislator is assumed to be motivated solely to stay in office; and it is assumed that the sole factor affecting his tenure is his success in passing those bills of most interest to his constituents. Neither of these assumptions corresponds directly to reality, though both factors are present in concrete legislatures. In order to simulate this process, all other elements are suppressed, and the single process abstracted from reality. The resulting simulation hardly mirrors reality, but instead mirrors only one component of it. In the second level of the game, a second source of motivation is assumed for the legislator: his own position, taken prior to knowledge of his constituents' interests, on each bill. Winning depends both on re-election by his constituents and on his voting in accord with his own beliefs. This introduces merely one more element into the simulation, which remains far from the reality of actual legislatures, but is instead merely an abstraction of certain important processes from them. In the COMMUNITY RESPONSE GAME, the appropriate balance between orientation to self-interests and to those of the community is important, yet difficult to obtain. In part, this is obtained by a balance between the anxiety elicited by failure to solve individual problems and failure to aid in solution of the community problems. But upon further reflection, it appears that in addition to mere anxiety reduction in a disaster, individuals are to some degree motivated by their conception of the regard in which they will be held by their neighbors, among whom they must live in the future. Consequently, among the three players who have accomplished the greatest anxiety reduction, the players vote for the one who has contributed most to the community, as the overall winner of the game.

Altogether, specification of the goal for each type of player in a game is the principal means by which theoretical assumptions are introduced into the game. If incorrect goals are introduced, then the behavior of the players will deviate from the behavior that it is intended to simulate, because the players are incorrectly motivated. In most simulations, the goals introduced are only partial goals, because the game, like a social theory, is an abstraction from reality, and should contain only those motivating elements that produce the aspect of behavior of the processes being simulated.

In addition to the goals of the game, the procedure rules and the behavior constraints are also partly determined by theoretical assumptions. In the PARENT-CHILD GAME, the procedural steps used in the game are, as indicated earlier, an expression of assumptions about the activities that occur in the determination of adolescents' behavior. In every game, certain of the behavior constraint rules correspond to role obligations of the individual being simulated. Sometimes, these are directly observable in the situation being simulated; sometimes they are not. In the PARENT-CHILD GAME, it is assumed that the adolescent is free to behave as he wishes, subject to possible parental punishment. But in reality, this is so only in some areas, such as staying out at night. In doing homework, or other activities carried out at home, the parents' supervision may come not merely after the behavior, but during the activity itself, to ensure its completion.

THE KINDS OF PROCESSES SIMULATED
AND THE MEANS OF DOING SO

It is difficult and perhaps unwise to make any general statements about what social processes most easily lend themselves to game simulation, and what is the appropriate means to mirror them. For, obviously, judgments about these matters derive from what has been done in very limited experience. Consequently, what I shall attempt here is to make some generalizations about the types of processes the Hopkins group have so far found it possible to simulate in games, and the kinds of devices members of this group have used in so doing. This may then give some insight into one general style in the development of social simulation games.

First, it is striking that in nearly all the Hopkins games, the player's goal achievement is measured by his achievement of "satisfaction" points, or some variation thereof. In the COMMUNITY RESPONSE GAME, it is the complement of this—reduction of "anxiety points." In the HIGH SCHOOL GAME, it is units of "self-esteem" that the player tries to gain. Only in the Legislature game is there any real deviation from this approach, for the legislator attempts to gain votes from constituents.

Even here, however, if the game were made more complex through introducing other sources of motivation for the legislator, one way of integrating these various sources of motivation to provide a single measure of goal achievement would be to calibrate all the objective measures of achievement (such as re-election, chairmanship on committees, voting in accord with prestated beliefs, etc.) onto a single scale of satisfaction. In fact, it appears likely that this is the source of the widespread use in these games of "satisfaction units" as measures of goal achievement: as the one common denominator against which otherwise incommensurable objective achievements can be scaled.

In relating these objective achievements to subjective satisfaction, two quite different approaches have been used: to fix in advance, as part of the goal achievement rules, the conversion ratios between each kind of objective achievement and subjective satisfaction; and to allow the player himself to fix these ratios. Most of the games use the fixed-conversion approach, but in nearly all the games, a more advanced form can be developed in which the player himself sets these (subject to constraints that prevent him from gaining advantage in later play by strategically set conversion ratios). In a second-level form of the LIFE CAREER GAME, the player decides the relative importance of each of four areas of life activity, thus fixing his own conversion ratios for satisfaction. In a form of the COMMUNITY RESPONSE GAME, used experimentally, each player was allowed to distribute his initial anxiety points among the different sources of anxiety in a way that corresponded to the relative anxiety he believed he would feel in each area. Similar variations have been developed in the PARENT-CHILD GAME, the HIGH SCHOOL GAME, and the DEMOCRACY GAME. In all but one of these the setting was based upon the player's own preconceived estimates of the satisfaction involved. But in one game, the HIGH SCHOOL GAME, the conversion ratio was determined by the player as a result of his experience in the game: he decided what proportion of his "attention" he should pay to esteem he received from other students and what proportion to esteem from parents. This relative attention then becomes the weighting factor in converting esteem from others to self-esteem. This approach, also used in the

LIFE CAREER GAME, is the most theoretically advanced of the approaches discussed above, for it introduces as one of the processes being simulated the selection and modification of goals contingent upon the consequence of the player's actions.

Exchange processes and means of their simulation: In sociology and social psychology, the recent theoretical development most akin to this utilitarian, purposive approach used in games depend greatly upon the idea of exchange. This is evident in the work of its principal exponents, Thibaut and Kelley, Homans, and Blau. These theoretical developments, the idea of exchange of intangibles such as deference, acceptance, autonomy, aid, and similar quantities, constitute the foundation of the approach. Each party to the exchange engages in it because of a gain that he expects to experience from it. Thus, the question of how social simulation games express the processes of exchange of such intangibles naturally arises.

In the Legislature game, there are two types of exchange, simulated in quite different ways. One is the exchange between legislators of votes, or control of issues. This exchange is not incorporated in the rules of the game, but arises from the motivations induced by the players' goals, together with the fact that no constraints *against* such exchange exist. The exchange is not expressed by a tangible or physical exchange (as it would be, for example, if pieces of paper representing votes were physically exchanged). This has certain consequences for the functioning of the system: the exchanges are merely "promises to pay" a vote or unit of control of an issue, and the promise may not be honored; nor is the exchanged quantity easily negotiable by the receiving party.

A second type of exchange in this game is the fundamental exchange of representative democracy: continued support of the legislator by constituents, in exchange for the legislator's pursuit and realization of the constituent's interest in legislation. This exchange is simulated through the environmental response rules: the legislator's score is dependent on cards he receives showing the interests of his (hypothetical) constituents.

These two examples from the game of Legislature illustrate that a social exchange process may be mirrored in games either by an exchange between two players, or by an exchange between a player and the non-player social environment, according to the environmental response rules. Both of these cases present certain complications, and each will be examined in turn. For exchange between players, the most fundamental point, though it appears obvious, must be made: the exchange must be motivated for both parties. The exchange must contribute to both players' goals. In the Legislature game, for example, an exchange of control occurs between legislators not because it is prescribed by the rules, for the rules make no mention of such an exchange. It occurs because it is to the interest of each to concentrate his power on those issues that will contribute most to his re-election or defeat. As a consequence, the exchange occurs only when two legislators see a mutually advantageous exchange of control.

Exchange between players can be either between persons in the same role, such as legislators in the Legislature game, or between persons in different roles, as parent and child in the PARENT-CHILD GAME, consumer and finance officer in the CONSUMER GAME, worker and manufacturer in the ECONOMIC SYSTEM GAME. When exchange is between persons in the same role, as two legislators, only one motivation need be supplied by the theory on which the game is based, for it serves both players. When the exchange is between persons in different roles, two different motivations must be supplied from a more complex theoretical base. For example, in the two-stage form of the DEMOC-RACY GAME, each citizen-constituent is attempting to maximize his satisfaction from the collectivity's legislation. He does this through exercising his power in a collective decision (a community action meeting) that determines the legislation that the representative-legislator must obtain in exchange for the constituency's support in re-electing him. Thus, for this exchange to take place, two different sources of its motivation must be inherent in the goal achievement rules: the representative-legislator must have control of some actions (in this case, legislation) that contribute to the citizen-constituent's goal achievement; and the citizen-constituent must have control of some actions (in this case, votes for re-election) that contribute to the representative-legislator's goal achievement.

In the PARENT-CHILD GAME, the parent and child negotiate over each of five areas of the child's behavior, and the implict starting-point is that each has partial control over this behavior. Thus, in the negotiation, there is an exchange of control, with each being motivated to gain control of those activities most important to him (i.e., those areas where the child's behavior affects his level of satisfaction more), in return giving up control over those activities of less importance to him. Obviously, it is only in those families where different activities have different relative importance to the child and parent that a mutually profitable exchange can take place.

The initial structure of control in this game is probably not in accord with reality (although the deviation may not greatly affect the functioning of the game). It seems rather that the adolescent has full control over some actions, the parent has full control over others, and for some, both parties have a veto power. However, if the initial structure were changed in this way, the basic commodity exchanged, control over the child's activities, would remain as it is.

There is a more subtle process that develops over time in this game, akin to exchange, but somewhat different. If the parent supervises the child's actual behavior, in a later stage of the game, and punishes the child for deviations from previous agreements, then both parent and child stand to lose in "family happiness." Thus, it is to the long-term advantage of the parent to make an investment of trust in the child, if the child generally honors this trust. (If the child does not, the parent loses satisfaction.) Similarly, it is to the child's long-term advantage to honor the trust, though he may make short-term gains in satisfaction from behaving in ways that give him most satisfaction, regardless of previous agreements. (This investment may also be described as an exchange, with the parent giving up the activity of supervision and punishment in return for the child's giving up immediate gratifications. However, because the returns to the parent are long-term, it appears more useful to describe it as an investment of trust.)

Whenever, as in the exchange between legislators or the agreement between parent and child, there is no exchange of a physical commodity, but merely a promise to perform, the exchange can be considered an investment of trust by the party whose return is most delayed (e.g., in the case of legislators, the legislator whose issue of interest, on which a vote is promised to him, comes up for a vote after he himself has delivered his promise). It is seldom, in areas of social behavior, just as in areas of economic activity, that two activities on which an exchange is made occur simultaneously. Thus is is almost always true that one party must make an investment of trust. In economic exchange, one of the principal functions of money is to facilitate exchange by transferring this trust from the person engaged in exchange to a central authority, whose "promise to pay" will be accepted by all as a trustworthy promise.

In games where the action of other persons or organizations is incorporated in the environmental response rules, such as the LIFE CAREER GAME and the HIGH SCHOOL GAME, a different and less explicit approach to exchange exists. The player acts, and the environment responds with either rewards or punishments, depending upon his action. As indicated earlier, the environmental response rules need not show how this response contributes to the goals of the person or organization whose response is simulated. In the HIGH SCHOOL GAME, the esteem from parents to the adolescent for his achievements is given by environmental response rules representing the parents, according to a schedule that corresponds roughly to empirical reality. It does not show how the exchange of esteem for achievement contributes to the parents' goal achievement, for since the parent is not a player, he need not be motivated to engage in the exchange. It is evident from this example and similar ones that the theoretical foundations of the game must become increasingly rich as the social environment is moved out of the environmental response rules and into the play by actual players.

The exchange of control over actions: It appears that exchange processes generally, in social simulation games and in reality, including economic exchange, can be usefully conceptualized as exchange of control over actions. Because of the interdependence of which society consists, actions taken by one person or collectivity have consequences for others as well. When these consequences for another are great enough, he will seek to influence the action, and often his most efficacious means of doing so is through offering control over another action in return. In economic exchange, where the exchange is ordinarily conceived as "exchange of desirable commodities," or exchange of a commodity for a promise to pay, in the form of money, the present framework would view the exchange as exchange of control over disposal of the commodity. This view accords with that of one of the most perceptive students of the nature of economic exchange, John R. Commons, who insisted that "exchange of goods" is not a fruitful way of describing economic exchange. Commons says, in describing exchange of economic goods, "Each owner alienates his ownership, and each owner acquires another ownership. Prices are paid, not for physical objects, but for ownership of those objects."[2]

In some cases, the control that is exchanged is full control over an individual action. In other cases, it is partial control over a collective action. Both processes are mirrored in the games described above. Apart from this distinction, there appear to be several other important structural differences in exchange processes, all of which have been exemplified above. One of these is the distinction between actual transfer of control and a promise to carry out the action under the other's direction. The former occurs in exchange of control over economic goods, while the latter is more frequent in social exchanges, which are ordinarily described as exchange of "intangibles." The so-called intangibles that are exchanged, such as deference, aid, acceptance, are in fact performance of actions in accord with the wishes of the recipient. A third distinction between exchanges are those in which both actions occur simultaneously, and those, far more numerous, in which one action occurs after the other, requiring an investment of trust by one player. Although the processes of trust investment do occur in the games described here, none of these games simulates extensive investments of trust, such as those that occur when a group allows its activity to be determined by a leader, or the investments which an organization manager makes in subordinates when giving them control over portions of the organization. Investments of trust such as these give rise to important social phenomena which can be simulated in games like the ones described above.

A final distinction in the structure of exchange processes is between those that involve only two parties and those that involve three or more. It may well be the case that player A has control over an action affecting B, B has control over an action affecting C, and C has control over an action affecting A, allowing a mutually profitable three-party exchange to occur where no two-party exchange could have taken place. Infrequently, such three-player exchanges occur among legislators in the Legislature game; but their relatively infrequent occurrence suggest some serious barriers in their way. One of these is the mere mechanical difficulty of discovering a profitable transaction and arranging it; another is the greater investment of trust, requiring each of two players to trust another to whom he may have no subsequent means of retribution. However, certain organizational structures are largely composed of exchanges involving three parties, one acting as a guarantor to one party, in much the same way as the government acts as a guarantor of the value of money exchanged in an economic transaction. For example, in a business organization, one employee performs services for another, and is not recompensed directly by the other, but by the overall management of the organization. It is likely that similar structures exist in a less economic framework.

The possibility of conceiving of all social interdependence in terms of inter-dependence of actions that can lead to mutually profitable exchange of control over actions suggests that all forms of social interdependence can be mirrored by social simulation games, limited only by the imagination, ingenuity, and theoretical acumen of the investigator.

Currency in the system: In economic systems, exchange ordinarily occurs through a physical transfer of goods, or by physical transfer of money. In non-economic exchange, there is seldom a physical transfer (though there are exceptions, such as assignment of a proxy to another person, giving him full control over the casting of the vote). Instead, each gives the other, or promises the other, effective control over an action by undertaking to act in the other's interest, while still retaining execution of the action himself. As a consequence, perhaps the fundamental difference between economic and social exchange is that nothing changes hands in the latter case. Indeed, it could hardly be otherwise, for in most cases of social exchange, it is intrinsically the *other's* action in one's interest that is the desired result. Constituents delegate their political authority to their representative; it is his action in their interest that they expect from him. He can carry out his part of the exchange only by acting in their interest—not by giving them physical control over anything.

The critical question, then, is whether in a social simulation game it is possible, and if possible, desirable, to represent such exchanges by physical transfer of something representing the "thing" that is being exchanged. It was indicated earlier that physical transfer does make an important difference, because it allows use of the thing received for further negotiability. Apart from this question, however, it appears unlikely that in most cases anything could be transferred physically, simply because it is "acting in the other's interest" that is being offered. There are exceptions, such as votes, which could be represented as a transferable commodity; but in general, it appears that the nature of most social exchanges does not allow such a transfer.

This is not to say that no elements in social exchange can be represented by a physical transfer. In nearly all exchanges, the action of one party in satisfying his part of the exchange occurs prior to that of the other. In some of these cases, the payment for the first party's action is not in terms of a specific action in return, but in terms of a kind of "social credit," which the second person can call upon when he needs it. This credit sometimes takes the form of status or reputation, and manifests itself in a variety of ways: deference, willingness to extend trust, and payment through specific actions. It is certainly possible that this "social credit" could be symbolized by a physical transfer of some paper units of account. However, this would be of use only it if served some function: if the notes thus transferred were useful to the recipient, either as negotiable property in further exchange (like the bills of exchange in Lanchashire before 1800, which were promises to pay that came to have the property of negotiability, and passed from hand to hand at face value, although they were private accounts between two parties), or as a debt for which the debtor could be held to account in the courts, i.e., in the rules of the game. Yet neither of these things is true of the social credit that is incurred in social exchange. Thus, provisionally at least, it appears questionable whether a representation of the conceptual quantities that arise in social exchange is possible even if it were desirable.

It may well be that the possibility of such representation merely waits upon the further development of ideas, to provide the basis for a scheme by which accounts are balanced, and also a unit of value in terms of which accounts may be kept. Certainly primitive systems of economic exchange have in early stages not had a unit of value, and have in many ways more nearly approximated social exchange than modern economic exchange. Yet the introduction of money as a unit of account, and strict balancing of accounts into such systems has changed their functioning, and it may well be that a simulation of social processes must not be based on a conceptual structure that consists of a tightly rational and fully accounted system. However, the idea of a conservative system, in which there is conservation of some quantity, such as energy in a physical system, is an attractive one. The issue must remain unresolved, awaiting further theoretical or game development. It may be noted, however, that if one abandons the idea of social simulation games as necessarily mirroring what *is*, he can devise games that represent innovations in social organization, just as the credit card is a recent innovation in economic systems and money is an early innovation, and as the bureaucratic organization is an early innovation in social organization.

SELECTED ISSUES IN THE CONSTRUCTION
OF SIMULATION GAMES

To this point, I have described the approach taken by the Hopkins group to mirroring social processes through games. I would like now to discuss certain issues that have been resolved differently in other games.

In all the games discussed above, the players receive a score, which most often is described as "satisfaction." In some other social simulation games, however, there is no final score at all. Instead, the players are assumed to measure their satisfaction by the events of the game. Two varieties of this approach can be distinguished. In one, the objective outcomes of the game clearly constitute a measure of winning and losing. Among the games discussed above, the Legislature game, in which each legislator receives votes for re-election rather than "satisfaction points" is closest to this. The votes are objective outcomes, and because they constitute a unidimensional measure, they can be used as a score for the game. Similarly, in the commercial game of DIPLOMACY, that nation which outlasts all others is the winner.

The use of such an objective criterion is an excellent measure of success in the game, so long as this single objective achievement is in fact the single objective goal of persons in those roles being mirrored by play of the game. This is most often the case in games which constitute a contest for political power or ascendancy. In such games, the final power positions constitute the outcome of the game. But more often, goals of persons in roles consist of a mixture of objective results, results of different types contributing to the person's satisfaction. When this is the case, it appears difficult to use as a measure of a player's success in the game the objective outcomes of any one of the activities that contribute to the goal. The economist's solution for a similar problem has been to devise a concept of "utility" as a way of giving subjective integration of the otherwise incommensurable objective things toward which the individual strives. Until another theoretical device accomplishing the same thing is discovered, some variant of the economist's solution is necessary.

A second variety of the no-final-score approach stems from quite different directions, from the game designer's distaste for competition, distaste for the idea of "winning" and "losing." It is a defect of social simulation games in general, whether there is an explicit winner and loser or not, that they motivate the players for success *relative* to others, while in some activities (but not all), his goal derives from the absolute levels of results. For example, in the CONSUMER GAME, although units of satisfaction accrue as a result of objective purchases, the player is motivated simply to do better than others, to maximize the positive difference between his satisfaction and that of others. Often, this gives behavior no different than would occur if the goal were in fact to maximize his absolute level of satisfaction; but in certain cases, such as those in which he might act to interfere with another's performance instead of implementing his own, it can be different. Yet, it is not clearly the case that in real life people strive to maximize the absolute level of achievement, rather than the relative one. The phenomenon of relative deprivation in social life attests to the fact that relative outcomes do play an important part in one's level of satisfaction.

The principal defect of the no-winner variety of the no-final-score approach is that it assumes what is hardly true: that the player can understand and internalize the goals of persons in the role he is playing in the game, and when those goals are not given to him by the rules of the game, and then evaluate his performance on the basis of those assumed goals. For if he cannot, his behavior will be aimless, that is without a goal, or will be directed toward incorrect goals, thus destroying the value of the simulation. Parenthetically, I should note that this anti-competitive view apparently is a misdirected generalization from the harm that punishment through low school grades, and punishment from adults generally, does to children. The idea of winning and losing in a game, and accepting defeat, is an early element in the socialization of a child. Children unable to accept defeat in a game are as Piaget's researches suggest, at a very early stage of socialization, approximately the four to six year age level.

A second issue that is sometimes resolved differently in social simulation games is the issue of abstract simplicity versus realistic complexity. Some games, in the area of international relations, legislatures, business, and others, have been developed as realistic and complex configurations of processes, attempting to simulate reality as well as possible. In contrast, the games discussed above are analytic abstractions from reality of single processes or delimited combinations of them. The virtues and defects of each approach as a learning device are not known. But it appears that as aids for theory, they are relevant to different aspects of theory-study of single processes or small combinations of processes. Yet, because they do not attempt to mirror the richness of reality, empirical tests against reality cannot be easily made. There is too little experience, however, to have a good assessment of the values of empirical richness and analytic abstraction in social simulation games.

THE USE OF GAMES AS INSTRUMENTS OF THEORY

The relation between purposive behavior theory and social simulation games is evident from the discussion in earlier sections. It remains here only to suggest the role that the construction and use of games can play in the development of behavior theory.

Social simulation games appear to be most useful in the intermediate stages of theory development—between verbal speculation and a formal abstract theory. For a simulation game appears to allow a way to translate a set of ideas into a system of action rather than a system of abstract concepts. The concept development is necessary (if the concept of money did not exist, it would be necessary to invent it in order for a system of economic exchange other than barter to exist) but what is necessary is not to specify "relations between concepts" in the usual way that theories are developed. Instead, it is necessary merely to embed the concept in the rules of the game.

In addition to those concepts and action principles that are part of the design of the game, additional phenomena arise which require further conceptualization, and extension of the theory. For example, in playing the Legislature game, exchange of votes occur, though this is not in the rules; and observation of this exchange led to: (a) conceptualizing the process as one of exchange of partial control over the collective action; (b) developing the concept of a player's interest in the action as the difference between the utility for one outcome and that for the other (i.e., re-election votes under one outcome and the other); and (c) the proposition that a player will exchange control so as to maximize his control over those actions that interest him most. Again, in the PARENT-CHILD GAME, although the concept of trust and investment of trust plays no part in the rules of the game, behavior arises during the game that suggests these concepts as ways of describing it.

It might almost be said that construction of a social simulation game constitutes a path toward formal theory that is an alternative to the usual development of concepts and relations in verbally stated theory. For rather than abstracting concepts and relations from the system of action observed in reality, the construction of a game abstracts instead a *behavior process,* describing through the rules the conditions that will generate that process. Then, after construction of the game and observation of its functioning, the concepts that adequately describe this process can be created, proceeding next to the development of formal theory. An important virtue of this path is that one learns, by malfunctions of the game, the defects and omissions in his abstraction of the behavior process. As a consequence, extensive corrections to the theory can be made in making the game function, even before the conceptualization that follows play of the functioning game.

NOTES

1. Descriptions of most of the games discussed in this chapter are included in other chapters in this book: the Legislature and Life Career Games in the Boocock chapter, in Part II; the Parent-Child Game, in the chapter by Schild, Part II; the Community Response or Disaster Game, in the Apprendix to the chapter by Inbar, Part III; and the Consumer Game in the Appendix to the chapter by Zaltman, Part III.

2. John R. Commons, *The Economics of Collective Action.* New York: Macmillan, 1950, p. 46.

SOCIOLOGICAL THEORY AND THE "MULTIPLE REALITY" GAME

CATHY S. GREENBLAT

Social simulation games have been described as languages of theory, explicit statements about what the designers believe about reality, and operating models of central features of systems or processes. All these definitions point to the intimate relationship between simulation games and social theory: the games are dynamic representations of theory. Simulation design demands a high level of explication and articulation of theory, for to develop a simulation we have to define the essential features of the referent system, abstract from reality, and develop a mode of presentation of both structural and functional relationships. Through the process of gaming or simulating, therefore, we refine our theoretical formulations (see Raser, 1969; Coleman, 1969; Stoll and Inbar, 1972).

Many of us also use simulations to teach theories to our students. We have our student-players operate a model, observe and analyze the way the system functions, and assess the extent to which it resembles the real-world correlates. We do this in the belief that this process makes it easier for them to grasp and assimilate systemic relationships which are difficult to comprehend through lectures or other linear presentations, and that it makes social science subject matter more real and vivid (see Boocock and Schild, 1968; Greenblat, 1973, 1971).

It is incumbent upon us, then, as designers and users, to examine the theories built into simulation games to see how accurate and inclusive such models are. My perusal of a number of games that are useful for teaching sociology recently led me to the belief that many social simulations have omitted a critical concept of sociological theory: the concept of differing definitions of reality or "multiple realities." The present paper is an attempt to review this aspect of theory and to show how it might be included in game models.

Reprinted from Simulation & Games, Volume 5, March 1974. ©1974 by Sage Publications, Inc.

AUTHOR'S NOTE: *I am very much in debt to Richard D. Duke for his encouragement and comments as I began work on this paper. Harry C. Bredemeier and Allan Feldt also provided helpful criticisms of earlier formulations. A grant from the Rutgers Research Council permitted me to deliver this paper at the Third Annual Symposium of the International Simulation and Gaming Association in Birmingham, England, July 5-9, 1972.*

DIFFERING DEFINITIONS OF REALITY

Men go through their daily lives trying to make sense of the world in which they live. Through the process of defining situations and constructing realities from shreds and patches of experience and memory, they impute meaning to persons and objects (see Berger and Luckmann, 1966; McHugh, 1968; Lyman and Scott, 1970; Cicourel, 1969; Douglass, 1971). The same events, persons, or objects, however, take on different meanings for different people, who thus create varying interpretations, or "multiple realities."[1] Literature and drama abound with examples—e.g., *Rashomon* (Kanin and Kanin, 1959) and *$100 Misunderstanding* (Gover, 1961), and all of us have seen personal examples or social science accounts of husbands' and wives', representatives of labor and management, or inmates' and guards' highly conflicting accounts of the same happenings (e.g., Manocchio and Dunn, 1970).

These definitions of reality tend not to be totally unique and individual. There are several sources of commonality and difference in people's perceptions and definitions, including (1) positions in the social structure, (2) goals which direct attention and order priorities, and (3) personal biographies. First, members of society are born into groups which have social heritages with preformulated systems of relevances and typifications, their own argots, and sets of values. Through the process of socialization, the newborn, like the newcomer to any group, is given a picture of reality as seen by his group. This picture includes definitions of himself, his group, and others, and of the appropriate degrees of contact with various others who have contrary or conflicting information or ideas. Hence, as his view of the world is developed, it tends to become somewhat insulated from challenges (see Allport, 1954). Position in the social structure accounts, then, not only for the extent of available resources, but for the rules, norms, and definitions one receives, for the degree of cross-group communication, and for the extent of challenge to certainty one encounters and accepts.

Multiple realities also stem from variability of goals. Different goals are salient for different people at any given time, and to the same person at varying points in his life. Depending upon one's goals, various aspects of a situation or event will be seen as relevant. Where men have different goals, we can at least expect that they will have different evaluations of things, if not different interpretations.

In addition, any social actor has a history, and, hence, definitions of a situation are partly biographically determined, affected by the individual's unique stock of previous experiences and recollections. Individual combinations of contacts and socialization experiences thus render differences between the definitions that might otherwise be expected of those with similar roles and goals (see Heeren, 1970). Thus, multiple realities often arise among those variously situated in the social structure with respect to the threats, dangers, or liabilities they are exposed or vulnerable to, as well as the opportunities and action alternatives open to them.

These multiple realities are more than just objects of curiosity for the analyst of social life, for differences in the meanings of events and persons produce differences in behavior. W. I. Thomas' famous aphorism, "If men define situations as real, they are real in their consequences" (Thomas, 1928: 584) offers the link between the social construction of reality and the encounters between men. To understand social behavior, one must understand not simply the social structure, but the images of reality that members carry and which make certain kinds of behavior appropriate and, more than appropriate, typical. Constraints and opportunities exists not simply as objective characteristics of social systems, but in the heads of participants. To understand a person's behavior, then, you have to understand what he *thinks* exists, not what "really or objectively exists." It is not, for example, the number of actual job opportunities that you must inquire about to comprehend the behavior of the ex-convict who feels pushed to return to crime as the "only way." It is not just the "real power" of a student body you must assess to understand the dynamics of campus politics at many colleges, but the students' *felt* powerlessness to alter existing conditions. Sociological literature contains increasing data attesting to the ways in which one belief may lead to another, affect later observations and behaviors, and thus create self-fulfilling prophesies (e.g., Merton, 1968; Lemert, 1962; Becker, 1964). Some of the ways in which our *expectations* may create what seem to be scientific findings have also been documented (Rosenthal, 1966).

For the sociologist, then, an important step in trying to comprehend a social system is to learn how the actors define their situations and the events that transpire. If, as game designers, we wish to create models that operate like these real systems, we must simulate both the structural elements and the differential perceptions of system participants.

Have we done this? Generally, I think not. Most games I have seen include two of the sources of differential definitions: different roles and different goals for players. Thus, over the course of play, participants may develop alternative perceptions of the same things and events. In most games, however, the constraints carried in the heads of the real-world system-participants—constraints developed through their prior socialization—are not given to the players. Hence, the multiple realities, conflicting views, and barriers to challenge of these views which are of critical importance *from the first moment of interaction* between the real-world system participants are absent. This frequently leads to game behavior quite different from behavior in the system simulated. We cannot and should not rely on game players who have not lived within the referent system to come to the interaction with the stereotypes, blinders, and prior ideas which affect the perceptions, attitudes, and behavior of their real-world counterparts. Yet this is what we have generally done.

What I am arguing, then, is that players should be given different and conflicting information, corresponding to the different and conflicting perceptions held by their real-world counterparts. All should not be presented with the same manual for play. It is not enough to provide *missing* information; we must provide *mis*information. We have to allow different views of reality, if not "corrected," to create differential opportunities and restrictions, and thus to provide the potentials for self-fulfilling prophesies.

Let me offer an example, utilizing a college setting, since that is presumably a familiar system to most of us. At hypothetical College X, if you asked the junior faculty about the distribution of influence among junior faculty, senior faculty, and administration, they might tell you that, of every ten "units" of influence, junior faculty have one unit, senior faculty have two units, and the administration has seven units. Further interviewing at College X might reveal that the senior faculty believe that junior faculty have three units, senior faculty have four units, and the administration has three units. The administrators, you might find, see the distribution yet a third way: junior faculty, two; senior faculty, five; and administration, three. A social scientist, coming to do a study of power and influence at College X, and thus assessing the relative influence of the three constituencies, might conclude that the distribution is really junior faculty, one unit; senior faculty, four units; and administration, three units.

Now we decide to build a simulation of the system. If we follow the usual procedure, we are likely to turn to the social scientist's "objective" description and write a game manual in which all players are informed that the junior faculty has one influence point, senior faculty has four influence points, and the administration has three influence points to allocate. Then we will give them the task of making decisions, voting on policies, negotiating, and so on. What we have thereby done, I believe, is to act as if the reality that motivates each real-world group and underlies their actions is the scientific reality rather than its own individual reality. But remember W. I. Thomas' statement, quoted earlier: "If men define situations as real, they are real in their consequences." College X operates as it does because of the separate realities that are brought together through the interaction of the constituencies; thus, the simulation must include the three realities. Rather than creating a shared reality that is nonexistent in the real-world college, we must tell each group of game players that the distribution is as their real-world counterparts see it. Each group should thus be given misconceptions about the influence possessed by other groups and about the proportion of influence each group—including itself—has, as shown in Table 1. Thus, we should tell junior faculty role players that the distribution is as the real junior faculty perceives it: one, four, five, and then give them their one point of influence. We should tell the senior faculty role players that the distribution of power is as senior faculty described it to us: three, four, three, and then give them their four influence points. Finally, we should tell the administration role players that the distribution is two, five, three, and give them three points. Each group would thus have the number of influence points they think should have—that is, one, four, and three, respectively—but they would be operating in terms of multiple realities. Under these conditions, I believe the behavior of the role-players would bear considerably more resemblance to the behavior of the real-world counterparts at College X than if the game were created as most seem to be, with all participants starting play with a shared reality. As in the real world, with this kind of model, players might think they are playing the same game, but in fact they are playing different versions of a game with the same name!

TABLE 1. Differences Between Perceived and Actual Influence at College X and in Simulation of College X

	Perceived influence	Perceived proportion of total influence	Actual influence	Actual proportion of total influence
Junior faculty beliefs:				
About junior faculty	1	1/10	1	1/8
About senior faculty	2	2/10	4	4/8
About administration	7	7/10	3	3/8
Senior faculty beliefs:				
About junior faculty	3	3/10	1	1/8
About senior faculty	4	4/10	4	4/8
About administration	3	3/10	3	3/8
Administration beliefs:				
About junior faculty	2	2/10	1	1/8
About senior faculty	5	5/10	4	4/8
About administration	3	3/10	3	3/8

METHODOLOGY

How, then, does the designer go about introducing a "multiple reality" component into his games? The methodology will vary depending upon the element or elements he wants to develop in this way: role definitions, resources, or rules of play.

The general procedure in constructing role profiles for a multiple reality game involves asking not, "What are the groups (constituencies) and how can they be characterized?" but rather, "What are the groups and how do they characterize themselves and one another?" Instead of deriving a linear set of "objective role definitions," the designer thus generates a matrix of the sort shown in Figure 1.

The cells of the matrix that are marked with an asterisk (*) contain data of *self-perceptions.* Row blocks indicate the multiple realities with respect to each group. For example, the row marked ⁄⁄⁄⁄⁄ contains data on different ways in which the As are defined—by the As themselves, by the Bs, and by the Cs. The column blocks present the specific, individual views of reality of each particular group. Hence, the column marked ＼＼＼＼ , for example, contains the data on how the As view themselves, the Bs, and the Cs. The information in the total matrix offers the multiple realities of role definitions as of the beginning of the simulation.

Group being
described,
defined:

Group describing or defining:

A   B   C

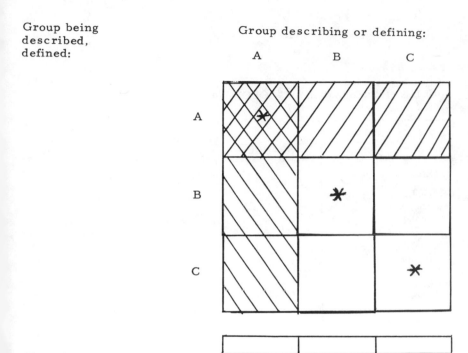

Shared information about
A, B, and C

FIGURE 1. Role Definition Matrix for a Multiple Reality Game

A matrix such as this serves as the basis for role definition in the players' manuals. The As would receive, in prose form, the information in column A; the Bs, the information in column B; the Cs the information in column C; all would receive the "shared information." Table 2 shows a hypothetical example of a filled-in matrix and Table 3 offers the corresponding manual pages.

Notice that only the game director is aware that each of the groups actually possesses the same amount of influence: medium. Each group misperceives the others and is in turn misperceived by them. While the Oranges, for example, see themselves as Liberal, they are seen by the Greens as Radical; likewise, while the Greens see themselves as Moderate-Conservative, the Oranges see them as Ultra-Conservative.

From this point, the design process is similar to the "regular" design process: goals must be spelled out, a scenario created, and rules of play determined.

TABLE 2. Role Definition for "COLOR WAR AND PEACE"

Group Being Defined:	Group Defining:		
	ORANGE	**BLUE**	**GREEN**
Aspect:			
ORANGE:	*		
Sociopolitical orientation	Liberal	Liberal-Radical	Radical
Power or influence	Medium	Low	Low
Definition of present situation	Terrible	Very bad	Bad
BLUE:		*	
Sociopolitical orientation	Moderate to Conservative	Moderate	Liberal
Power of influence	Low	Medium	High
Definition of present situation	Poor	Bad	Very Unpleasant
GREEN:			*
Sociopolitical orientation	Ultra-Conservative	Conservative	Conservative-Moderate
Power or influence	High	High	Medium
Definition of present situation	Good	OK	OK with some strains

SHARED INFORMATION: (i.e., all three groups know the following) Oranges see Blues as sympathetic to them but not very understanding, and see Greens as hostile and rigid. Blues see Oranges as unappreciative of their support, and see Greens as patronizing and suspicious. Greens see Oranges as highly agressive and immature, and see Blues as sympathetic but not very understanding.

TABLE 3. Role Profiles for "COLOR WAR AND PEACE"

As given to ORANGES:

"There are three groups in the game, as described below: Each player will be assigned to one of these groups, so read the descriptions carefully.

    (1) ORANGES

         The Oranges' sociopolitical position can be characterized as Liberal. They wield a medium amount of influence, and define the present situation as terrible.

    (2) BLUES

         The Blues' sociopolitical orientation is Moderate to Conservative. They wield a low amount of influence, and define the present situation as poor.

    (3) GREENS

         The Greens' sociopolitical orientation is Ultra-Conservative. They wield a high amount of influence, and see the present situation as good.

Past relations among the three groups have been strained. The Oranges feel the Blues have been sympathetic but not very understanding, and that the Greens have been hostile and rigid. The Blues feel the Oranges have been unappreciative of their support, and feel the Greens have acted in a patronizing and suspicious manner. Finally, the Greens feel the Oranges have been highly aggressive and immature, and that the Blues have been sympathetic but not very understanding."

As given to the BLUES:

"There are three groups in the game, as described below. Each player will be assigned to one of these groups, so read the descriptions carefully.

    (1) ORANGES

         The Oranges' sociopolitical orientation can be described as Liberal to Radical. They wield a low amount of influence, and they define the present situation as very bad.

    (2) BLUES

         The Blues' sociopolitical orientation is Moderate. They wield a medium amount of power and view the present situation as bad.

    (3) GREENS

         The Greens' sociopolitical orientation is Conservative. They wield a high amount of influence and see the present situation as OK.

Past relations among the three groups have been strained. The Oranges. . . .
(CONTINUES SAME AS FOR ORANGES)

As given to the GREENS:

"There are three groups in the game, as described below. Each player will be assigned to one of these groups, so read the descriptions carefully.

    (1) ORANGES

         The Oranges' sociopolitical orientation can be described as Radical. At the present time they wield a low amount of influence and define the present situation is very poor.

    (2) BLUES

         The Blues' sociopolitical position is Liberal. They wield a high amount of influence, and define the present situation as very unpleasant.

    (3) GREENS

         The Greens' sociopolitical position is Conservative to Moderate. They view the present situation as unpleasant, and wield a medium amount of influence.

Past relations among the three. . . . (CONTINUES SAME AS FOR ORANGES)

The multiple reality game need not, as in the preceding example, contain homogeneous subgroups. The matrix in Table 4 shows how the designer could set up differential resources based upon varying beliefs and expectations within and across groups. Note that this is quite different from saying that the groups have different *values*, which is commonly noted in games. Table 5 then offers a sample filled-in matrix.

Individual role profiles would then be constructed. So, for example, one Yellow player might be given beliefs 1b, 2a, and 3b: another might get 1c, 2b, and 3c; and so forth. The Purples' theory would also be given out. Distribution could be either through deliberate or random combination, depending upon the designer's beliefs about whether the beliefs tend to cluster.

Note also that, in this sample, the *actual* distribution is such that both groups have equal objective access to the resources. Yet differing proportions of the two groups perceive their resources as equal to those of the other group; some see their group as disadvantaged, and others see it as "advantaged." I believe that, as in the real world, these perceptions would be likely to persist and color the participants' views and interpretations of events that transpire. So, for example,

TABLE 4. Matrix for Creating Differential Resources Based upon Differential Beliefs and Expectations in a Multiple Reality Game

Belief sets and specific beliefs:	Group A	Group B	Designer's actual set-up
Set X:			
1 ⎧ Exhaustive,	____ %	____ %	
2 ⎨ mutually	____ %	____ %	
3 ⎩ exclusive	____ %	____ %	
beliefs			
	100%	100%	
Set Y;			
1 ⎧ Exhaustive,	____ %	____ %	
2 ⎨ mutually	____ %	____ %	
3 ⎩ exclusive	____ %	____ %	
beliefs			
	100%	100%	
Set Z:			
1 ⎧ Exhaustive,	____ %	____ %	
2 ⎨ mutually	____ %	____ %	
3 ⎩ exclusive	____ %	____ %	
beliefs			
	100%	100%	

TABLE 5. Resource Distribution in "COLLEGE COLOR WAR AND PEACE"

Belief sets and Specific beliefs:	% of YELLOWS with this belief	% of PURPLES with this belief	Actual distribution
1. Academic resources			
a. Yellows get preferential treatment from professors	20	25	
b. Yellows and purples get equal treatment from professors	20	50	Given equal time invested in studies, yellows and purples have equal proba-
c. Purples get preferential treatment from professors	60	25	bilities of good grades
2. Money resources (scholarships)			
a. Yellows have most of the scholarship funds, though purples have as much need	0	20	
b. In general, purples are from wealthier backgrounds than yellows; they have less need for scholarships and fewer have such funds	80	80	Scholarship funds are distributed on the basis of need. 40% of the Yellows have scholarships, and 20% of the purples have scholarships
c. Although purples are generally wealthy and have no need for assistance, they have most of the scholarship funds	20	0	
3. Potentialities for intergroup relations			
a. Yellows don't want to have relationships with purples (although they may pretend to)	0	40	100% of each group are told that inter-group relationships are desirable, but 60% of each group
b. Yellows and purples generally desire good relations with one another and can have them with a little effort	60	60	are told this is possible, and 40% of each are told it is impossible be-
c. Purples don't want to have relationships with yellows (though they may pretend to)	40	0	cause the other group doesn't want relationships

to determine any player's grades for a given round, the game coordinator might ask how many hours had been invested in studies, ask the player to roll a die, and, using something like the chart in Figure 2, tell him his grades. As was noted in Table 5, the same probabilities prevail for Yellows, and Purples; several, however, are not likely to believe this and will make decisions accordingly.

Time	Die	Grades		Time	Die	Grades		Time	Die	Grades
Low	1 —	F		Medium	1 —	D		High	1 —	C
	2				2				2	
	3 >	D			3 >	C			3 >	B
	4				4				4	
	5 >	C			5 >	B			5 >	A
	6 —	B			6 —	A			6	

FIGURE 2. Grades in "COLLEGE COLOR WAR AND PEACE" Game, Based upon Time Spent and Roll of Die

CONCLUSION: PROPOSALS FOR DEVELOPMENT AND USE OF THE CONCEPT

I am NOT, I wish to emphasize, urging that all games designed from now on include a multiple-reality component. Promising as I think the idea is, it carries with it a number of costs that must always be weighed against the benefits. I would like, then, to conclude by reviewing some of these factors, both negative and positive.

The costs include, first, the need for secrecy and careful planning to prevent players from immediately learning of the differential materials. All versions of the manual or other multiple-reality materials must be made to look the same to players, but be like a deck of "marked cards" so that the game director can distribute them easily and accurately. Ballotting or other procedures must be secret, so players cannot see that the actual number of votes cast in the resource distribution has been differentially defined. But these are not brand new problems: in CRISIS, the director must be careful in the distribution of the two newspapers; in PLANS and METROPOLITICS, votes must be privately turned in to the coordinator, albeit for different reasons than those in the procedure advocated here.

A second cost is closely related to this: such materials would be more expensive to produce, as several manuals must be provided, and the packaging is more complex.

A third cost also derives from the need for secrecy: the game would only "work" with players who did not know of the built-in differences in perceptions. Thus, the game director would have to urge those who have played, at the conclusion of the postgame discussion, not to reveal the "secret" of the game. Much as theatre producers of plays such as *Sleuth* enjoin such silence on the part of members of the audience, so game directors would have to urge players to say nothing. To the extent that the game was enjoyable and worthwhile, and players understood the functions of the not-to-be-revealed components, their cooperation could probably reasonably be expected. Again, this is not a totally new demand on the game director, who has already made such pleas with such games as STARPOWER. The problem, however, would be vastly compounded if a high proportion of new games were to incorporate my multiple-reality suggestions, for now we would have the problem of players who had heard nothing about the *specific* game they were about to play, but suspected the multiple-reality "fix" because they had encountered it in several other games. How much of a problem versus a gain this is depends upon whether such "game sophistication" is paralleled by the sociological sophistication of realization that the "trick" is really not a "trick," but rather a simulation of a critical dimension of each of the systems.

This leads me to the fourth potential "cost" of the multiple-reality game: the chance that participants may become hostile to the game director on discovering the existence of different manuals with different information. I firmly believe that the game designer bears responsibility for presenting the compound information in a way that minimizes such reactions, preparing the game director for this possibility, and providing him with the background information and rationale for the multiple-reality inclusions so that he can make the experience a fruitful one (even for those who react adversely) by accurately explaining them in the postplay discussion.

Fifth, and finally, I believe that the benefits will often not outweigh the costs where the game is designed to simulate a social system or process in a very abstract way. For example, in SIMSOC, players begin with minimal structural arrangements and confront the need to develop a social organization. The seven basic groups and four regions have no direct, exact counterparts in any real-world society. Hence, no "prehistory" vis-à-vis one another is necessary to avoid artificial and often deceptively facile early interactions although it is desirable, as I have argued, in games that purport to simulate more specific systems or processes, such as negotiations between school board members or between city residents confronted with referenda on government reform. Whereas I believe that the latter game is strengthened by my inclusion, the former, highly abstract game, I believe, teaches better by allowing the multiple realities to be emergent rather than supplied.

What, then, are the benefits to be weighed against these increased costs to both the designer and the user? The discussion in the body of this paper has offered simply the embryonic formulations of my idea. A wide variety of elaborations is possible, from the simple to the highly complex. For example, one could vary not simply the agreement-disagreement dimension of consensus, as was done in the examples, but also could vary the knowledge or understanding of this disagreement, and even the realization of the understanding or misunderstanding (see Scheff, 1967). Many of the complexities of differential perceptions, as they affect social behavior, therefore, could be built into simulations by the ambitious designer. Such enterprises should bring the rewards indicated at the beginning of the paper: higher verisimilitude of participant-real-world behavior, and, hence, a better teaching tool or research device. Rather than hoping some player will come up with a bluff, that another will spread rumors or scandalous reports, and that yet another, like Hamlet, will feign madness, we build in the bluffs, rumors, scandalous reports, and charges of madness that we know play a significant part. Rather than waiting for the "equiprobability through ignorance" principle to operate to lead players to reduce ambiguity through creating interpretations, we provide the definitions known to characterize system participants (Brim, 1955). Particularly where players are naive about the nature of the system and the types of attitudes and behaviors characteristic of those whose roles they are playing, or where the backgrounds of participants are quite different from those system members or are generally homogenous, the supplying of those perceptions, stereotypes, and blinders that guide and influence the real-world system participants should contribute to simulation activity and attitude closer to actual behavior. Hence, observation and analysis are likely to generate better data for testing of model versus real world, and thus to provide a more fruitful learning experience for students and for those of us who hope to utilize simulations for the construction and refinement of theory.

NOTES

1. This is a somewhat different usage of the term than that of Schutz (1962).

REFERENCES

Allport, G.
    1954 The Nature of Prejudice. Reading, Mass.: Addison-Wesley.

Becker, H.
    1964 The Other Side: Perspectives on Deviance. New York: Free Press.

Berger, P. and T. Luckmann
    1966 The Social Construction of Reality. Garden City, N.Y.: Doubleday.

Boocock, S. S. and E. O. Schild
    1968 Simulation Games in Learning. Beverly Hills: Sage Pubns.

Brim, O. G., Jr.
    1955 "Attitude content-intensity and probability expectations." *Amer. Soc. Rev.* 20 (February): 68-76.

Cicourel, A. V.
    1969 "Basic and normative rules in the negotiation of status and role," in D. Sudnow (ed.) *Studies in Interaction.* New York: Free Press.

Coleman, J.
    1969 "Games as vehicles for social theory." *Amer. Behavioral Scientist* 12 (July/ August): 2-6.

Douglass, J.
    1971 American Social Order: Social Rules in a Pluralistic Society: New York: Free Press.

Gover, R.
    1961 *The $100 Misunderstanding: A Novel.* New York: Grove.

Greenblat, C. S.
1973 "Teaching with simulation games: a review of claims and evidence." *Teaching Sociology* 1 (October).

———

1971 "Simulations, games and the sociologist." *Amer. Sociologist* 6 (May): 161-164.

Heeren, J.
1970 "Alfred Schutz and the sociology of common-sense knowledge," pp. 45-56 in J. Douglass (ed.) *Understanding Everyday Life: Toward the Reconstruction of Sociological Knowledge.*

Kanin, F. and M. Kanin
1959 "Rashomon." (adapted from the stories of Akutagawa) New York: Samuel French.

Lemert, E. M.
1962 "Paranoia and the dynamics of exclusion." *Sociometry* 25: 2-20.

Lyman, S. M. and M. B. Scott
1970 A Sociology of the Absurd. New York: Appleton-Century-Crofts.

McHugh, P.
1968 Defining the Situation. New York: Bobbs-Merrill.

Manocchio, A. J. and J. Dunn
1970 The Time Game: Two Views of a Prison. Beverly Hills; Sage Pubns.

Merton, R.
1968 Social Theory and Social Structure. New York: Free Press.

Raser, J.
1969 Simulation and Society: An Exploration of Scientific Gaming. Boston: Allyn & Bacon.

Rosenthal, R.
1966 Experimenter Effects in Behavioral Research. New York: Appleton-Century-Crofts.

Scheff, T. J.
1967 "Toward a sociological model of consensus." *Amer. Soc. Rev.* 32 (February): 32-46.

Schultz, A.
1962 "On multiple realities," pp. 207-259 in Collected Papers I: The Problem of Social Reality. The Hague: Martinus Nijhoff.

Stoll, C. and M. Inbar
1972 Simulation and Gaming in Social Science. New York: Free Press.

Thomas, W. I.
1928 The Child in America. New York: Alfred A. Knopf.

SPECIFICATIONS FOR GAME DESIGN

RICHARD D. DUKE

The following list is a series of questions for which parameters must be provided *before* game construction begins. The responses, if carefully delineated, provide detailed specifications at the outset of game construction against which the final product can be evaluated.

IS GAMING APPROPRIATE?

Define the Problem

What is the need, conditions, or circumstance that prompted consideration of a game? This statement should be brief and convincing to a neutral observer. The problem statement must be sufficiently detailed to permit evaluation of the success of the game when completed.

(1) Client: Who is the responsible agent? Who is authorized to approve the detailed specifications prior to construction of the game? Who will evaluate for successful completion? *If no client* (as when the game is spontaneous by the designer), an imaginary client should be conjured to ensure thoughtful review of the specifications.

(2) Purpose: What is the primary purpose to be achieved through the game? To transmit information to an audience? As a questionnaire to extract information or opinion from the players? To establish dialogue between players (for example, as a research team)? To motivate players or to prime them for some related experience? To provide an environment in which creative ideas will spontaneously occur? If more than one of these purposes applies, each should be stated explicitly, and they should be clearly placed in order of priority.

(3) Subject matter: The substantive material which is to be dealt with by the game should be defined as explicitly as possible. If no specific subject matter is implied (e.g., "frame" games like POLICY NEGOTIATIONS, NEXUS, or IMPASSE?) this should be stated and typical substantive example(s) cited.

Reprinted from Gaming: The Future's Language, by Richard D. Duke. ©1974 by Sage Publications, Inc.

(4) Intended audience: For whom is the game intended? Will the players be homogeneous or heterogeneous? What age? Motivation to participate? Sex? Size of group? If several audience profiles are anticipated, define each category. List in order of priority to indicate which group is prime if design considerations require a tradeoff.

(5) Context of use: Under what conditions will the game normally be used? Will it be free-standing or part of a series? Or in conjunction with an academic course or training program? What follow-up circumstances are anticipated? Will the same group run the same exercise repeatedly?

Practical Constraints

What mechanical, political, financial, or other considerations are anticipated relative to constraining the use of the product?

(1) Resources: What financial resources are available for game design, construction, and testing? Is there flexibility to permit alternative designs or cost overruns? How much? Under what circumstances? What financial constraints govern the use of the game? Is the typical cost of play to be within certain limits? What are they? How much time is available for game design? Construction? Testing? Is evidence of productivity to be demanded? At which stages? What evidence? What are the time constraints during the *use* of the game? May it be discontinuous? Will evening or weekend sessions be a normal operational style?

(2) Paraphenalia: Are any constraints to be imposed on materials used by the game? Must it be portable? Under what conditions? Are storage requirements to be specified? What are they? What demands for reproduceability are to be specified? Must the kit use standardized materials? What are they? Is do-it-yourself reproduction permitted? Required? Are instructions for reproducing the kit to be included?

May a computer be used? Which one(s)? What programming language(s) are permitted? Required? Must it be batch-processed? Run on terminals? Are special peripherals permitted (X-Y plotter, C.R.T. tube)? What are the limits to computer use in dollars or hours? Per cycle? Per run?

Medium Selection

State the reasons for not using other forms of communication (games are generally the most specific and therefore the most costly medium). What are the particular characteristics of the message to be conveyed? May other media be employed with the game? Required? Permitted? Under what constraints (duration, timing, frequency, structured or spontaneous)?

DESIGN

Conceptual Map

Games are a communications medium. What is to be conveyed by this game? Define the system, its components, characteristics, roles, linkages (component to component, role to role, component to role). What considerations are to be emphasized? What themes, issues, or problems are to be stressed?

How is this "message" to be transmitted to the player? Is it to be implicit, buried in the game? Explicit as graphic material or text? Integral in that play of the game requires confrontation with it?

If the message to be transmitted (reality system) is unknown, the purpose of the game being to deduce or extract some perception of reality, system, or message, in what format should the game produce it? Only as a perception retained by the players? As a written record through text and graphics (player-generated)? As a physical representation (iconic model) created by the players? What roles govern the retention, release, or transmission to sponsor of results of play?

Gaming Considerations

Gaming is characterized by an incredibly diverse set of techniques. Which of these are desirable? Required? Prohibited? Are parallels with particular existing games sought? Which ones? Any to be omitted? Why have they been rejected?

(1) Repertoire: What style of game is sought? Emphasis on group dynamics? Or more intellectualized emphasis? Should allocation of scarce resources be central? Or is emphasis to be placed on formulating or conveying a system's gestalt? What basic style or character is desired?

Is the material (message, conceptual map) to be expressed from a particular orientation (field or discipline—e.g., psychology, geography, etc.) implying both the perspective and jargon?

What level of abstraction is desired? Geographic or social scale? What time frame or horizon? What scale of reduction in time? How are power or finances to be conveyed?

Are game management constraints to be specified? Operator skill, quantity, or training? What level and style of protocols and administrative forms are desired?

How are players to be organized? By group? Individually? Are coalitions to be permitted? Encouraged? Are mult. e roles for one player appropriate?

Are particular analogies to be encouraged or omitted by the designer (some known system or concept)?

What structure is to be incorporated for issue generation within game play?

Are they to be predetermined or player-generated? Random or preset? Linked? Under what conditions?

Is a physical board to be used? Are constraints of size or complexity required?

(2) Art: "Gaminess" in the final analysis will depend on the skill of the designer. Some considerations, however, may be specified by the client which limit or direct this somewhat elusive characteristic. What degree of player involvement is desired? Large emotional content or emphasis? Or a more deliberate or intellectualized character? Ideally periods of intense involvement will be interspersed with more detached or analytic sessions.

How is the game to be staged? Are particular room arrangements or player configurations to be specified? Are constraints to be placed on pre-game activity? Are warm-up exercises to be considered? Under what conditions are critiques to be held? What duration? Is the style to be free (a "happening") or more controlled?

Are particular learning principles to be emphasized? Is an iterative approach required? How many cycles minimum? Are they to grow more complex? Can level of complexity be stipulated or guided?

Games are best perceived as environments for self-instruction. Are players to have complete freedom of movement within the game or is it to be guided? By the designer? By the operator? By the conscious choice of the players? Are the players to be permitted to invent their own rules? May they alter game procedures? Under what circumstance and frequency?

(3) Design principles: Several elements of design are common to all (or most) games; it may be desirable to specify conditions or characteristics of some of these as design requirements. Is a scenario to be specified? Are alternative scenarios desired? Is the scenario to be explicit, detailed? Highly abstract? To mimic some existing source document? Is it to be capable of modification by the players? The operator? How much resources (time, money) may be permitted for modification during a typical use of the game?

A game is a medium which employs its own distinct "language," and therefore each game requires a unique symbolic structure. The character of this structure should be specified. Is it to be physical? Three-dimensional? Should its complexity be restricted to permit being learned in a specified time by new players? Are symbols to be of a commercial source (such as Lego)? Are they to be presented initially in a codified form? A glossary? A visual aid for continuous display during play? Is a board to be employed? A map or other visual(s)? What degree of complexity is permitted (quantity of new symbols employed, visual details of maps or boards)? Are there any constraints or requirements as to conventional languages or media to be employed (Math? Statistics? Fortran? Flow charting)?

The character and utility of a game are heavily influenced by the rules and procedures they employ. These may be very rigid (as in games of logic);

procedural in that they are specified as requirements to orderly play (as in most social science games); or only partially existent (as in situations mimicking a social dynamic) where players are encouraged to develop their own, or to modify a starting set. What is the character sought in this game? Are circumstances governing the basic accounting system to be treated differently from rules governing player behavior? Is the accounting system to be expressed in a computer language? If not, may it be so complex as to require a calculator? How much operator training for use? Time for routine processing?

Basic complexity of play of any given game is probably best revealed by inspecting the list of "steps of play." Are these, or any subset, to be pre-specified (e.g., "Prepare a budget," or "Make written estimate of consequence of . . . before viewing next output," etc.)? How lengthy or involved is this to be? Can the desired effect be described as advice to the designer? How are they to be presented to the players? Initially? Each cycle?

What information flows are to be provided for? Denied? Emphasized? (Player to player, player to component, component to component.) Are they to be monitored? Recorded? Under what circumstances? By whom? Are they to be preserved beyond the game?

What time scale is to be employed? How is the game to be paced? What duration to a cycle? A game? Is timing to be truncated with successive cycles?

Concept Report

Games are notoriously hard to evaluate. This can be improved upon if a concept report is required of the designer *prior* to game construction. This report should be a synthesis of those considerations reviewed under "Is Gaming Appropriate" and "Design." It should be a statement of the reality to be conveyed or objectives to be achieved through the game. Is the concept report to be reviewed? By whom? What time intervals for review? What procedures apply to resolve disputes? Is the final product to conform to the concept report? What latitude of difference is acceptable? What penalty (ies) apply in case of failure? Is construction to begin before the detailed concept report is approved by the sponsor?

CONSTRUCTION

If the concept report is carefully prepared and reviewed, construction should be routine and uneventful. However, if the project is large, some particulars may be pre-specified by the client.

Pre-Player Activities

Are project management procedures to be specified? Are reports to be prepared which document the various components of the game? If so, how circulated? Must there be client approval? Most games deal with systems that are complex and nonlinear; yet game processing is usually rigidly sequential because of mechanical constraints. The order of processing of various models, components, or decisions must in some sense be artificial. Is it to be pre-specified to the designer? In games of any complexity, there are numerous components (roles, models or simulations, paraphernalia, accounting systems, etc.) which must be separately designed and/or constructed. These are then assembled and tested. Usually many modifications are necessary to bring a good game together, often necessitating compromise with original specifications or objectives. Are such changes to be reported to the client? Is the concept report to be altered to reflect them? Is approval required? Is data collection, reduction, and loading subject to any quality constraints? Which data items? What degree of accuracy during the assembly and linkage? The entire model must be calibrated to reflect accurate response. Fudge factors (arbitrary values to scale some component up or down) are often used to correct errors. Are those to be reported and appended?

Testing the Game with Players

Most games cannot be claimed as valid unless they have met the "Rule of 10"—at least ten "live" games, the last three of which required no changes (other than perhaps cosmetic or superficial adjustments). Is the game to be tested? With what conditions? Are written evaluations to be required of players? Is the game to be demonstrated for the client?

USE

The game, when complete, is no longer the responsibility of the designer; however, certain considerations incidental to use may be pre-specified.

Dissemination

(1) Ethics: Is the product to be private property (client's or designer's) or in the public domain? What are the designer's obligations to the sponsor? To users? If private property, who holds the copyright or patents, if any? Are royalties to be paid?

What user-related ethics must be considered? Is the design nonmanipulative? Is related literature accurate in describing the game objective and particulars? Are certain debriefing considerations to be mandatory for player protection? Is the game to be used in some larger training context? Are "demonstration" runs potentially a waste of player time which they must be appraised of?

(2) Mechanics: What requirements are to be established for packaging, distribution and operator training? Is the game to be modified and updated? How will these changes be released? Who will distribute the game? Are fees or charges

to be permitted for game materials? Are these to be limited to a particular dollar amount? Are they to be waived for certain users? Is the client entitled to multiple sets? How many?

Interpretive Criteria

Games, unlike other media (books, films) are very difficult to judge unless they are actually played. While this may be an inherent difficulty, some steps can be taken to lessen the problem for perspective users.

Is a classification number of some type to be assigned and included in the documentation (e.g., Dewey Decimal or Library of Congress)? Is a description to be included integrally to the kit which addresses the major points under "Use," or is an abstraction of the concept report required? Is this to be pretested with potential users and reevaluated after they have used the game to ensure accuracy and clarity?

Evaluation

Is the designer-builder required to initiate a formal evaluation of the utility of the product? Is validity to be measured against the reality it purports to address? Against the independent judgment of professionals who know the intended context of use? Against the independent judgment of professional gamers? Against the reaction of players? Other criteria? All of the above? If the game is found wanting, what procedures apply for its modification?

To conclude, these notions are intended to suggest a reasonably comprehensive set of questions a client (and/or designer) may wish to specify before a game is commissioned. Since each game is to fill a specific need, these can only be used to prompt a careful search of conditions appropriate in your particular context.

OBSERVATIONS ON THE DESIGN OF SIMULATIONS AND GAMES

ALLAN G. FELDT and FREDERICK GOODMAN

University of Michigan

On several occasions the two authors have served together as panelists discussing the design of games. Through repetition and argument we have found ourselves arriving at a reasonable consensus on a number of points regarding game design which seemed sufficiently insightful and potentially helpful that we felt they should be shared with a wider audience; hence, this paper. Before beginning the enumeration of our joint observations, we have usually found it appropriate to point out that neither of us feels we can teach others how to design games in any formal sense. To us, the question of design seems more an art form than a definable and reproduceable methodology. Nonetheless, this art form has a number of fairly consistent "rules of thumb" which we have attempted to embody in the following observations.

The fundamental decision in designing a game or simulation is specification of objectives—i.e., what the system and set of interactions you wish to represent are. Perhaps even more important is the specification of those systems and actions you explicitly do not wish to represent. While the advice may seem prosaic, this problem is the single biggest hurdle in designing a workable game. It is common for persons to seek to design a game on "politics," "criminal justice," and even "love and marriage." While these are admittedly interesting and important topics for game design, they do not clearly specify a system of actors and interactions to be simulated. Attempts to proceed with design on the basis of such specification alone are unlikely to succeed. The real specification occurs, for example, when the designer decides that the game on politics will deal with the local electoral process and the campaign strategies for winning local voters to the side of one or another candidate. Similarly, a game on criminal justice might deal with the interrelationship between police procedures in apprehension and arrest coupled with the flow of activities following through pretrial hearings, trial, and sentencing. Again, a game on love and marriage may specify that it deals with male-female courtship, marriage, and family relationships. While games may deal with abstract concepts such as those initially cited, the design of the game requires a clear specification of the setting and actors within which those concepts occur.

As Armstrong and Hobson (1969) have pointed out, most games appear to be made up of three major components: a set of roles, a scenario, and a set of accounts. Attempting to design the game around specification and elaboration of these three principal components often proves very helpful, especially in the initial stages. Any given game may tend to emphasize one or another of these three components much more heavily than the others, and, indeed, some games in their final form do not always have all three present. In early stages of design, however, considerable progress can be achieved by attempting to use these three elements as an ordering principle in the specifications to be attempted. In this context, the roles define the set of actors to be represented in the game, the scenario specifies the historical, cultural, or geographic setting of the game—including the possibility of fairly elaborate data sets in some type of games—and the accounts specify the manner in which the results of decisions by players will affect the status of other players and modify the setting or data set within which the game occurs. A game with very strong roles and a relatively weak scenario and accounts seems to approximate what is often called a "role-playing exercise." A game strong on the scenario aspects often approximates a case study. A game with major emphasis on the accounting procedures tends to approximate what is sometimes called a "pure" simulation.

In order to make a game a good teaching and communication instrument, the designer should ensure that at every important decision point in the game there are at least two and preferably three persons involved in making the decision. In this manner, players are forced to communicate among themselves concerning their understanding of the problem posed and to try to justify to each other the decision they prefer. Thus, the players will inevitably teach each other what they know about the subject matter relevant to play of the game. A single person making important decisions without the necessity of justifying his actions can and often does behave in a whimsical or abitrary fashion. Logic, argument, and understanding are the more likely ingredients of decisions among a small group. Three persons are preferable to two simply because this decreases the likelihood of a single strong personality dominating all decisions. More than three persons is generally not desirable unless extended debate and discussion on the issue are more important than getting a decision out in a reasonable length of time. In some games, it might be desirable to teach a player something about loneliness or isolation and, in such circumstances, this rule would, of course, not apply. Similarly, if the information or learning potential of certain decisions is not important to the game, the designer should assign these decisions to a single person in order to minimize the amount of time and energy spent in making them.

All games consists of some combination of personal choice and chance, with learning games usually involving a high proportion of decisions or plays based on personal choice and a relatively small number based on chance. Most moves involve some combination of both elements and may be called moves based on "partial chance," where the player weighs his understanding of the process against his judgment of his own skills in the process and against the possibility of something happening in a more or less random fashion which would improve or damage his position. A game based entirely on chance seems unlikely to produce much learning since there is no payoff from increasing one's understanding of the nature of the game, beyond learning something of the laws of probability. A game of pure personal choice (which is similar to Luce and Raiffa's [1957: 44-47] game of pure strategy, GOPS) may be intellectually stimulating but may not be capable of sustaining play for long periods of time. This occurs when one player realizes that he understands less than his opponent and has no chance of winning against him due to the absence of any "lucky breaks" or random events which might otherwise reverse their positions. The only reason for the inevitable loser to continue such a game would be to learn as much from the winner as possible in order to be able to win against others at some other time. The introduction of a small amount of chance in such a game will often substantially increase the ability of the game to sustain the interest of less well-informed players and allow them to continue to learn from the game.

All successful game designers are very much aware of the necessity of allowing five to ten trial runs of their game before it can be said to be ready for "production runs." Like all design processes, game design is inherently itera-tive. A draft version is prepared, criticized, revised, criticized and reprepared. Then the game is played among close and sympathetic friends, revised, played again, and revised again. Through successive revisions, the game improves, and less sympathetic audiences may be used for trials, including, in the later trials, persons totally unfamiliar with either the subject matter or games. Throughout this process, the game will continue to grow and change. Preparing final copy of the rules and procedures should not be attempted until these "shake down" runs are completed. The designer should know beforehand that his first trial run of the game will necessarily be a disaster and should choose his audience and their refreshments accordingly. Many very promising games have been lost or severely damaged by presenting them to a body of unsympathetic critics too early in their design. Regardless of production schedules, deadlines set by granting agencies, and other political pressures, a good designer must allow himself enough time to break in the game and improve it through trial runs before presenting it as a finished product.

When possible, the game designer should try to present the rules of the game in a format such that players themselves will be encouraged to criticize and possibly change the rules themselves. This allows the possibility for the game to be improved over time by players who may have greater understanding or insight than the designer himself. It also allows players to become involved to some degree in the design and redesign process of the game by encouraging them to speculate on the purpose of a particular rule and the nature of change to the game which might result from its modification or elimination. The game THEY SHOOT MARBLES, DON'T THEY? (Goodman, 1970) illustrates this type of design by providing an absolute minimum of rules at the beginning of the game.

Players find themselves in need of additional rules during play of the game, and they themselves develop the rules needed to keep the game in operation. Another variation is found in WALRUS I (Feldt et al., 1972), where all the game rules are broken down into man-made rules and natural laws. Players may change man-made rules during play through representations of standard legislative processes. They are encouraged to seek to change natural laws as well by allowing them a period at the end of each round during which they may challenge their accuracy, workability, or relevance to the game. Such a challenge session within each round offers a form of continuous debriefing and criticism of the game which seems very promising in its own right.

Most novice game designers become seduced by the lure of adding detail and realistic interest to a game until it becomes overburdened with details which make it largely incomprehensible. All games should be designed in the simplest possible format for the problem at hand. The important principle in good design is how to keep things out of games, not how to put them in. It seems possible that many games fall into one of three classes of design and use, reflecting their level of complexity and the number of "variables" represented within them. Games containing less than ten variables seem amenable to play in periods of one or two hours, although more extended use can be made of them. MARBLES is a good example of such a game. Other games which deal with 25-50 variables seem to require four or five hours at a minimum for adequate presentation and play. They are less conveniently used in a simple classroom setting and often require more elaborate preparation and assistance in their conduct. CLUG (Feldt, 1972) is a good example of such games. Substantially more elaborate games generally require computerized assistance or very large staffs to assist in their management and frequently require several days or weeks to play. METRO-APEX (Duke et al., 1971), the CITY and RIVER BASIN GAMES (House and Patterson, 1972), and other quite elaborate game/simulation models are examples of such games. Within the constraints of the purpose and necessary levels of detail which a particular purpose may require, the game designer is urged to follow Thoreau's advice and "simplify, simplify."

The utility and power of any game is significantly enhanced if the designer makes creative and imaginative use of graphic display and mnemonic materials. Game boards, counters, color coding of objects, creation of two- and three-dimensional composite representations of game components and their status, pins, strings, badges, and hats are all important parts of the creative and utilitarian design of games. Such objects are not simply "gimmicks" to amuse and engross potential players, however. Chosen and used effectively, they are important assests to the understanding and insight of the players as to ways in which the game is progressing and useful accounting mechanisms for keeping track of various accounting processes operating within the game.

An aspect of game design related to the creative use of graphical displays is the possibility of representing some of the game rules by placing limitations on the ways in which graphics may be used. A fairly complex rule which would be difficult to describe in written form or difficult to enforce may then be codified and even enforced quite simply by a simple statement such as, "players must always have at least one hand on the desk." A good illustration of such use of graphics to explain and enforce otherwise complex rules is found in END OF THE LINE, a game dealing with the process of aging (Goodman, 1973). In this game, it is important that the players experience the gradual decrease in physical mobility and increasing dependence on help from others which often accompanies aging. Verbal rules to accomplish this might have become quite complex, instructing players to limit the number of steps they may take in any given round to some number which would systematically or randomly decrease as the game progressed and players became "older." Such a written rule would also be very difficult to enforce and would require elaborate and careful checking mechanisms to guard against a very strong need for mobility and a potentially overpowering incentive to cheat a few steps on the part of most players. Goodman solves this potentially insuperable design problem by the simple expedient of requiring that each player must always be connected with his home position by a piece of rope as he moves about the room. A series of systematic and random procedures for shortening the rope each round then provides the desired decreases in mobility being sought and the enforcement problem becomes negligible from the game director's standpoint. Indeed, continued shortening of the rope becomes the single most powerful graphic illustration of the process the entire game is seeking to illustrate and deal with. Similarly, in a game dealing with land uses where blocks are used to represent different types of land uses, certain combinations of land uses may be simply prohibited by creating surfaces on the blocks such that they may not be physically stacked one upon the other without falling down. Rules prohibiting such combinations then become unnecessary since it is physically impossible to make the combination.

During the design of some games, there occurs a time when the overall form of the game begins to crystallize. At this time, the game begins to define its own requirements over and above those originally envisioned by the designer. This might be called the "jigsaw puzzle effect." At this point, the system or process being represented becomes sufficiently clearly defined that components or concepts not envisioned by the designer begin to become apparent as necessary ingredients in order to allow the final representation to work properly. At the same time, some of the concepts and components originally introduced become surplus to a discerning designer when he realizes that they are not really essential to the proper functioning of the game or simulation. When this phenomenon does occur, an experienced game designer recognizes it as a turning point and a first assurance that his representation of reality possibly has some real validity in terms of its internal structure and its ability to represent the problem originally defined. The phenomenon does not always occur or become obvious in all designs, and its absence does not necessarily indicate an inadequately designed game. Its occurence, however, is a strong indication of a successful game.

The assignment of roles in games is often a point of some confusion but actually follows relatively straightforward and traditional rules. Generally, when a game is intended for teaching purposes, players are assigned to roles either randomly or in some deliberate fashion so that they are unlikely to assume roles in the game similar to their roles in real life. Through such a procedure of role assignment, the players begin to gain greater understanding of the totality of the system by putting themselves in the place of other persons in positions with which they are not familiar. They then gain some insight into the operation of their roles and some empathy for their problems within the operation of the system.

If the purpose of the game is to generate predictions or possible futures or to engage in some form of policy-making exercise rather than strictly teaching, the role assignments are generally made as closely as possible to the actual role players hold in the real world. In this case, the outcome of the game and the decisions made generally are considered to be more important than the actual learning which might take place by individual players within the game. An interesting variation on the theme of role assignment and the standard choice of mixing or matching role assignments is the point made by Dr. Greenblat of Rutgers. Greenblat offers the observation that inadequate information on the role characteristics of different persons and even misinformation is an important part of social system operation and that in some games it may be desirable to misinform players as to the characteristics of certain roles, possibly including their own. In this way, the game can be used to provide insight and under-standing as well as possibly research on the problems of inadequate role defini-tion within society.

Some game designers have noticed that the degree of equality or inequality among players at the beginning of the game may produce a marked effect upon the way in which the game is played. At the extreme, significant changes in the initial status of players may completely alter the basic character of the game. Thus, a game which was initially designed to provide a balanced start among players becomes transformed into the same game coupled with questions of social justice and inequality upon giving a significant advantage to some players at the start. Games which are initially designed with heavily imbalanced begin-ning statuses usually are intended to focus on these same questions. Changing their initial status toward higher equality among players does not usually work since it effectively eliminates the major object and focus of the game. Provision of initial disadvantage to one group of players or another can be accomplished by different distributions of wealth or property, in games dealing with these types of artifacts.

More significant and long-term disadvantages may be developed in some cases, however, by simply requiring one group of players to wear gloves when the game requires some degree of manual dexterity, making them write or shoot left-handed, not allowing them pencils or erasers for keeping records, and so on. Another variation which offers promise on some occasions is to have some players enter the game an hour or two after others have begun with nc opportunity to have the rules explained to them. In such cases, a group of immigrants or a younger generation is created which becomes quite dependent upon the older inhabitants for information and socialization into the mechanics of play. Coupling these innovations with aging and removal of older players raises an entirely new set of concerns over intergenerational conflicts and educational processes, all within a game whose initial design may have been directed toward some other subject.

The importance of providing an adequate debriefing to a game is frequently made and bears most directly on questions of the use of games rather than their design. The importance of this function can be enhanced, however, through some steps in careful design. Innovations such as those described earlier providing for round-by-round challenges and discussions of the game format offer one way to improve the debriefing of games by building it into the game itself. If such mechanisms are not easily developable within a game, the designer should then maximize the quality of the debriefing his game is likely to receive by developing materials and suggestions for game users which can be used in the debriefing. Game designers as well as game users have a responsibility to ensure that all games are adequately debriefed.

In designing games, processes and components are often selected because they provide clear and workable analogues to some more complicated aspects of the real world. Such components or processes behave in substantially the same way as their real-world counterparts for substantially the same reasons, although in a more simple or more obvious way. Occasionally, however, the game designer will stumble across a homologue which may be equally useful, although initially confusing. A homologue is like an analogue, except that the reasons it behaves the same as its real-world counterpart are absolutely irrelevant or even wrong. Good, usable homologues are rare and must be examined carefully before being put into use. Well-chosen ones are useful, however, and they may be legitimately employed in game design in the same manner as analogues. A simple example is the example cited earlier, where the shape of certain blocks makes it impossible for them to be piled upon each other, thereby making certain combinations of land uses physically impossible in the game when in fact those same combinations should be prevented from occurring for some entirely different reason. Similarly, players seated in the back of the room may have problems in seeing or hearing all that is occurring in the front of the room. Instead of writing larger or speaking more loudly to correct the problem, it might be desirable to use their disadvantage as a surrogate for cultural or geographic isolation in the game for some particular set of players. A creative and imaginative turn of mind helps in seeing and utilizing homologues, and game designers should not be shy of exercising poetic license in choices and utilization of such opportunities. In related fashion, the meaning of colors or shapes of graphic materials used in games may be more easily remembered by players if the designer takes advantage of either analogous or homologous properties between the color or shape and the concept it is to represent. Thus, paper clips may be used to measure the strength or existence of a relationship by virtue of the fact that paper clips are used normally to establish a "relationship" between two pieces of paper. The color red may be used to measure some property of an area because, when the amount present gets very large, the area becomes dangerous and red is associated with danger. Round shapes may be used to count the number of females present and square shapes to count the number of males because of the male chauvinistic idea of the roundedness of the female form. The difficulty of recomputing and realigning a particular relationship within a game may inhibit players from making some changes, thereby producing a form of social inertia and habit in the game which is desired but obtained for the wrong reasons, and so on. Good designers learn to watch for and use such opportunities. Bad designers see these same opportunities as problems.

In designing a game, some care must be exercised to provide a game which can be run within the kinds of time periods which are normally available to the intended audience. In secondary schools, this may require a fifty-minute playing period for either a round or the entire game. In adult education courses or citizens' groups, periods of two or three hours in an evening session may be available. If possible, the cycling or total playing time of a game should be controlled to make it more usable and flexible for the kinds of playing situations expected. One way of helping to develop maximum flexibility is to make provision for periodic discontinuity within the game such that the game may be stopped at some point and begun again later with no serious loss of purpose or direction. Such discontinuity may reflect annual cycles or other temporal breaks and may be used to either allow players a chance to confer between sessions or possibly have them forget some of the important things they need to know in order to operate effectively. Recollection of previous states may be enhanced or hindered as desired through the provision of more or less written and recorded information on the status of the game at the end of each such cycle.

In designing and seeking to control the length of time a game requires, the designer should bear in mind that it may be perfectly legitimate and even desirable for the game to distort certain time frames from reality. Thus, it may be very useful for the purpose of the game to get players to spend an hour or more on one aspect which in the real world might only take a few hours, while on some other aspect or decision players might spend five minutes when in the real world the analogous situation would have consumed months of effort and time. The distortion of time to help to emphasize and control the attention and focus of the players is a useful and highly productive way in which the nature of the learning and communication process may be enhanced.

A final observation and suggestion for assistance in designing games is to take advantage of the fact that most people already know a number of more conventional games such as card games and parlor games. This familiarity may be used to advantage both in designing and in explaining a game of your own design in any of several ways. First, the game as a whole may be seen to be vaguely analogous to some more common game, and its overall operation and mechanics may be introduced in reference to its common counterpart. It may even be that the game was initially designed around the more common game, and this fact in itself will help players to understand it better. Alternatively, a more common game which has some properties in common with all or part of the more complex, designed game may be used as a "priming game" to get the players to think about some of the aspects of play which they will be undertaking. Finally, the play of other games within a game can often be used very constructively either to seek a linkage between two educational games or simply to provide a meaningfully defined task or activity to some group of players within a larger game. Thus, playing dominoes or completing jigsaw puzzles may be part of the role performance of some groups within a game where others are involved in other activities which may seek to hinder or help the completion of the puzzles or domino strings, according to the purpose of the designer. Similarly, a card game of rummy might be used to introduce a game dealing with problems of classification and taxonomy in some subject matter area. Only the imagination of the designer limits the possibilities available.

REFERENCES

Robert H. Armstrong and Margaret Hobson, "Gaming/Simulation Techniques: An Introductory Exercise, Management by Objectives." Birmingham, U.K.: Institute of Local Government Studies, University of Birmingham, 1969.

Richard D. Duke et al. *METRO-APEX*. Ann Arbor, Mich.: Environmental Simulation Laboratory, University of Michigan, 1971.

Allan G. Feldt, *CLUG, Community Land Use Game*. New York: Free Press, 1972.

Allan G. Feldt, et al. "WALRUS I, Water and Land Resource Utilization Simulation," Ann Arbor, Mich.: Sea Grant Program, University of Michigan, 1972, Technical Report No. 28.

Frederick L. Goodman, "End of the Line." Ann Arbor, Mich.: Univ. of Michigan School of Education, 1973.

Frederick L. Goodman, "They Shoot Marbles Don't They?" Ann Arbor, Mich.: Univ. of Michigan School of Education, 1970.

Cathy S. Greenblat, "Sociological Theory and the 'Multiple Reality Game'," Simulation and Games, 5, (March 1974).

P. House and P. Patterson, *An Environmental Laboratory for the Social Sciences*. Washington, D.C.: Environmental Protection Agency, 1972.

R. Duncan Luce and Howard Raiffa, *Games and Decisions*. New York: John Wiley & Sons, Inc., 1957. pages 44-47.

Part III

GAMING-SIMULATION FOR TEACHING AND TRAINING

GAMING-SIMULATIONS FOR TEACHING AND TRAINING:
An Overview

CATHY S. GREENBLAT

> By and large, however, teachers do not think in terms of how a group can
> be organized and utilized so that as a group it plays a role in relation to the
> issues and problems that confront the group. . . . In their training teachers
> have been exposed, almost exclusively, to a psychology of learning that
> has one past and one present characteristic: the latter is its emphasis on
> how an individual organism learns, and the former is that the major
> learning theories were based on studies of the individual Norway rat. If
> instead of putting one rat in the maze they had put two or more in the
> maze, the history of American psychology would have been quite dif-
> ferent. Conceivably, the social nature of learning might not need to be
> rediscovered.
>
> Seymour Sarason (1971: 190).

THE RANGE OF EDUCATIONAL APPLICATIONS

In a number of elementary schools in the Detroit area, elementary school
children are learning and practicing mathematics by competing in tournaments
based on Layman Allen's EQUATIONS game. Based on prior math knowledge,
students are put into teams of three competitors. They play a tournament,
practicing the manipulation of numbers and signs in an EQUATIONS game as
challenging as their abilities warrant (see Allen, 1972). The winner of each such
three-person tournament "moves up" a table to tougher competition; the player
with the middle score stays at the table, and the player with the lowest score
"moves down" a table. In such a fashion, each student is always matched with
those at a similar learning level, and he or she has fun while developing math
skills.

"What did you think their culture was like? How did they seem to you when
you saw them in their *own* territory? What did they seem like when they came
to visit your area? How did *you* feel as a visitor to a strange culture? Which of
the two cultures—yours or the other—would you prefer to live in if you had a
choice?" I listened as these and other questions were discussed at length and
with considerable vehemence by a group of high school students last week. Their
common experience was a one and a half hour play of BAFÁ BAFÁ, Garry
Shirts' newest game, which, like his earlier STARPOWER, is a huge success with
high school and college students and adults as well. We videotaped this class, and

the expressions on their faces as they entered the "other culture" and felt frustrated at not being able to figure out what was going on, or as they hosted visitors from the other culture and found themselves amused at the difficulties the others had in coping with their culture were also effective for teaching. Through the game, students gained new insights into the feelings, anxieties, misperceptions, and counter-productive attitudes of people who, by choice or circumstance, interact with members of another culture, and into the meaning and significance of "culture" as an explanatory factor in behavioral analysis.

At the University of Maryland; California State College, Stanislaus; California State College, San Bernardino; the University of California (Santa Barbara); the University of Wisconsin; and Rutgers University last fall, teams of four to twelve undergraduate students participated in the same foreign policy exercise (Noel, 1971). It was the *same* exercise in the sense that each class represented a different nation in the same "world"—a world coordinated by Robert Noel and his staff at the POLIS Laboratory of the University of California (Santa Barbara). According to director Noel, in a personal letter to the author:

> Our experience thus far encourages us to believe that network gaming and simulation has great potential in several aspects of instruction: the opportunity for an instructor to incorporate into a course complex exercises without having to assume the heavy administrative burden such exercises entail; the ability to focus on international relations and foreign policies of a particular nation or subset of nations, without having to form locally the number of nation-teams necessary for an entire game; high student motivation and involvement produced by the inter-campus aspect of the games; and a cost-per-student that promises to decline as our experience increases and as hardware and communications costs continue to decrease.

Plans for 1974 call for participation from students in England and Japan as well.

At Virginia Commonwealth University, 109 graduate students—the entire first-year class of the Graduate School of Social Work—and about 10 faculty members spent the first two days of their first semester participating in one of two simultaneous SIMSOC games. The enterprise was designed to expose them to problems of social organization, conflict, and control; and to give them a set of provocative questions as they began their course in "Community Study." Following the game and some regular course work, groups of twelve to fifteen received field assignments to neighborhoods in Richmond, where they had to cope with problems of social organization, conflict, and control. The simulated experience thus served as an initial "laboratory" to sensitize them to questions and issues and to help them develop a general view of community problems.

A group of five adults who regularly work as a team sat around a desk one day, moving little pieces of colored wood in a plastic tray. The game was . . .ET ALIA. . ., and the required paraphernalia consisted of a "nine-piece puzzle" available for about $.79, and a set of rules by Bob Armstrong, Margaret Hobson, and Jim Hunter. Each player was represented by a color, and the rules regarding "moves," "winning," "authority," and communications were deliberately ambiguous. When I halted play after a half hour or so, we began talking together about how they had dealt with problems that arose in the play session. Suddenly one person said, "Yes, but I think that's something we do *all the time*—we don't talk about goals in advance and see where there are conflicts between the things the five of us are trying to achieve. Remember the time that. . . ." A new sense of the nature and value of the game emerged as the rest of the discussion centered on parallels between game problems and real-world problems of small group dynamics. Through this training exercise, the group gained insights into patterns of behavior that aided and impeded their successful functioning.

SIMULATIONS, GAMES, AND TEACHING TECHNIQUES

As the above examples may begin to indicate, the teaching and training uses of gaming-simulations are legion, and the newcomer to the field may well feel inundated by the vast quantity of games and the wide range of applications. In addition, the new user will encounter in early explorations a morass of terms, often, but not always, used interchangeably: "simulation games," "game-simulations," "gaming simulations," "games with simulated environments," "teaching games," "learning games," "instructional games," "educational games."

How can these be sorted out? Some preliminary answers have already been given in our discussion of the terms "simulation," "game," and "gaming-simulation." Now another distinction must be offered as we introduce the idea of teaching or training. Figure 1 represents the different combinations and types that emerge when these three ideas are put together.

(A) In Category A are all sorts of teaching-training techniques and materials with which we are familiar: lectures, case studies, discussions, films, audio-visual materials, field study, etc. These are included here to indicate the broad range of other teaching devices.

(B) Sometimes simulations are used for teaching. In category B are those non-game simulations such as flight simulators for pilots and programs in which disasters are simulated to teach or train medical and paramedical personnel to deal with real-world disasters when they happen.

(C) Other simulations, as indicated earlier, are designed for research purposes. In category C we find these non-game, non-teaching simulations, such as the Simulmatics project, undertaken to attempt to predict the outcome of the 1960 presidential election via computer simulation techniques (Pool, 1965).

(D) There are some game-simulations played and enjoyed by many people at home, but generally unused for teaching purposes. These are in category D, represented by DIPLOMACY and GETTYSBURG, two gaming-simulations of military operations.

(E) Some games (category E) do not simulate a social system or process, and are played for amusement, fun, to pass the time, develop strategy, develop critical thinking, or any of a number of purposes. Though some are quite intellectually challenging, they generally have not been adopted for teaching purposes. Included here in

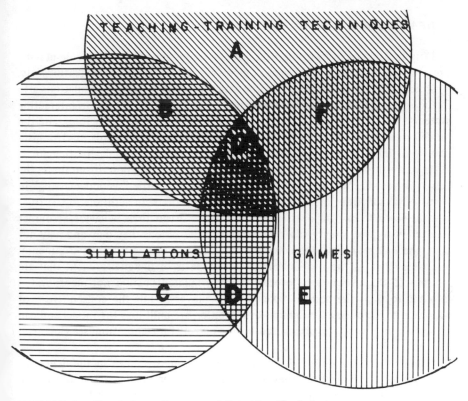

FIGURE 1. Simulations, Games, and Teaching Techniques

Category E are such games as Scrabble, Go, and Monopoly. (Monopoly is not clearly in Category E. Some argue, for example, that it is a gaming-simulation of the real estate system and hence belongs in D. My position is that having a social *setting* is not enough to qualify a game to be counted as a gaming-simulation; it must also contain a reasonably accurate representation of the goals, resources, constraints, and consequences of the real-world system. As such, I don't believe Monopoly can be counted as a gaming-simulation.)

(F) Recent years have witnessed the tremendous growth of games for teaching purposes. Some of the former, shown in Category F, include WFF 'N PROOF, EQUATIONS, ON-WORDS, and others in that series; IMPASSE?, a paper and pencil game for individual or group assessment of impacts of policies; and such group dynamics games as . . .ET ALIA. . . .

(G) Finally, in Category G, we find those gaming-simulations employed for teaching and training (and sometimes for research and public policy) including METRO-APEX; CLUG; the COMMUNITY LAND USE GAME; THE MARRIAGE GAME: UNDERSTANDING MARITAL DECISION-MAKING; GHETTO; STARPOWER; POLICY NEGOTIATIONS; SITTE; and many others.

When we speak of games and gaming-simulations for teaching and training, then, we are speaking of subgroups of games, simulations, and teaching techniques.

GAMING-SIMULATION AND ROLE-PLAYING

Despite some obvious similarities, gaming-simulations differ from role-playing exercises and from such endeavors as mock congresses, mock United Nations, etc.

Role-playing is an element of gaming-simulations, but gaming-simulations also include other components. In gaming-simulations, roles are defined in interacting systems; that is, emphasis is on the role as it interacts with other roles, and the model creates the dynamic interaction, the constraints, rewards, and punishments. In most role-playing exercises, on the other hand, the participant is assigned a role and given only the general outline of a situation. From there, the action is freewheeling. In gaming-simulations there is little "second-guessing" in terms of personalities of particular people or positions; in addition to the roles and the scenario, participants are given goals to orient their behavior, resources to attempt to meet their goals, rules to govern the actions they may and may not take, the order of play, the consequences of violations, and the environmental responses—that is, the probable response of parts of the environment that are relevant to the structure or process being simulated but are not incorporated into the roles or the actions taken by the players. Thus the consequences of actions in terms of goal attainment or inhibition are built into the model and can be known in advance or discovered by the player (for elaboration, see Coleman, 1968).

Gaming-simulations thus differ from role-playing exercises in the degree of structure or formalization they entail and in their emphasis on interaction processes rather than on the playing of individual roles. Further, in many instances of classroom role-playing, several students participate while the remainder of the class watches. In a gaming-simulation, all participate; none is a passive observer.

DIVERSITY OF MATERIALS FOR TEACHING

Even the reduction or limitation of our discussion to games and gaming-simulations used for teaching and training purposes leaves us with a great deal of variation. The several thousand teaching-training games known to be available (cf. Zuckerman and Horn, 1973; Belch, 1973; Gibbs, 1974) vary first of all in the form and complexity of their paraphernalia, the subsequent time, money, and personnel required for their operation, and in the types of learning for which they are therefore appropriate. They differ second in the form of their publication. Some are available in standard book format with manuals to be purchased by each player (e.g., SIMSOC, THE MARRIAGE GAME, METROP-OLIS, CLUG). Others come in boxed kits containing reusable sets of instructions, forms, and any additional materials needed for play (e.g., the games from SIMILE II and from Academic Games Associates; the boxed version of CLUG;

etc.). Yet others include manuals, programs, lengthy instructions for operation, etc. (e.g., METRO-APEX).

The available games and gaming-simulations differ also in the degree of specificity-abstraction of components: roles, goals, rules, resources, accounting system, etc., and in whether they are externally parametered (i.e., by the designer) or internally parametered (by the participants). Some of the differences this entails are indicated by Brent Ruben in Figure 2.

	Externally Parametered	Internally Parametered
Roles	assigned rigid low ambiguity simple	emergent flexible high ambiguity complex
Interactions	channels specified patterns prescribed low ambiguity few channels available predictable	channels emerge patterns emerge high ambiguity multiple channels available unpredictable
Rules	prescriptive fixed constant low ambiguity specified	prescriptive flexible changing high ambiguity emergent
Goals	imposed uniform single clearly defined	emergent individual multiple ambiguously defined
Criteria	predictable uniform single likely to involve "winning" clearly defined	unpredictable individual multiple unlikely to involve "winning" ambiguously defined

SOURCE: Brent Ruben, 1973:

FIGURE 2. Structural Elements and Design Characteristics of Experience-Based Systems

Additionally, games and gaming-simulations used in teaching embody different types of models: resource allocation, group dynamics, and system specification. Resource allocation models are structured around concepts of competition for scarce resources (e.g., limited availability of municipal funds, competition by family members for budget, competitive bidding for land ownership). Group dynamics models emphasize role-playing and/or interpersonal relationships. System specification models entail the explicit expression of a complex system, its roles, components, and linkages. Most gaming-simulations reflect all three of these considerations; however, typically one theme will be dominant. This does not mean to suggest that other types of models cannot be discerned in the broad array of gaming-simulations now available. These three types, however, encompass most.

Three readings offered in this section of the book present an overview of the range of differences in gaming-simulations used for teaching. James Heap describes uses of minimum-structure games, and Degnan and Harr describe the richness of material and teaching possibilities they found in using one of the most complex gaming-simulations available: METRO-APEX. Cohen and Rhenman's well-known article (1961) deals in depth with the use of games in business and management and the general lessons to be derived from that extensive experience.

THEORETICAL SUPPORTS

How can we account for the current popularity of games in the academic arena? Despite rises and falls in user enthusiasm (cf. Boocock and Schild, 1968: 13-18), games have been in the schools since the early 1960s, and there is evidence of their continuing and even growing foothold there (cf. James Coleman, 1971). There seem to be several theoretical bases upon which such usage rests.

First, as the Anderson-Moore and Watson papers in the last section indicate, games and play have long served as modes of socialization. Jerome Bruner, in his *Toward a Theory of Instruction,* remarks (1966: 134-135):

> It was a standard line of nineteenth-century evolutionary thinking that the function of play was to permit the organism to try out his repertory of response in preparation for the later serious business of surviving against the pressures of his habitat. It can be argued equally well that, for human beings at least, play serves the function of reducing the pressures of impulse and incentive and making it possible thereby for intrinsic learning to begin; for if ever there is self-reward in process it is in the sphere of 'doing things for merriment', particularly things that might otherwise be too serious in Niels Bohr's sense ("But there are some things so important that we can only joke about them.").

A second set of theoretical ideas conducive to the enlargement of the role of gaming in the schools embodies notions of the importance of group dynamics, peer learning, and peer pressure in the learning process. Much of this seems to have gained impetus and added credence from the findings of James Coleman's study *The Adolescent Society* (1961) and the now famous "Coleman Report" on the consequences of inequality of educational opportunities (1966). In the former document, Coleman urged the development of competitive games to be used as a learning method and as a means of improving adolescent regard for learning. It is this idea of the importance of paying attention to and capitalizing upon our understanding of the potentials of interaction between students that Sarason affirms in the quote at the start of this chapter.

Several theories that stress the importance of *activity* in the learning process have been brought to bear on the question of why games have potential educative powers, but fundamental questions remain to be answered about the relationship between cognition and action in learning. In our "ordinary teaching," we transmit information, hope it will be integrated, and that eventually it will be applied to shape action. We "test" after the information transmission, but often discover later that the succeeding stages of the process do not transpire.

Teaching via gaming-simulations entails quite a different sequence. As Coleman (1972) notes, here we use action and hope it will lead to cognition and then to generalization to other actions as it is applied. Early evidence indicates that the end result may in fact come about—but without the middle one. Further research on the process is surely needed.

EVALUATION

What happens when students are taught with games? Game designers and teachers have posed the question or offered speculative answers far more frequently than educational researchers have tried to answer it. The files of anyone who has taught extensively with games are likely to be filled with notes such as the following three letters, attesting to the intensity and duration of excitement generated in some student participants:

Because your course in Simulations of Social Relations was so enjoyable and educationally profitable, I have strongly advocated that method of teaching. Last year I ran "STARPOWER" in an undergraduate class, which I believe was beneficial to the class, the professor and myself. This semester I plan to run it in a graduate class concerning minorities and power structure. I am writing to you in hopes that you might be able to give me some tips.

I find myself in medical school taking a course called "The Patient, the Physician, and Society" and the last element of course brings you to mind. The course is a real bore and so I've decided to take it upon myself to liven it up a little. So I am writing you this letter to ask a favor. I would like to introduce my colleagues to the concept of learning through games. Two games come to mind as particularly applicable in this course: they are the Values Game, and the Doctor Game which your graduate students created last spring. I was hoping that you could xerox a copy of the questions in the Values Game and send along the instructions on how to play the Doctor Game. I would really appreciate this and my classmates will too.

Well, in two months, I will have completed my first year of medical school. . . . The responses to the Values Game and the Doctor Game were outstanding. With the Values Game I rewrote about 10 vignettes which were more applicable to value judgements which doctors face. The group just loved it. They found the idea of gaming an intriguing one and the vignettes most thought provoking. The Doctor Game was a game on Death and Dying, and the principles of the game came practically from the book of the same title by Kubler Ross. We have all read that book. So I modified that game so it dealt with Doctor-Patient communications more generally. Once again the response was fantastic. Young prospective doctors like to be put in their future roles and this game provided that opportunity, consequently they just couldn't praise enough. Congratulations to the authors of these games; they certainly have a place in the medical school curriculum if properly modified.

"Heightened motivation and interest" are not the only outcomes suggested by afficionados, nor, happily, are anecdotal reports the only type of data extant (though anecdotes are offered as evidence far more frequently than one would hope). The major types of claims about what happens when one teaches with games are outlined in the paper in the readings section entitled "Teaching with Simulation Games: A Review of Claims and Evidence." The claims are there organized into six major categories:

(1) motivation and interest;

(2) cognitive learning;

(3) changes in the character of later course work;

(4) affective learning re: subject matter;

(5) general affective learning; and

(6) changes in classroom structure and relations.

The available evidence as of September 1971 is summarized in that paper, and Frank Rosenfeld brings this up to date through his summary of evidence reported from 1971-1973. Data on evaluation is therefore not dealt with in the present paper. Several ideas presented there, however, warrant highlighting here:

(1) There is an increasing amount of positive data on the effects of teaching with games;

(2) where the evidence does not reveal benefits of gaming techniques over other modes of teaching, neither does it show the reverse; that is, those taught with games do not prove to have learned *less* than those taught in traditional ways;

(3) the time dimension continues to be inadequately dealt with; that is, questions of retention and application have received scanty attention;

(4) results from studies with particular games cannot be generalized to "learning with games in general";

(5) the quality of evaluative research seems to be improving as researchers become more sensitive to the methodological conditions that yield valid and reliable results. Nonetheless, problems persist by virtue of the nature of the games and the conditions of their operation. As Taylor notes, for example, "It should be pointed out that many users of simulation would not wish to evaluate its learning possibilities separately from the other strategies which make up the teaching UNIT in which it is included. They would argue that the simulation acts as a stimulus to subsequent learning and that this spin-off interest can properly be considered as part of the benefit of the technique, even though it may be developed through more traditional methods of learning" (Taylor, 1972: 39). Fruitful suggestions for those interested in research can be found in Jerry Fletcher's paper, "The Effectiveness of Simulation Games as Learning Environments: A Proposed Program of Research" (1971).

UNANSWERED QUESTIONS

Articles on gaming seem to have been written almost exclusively by those who are "believers"; surely critics and skeptics exist, but they tend not to have taken to paper with their doubts and criticisms. I, too, obviously am in the camp of those favorable to gaming, but I believe there are some important questions and reservations that the gaming community must confront in future research and conferences—questions that have not been publicly raised or adequately addressed. Some of these are outlined above and in the end of the review of claims and evidence; in addition are those following.

What Harm Is Done by Bad Simulations?

In Marshall McLuhan's book *The Medium is the Massage* there is a cartoon in which a man and a woman sit in front of the TV and one says to the other "When you consider TV's awesome power to educate, aren't you glad it doesn't?" Laments about the paucity of data showing students learning from gaming-simulations are predicated upon the belief that the games have something worthwhile to teach. The hope rests on the belief that the models are valid. But what if the model *distorts* reality?

Unfortunately, there are a number of *bad* simulations, and many of them are at least as seductive as the good ones. Though a simulation distorts, it may be very believable, especially to those unfamiliar with the subject. Such gaming-simulations may not only be bad, but may be dangerous, for they convey misinformation and perpetuate ignorance, stereotypes, etc. As a sociologist I have serious reservations, for example, about the BLACKS AND WHITES game. It is based upon a model which includes extremely high mobility for Blacks despite initial handicaps, very high probabilities of problems for whites, and allows Blacks and whites equal chances to "win" or "make it," though by different routes. The student player who "learns" that Blacks have equal chances to win if only they learn to play right may generalize this to the "real world" and conclude that since they're obviously *not* winning it's because they don't play right (that is, their failure is due to their ignorance rather than to a system in which equal roles and resources are not allocated to those with stigmatized identities) (see Schild, 1971, for further comments on this model).

Garry Shirts (1970) has discussed the same type of problem in terms of war games and some games dealing with social problems:

> Games are vulnerable in a way that textbooks aren't. Because the interaction between participants is genuine, there is a temptation to conclude that the model, the facts, and everything about the game are also genuine . . . there is a real danger that games about the black community, which are written generally by persons from the suburbs are based on a series of unfounded cliches about what it is like to be black, not only encourage stereotyping but create an attitude of condescension towards blacks. More

importantly, they can give the students a false feeling that they actually know what it is like to be discriminated against or what it is like to be black. Such games should not be played unless there is extensive input from the black community through talks, films, literature, personal confrontations and discussion.

What Harm Is Done by Bad Games?

A gaming-simulation may be bad because the simulation is bad (as described above) or because the *gaming* elements are poorly designed. How can these problems be recognized and dealt with? This problem is perhaps less serious than the one above, since poor gaming components will probably lead at worst to *no* learning rather than *mis*learning, but the problem is nonetheless one to be reckoned with. Absence of feedback to players, high role-play leading to over-involvement and limitation of analytic abilities, high "win-focus" leading to limited discussion, and other such factors may substantially reduce the efficacy of the experience. While factual information about gaming-simulations is now available, there is still little in the way of carefully done, widely disseminated evaluations of the quality of various games, and users are highly dependent upon word of mouth.

How Do Variations in Teacher Behavior and Attitude Affect Game Operation and Learning from the Experience?

Supposing that we have a *good* gaming-simulation to work with, a critical set of questions deals with the role of the teacher. We know that the teacher is an important variable in the game process, but it is unclear what elements of teacher behavior affect the success or failure of the game run. There are guidelines to successful game operation (see, for example, the second lead article in this section), but these are not always followed. What are the things that teachers do or fail to do that affect what happens and what is learned?

The role of the teacher is described in some of the literature as one of a "coach." Boocock and Coleman (1966) suggest that games teach because the teacher drops his or her role of judge and jury, but what if the teacher *does* interfere and criticize and suggest? I have heard stories of teachers running international relations games such as CRISIS and, because they felt the pace was too slow, precipitating a war. If the students are not aware that the resultant war was not a consequence of their decisions, what does this do to the learning of cause and effect in international relations? And if they *are* aware of the teacher's intervention and manipulation what attitudes does this lead to in terms of war, authority, manipulation, etc.?

What Kind of Game Is Useful for What Kind of Learning?

We know that games vary in their complexity, in their length, in their degree of abstraction, in the type of model employed, etc. But what are the consequences of these differences for learning? What are the relevant differences in parameters which would allow us to talk about *types* of games? And what hypotheses can be generated and tested about the impact of these differences in parameters for different types of learning and different types of learners? An example of such hypotheses is found in Brent Ruben's recent paper (1973), but research to test these and similar propositions has scarcely begun.

> The relative appropriateness of internal versus external parameters likewise would seem to depend upon the particular learning goals involved. External parameter simulations seem generally better suited for teaching specific content, while internal parameter games and simulations appear more appropriate for the learning of problem-solving processes.

> Where the learning goals suggest that what is to be learned is specific, prescribable, predictable, specifiable and determinant, external parameter environments are appropriate in that the designer or implementing instructor wants all students to learn the target facts, strategies and procedures in the same way. The structural elements of the game or simulation will therefore need to be controlled and manipulated intentionally, to insure that participants deal with specific issues, using specific interactional channels, according to prescribed rules, in order to achieve the designer's predetermined goals.

> External parameter activities are generally appropriate when the learning desired is in the form of answers or decisions which are fixed, constant over time and tend to be technical in nature.

> Internal parameters, on the other hand, are generally better suited than external parameters in instances where the learning is to be structural, general, not pre-scribable, unpredictable and indeterminant [Ruben, 1973: 14].

How Can We Understand the Asymmetrical Learning Experiences of Students Who Participate in the Same Game?

Discussions of the experience of being a player seem to treat participation as a constant factor. Yet in most gaming-simulations, students are cast in very different simulated roles (e.g., poor person, city administrator, clergyman, etc.) with different goals, constraints, resources, etc., attached to them. By internal definition, then, the experiences are different, and it could be anticipated that the nature and extent of learning by those in different roles would differ. In good post-game discussions there is of course a sharing of experiences and perceptions, but it is unlikely that this vicarious experiencing can fully equalize the differential learning. Attention thus must be directed to understanding and dealing with asymmetrical learning from the same gaming-simulation.

The problem of differential learning is compounded when we consider that the participant is really in three simultaneous roles: in addition to playing the *simulated role,* students are players in a game and students in a class. Those writing about games often ignore the latter two or assume that they are of no importance. Yet either of these may be dominant at any particular time and may seriously affect the learning that transpires. Add to this the effects of peer pressure, and the complications in understanding what is happening are magnified further. Teachers and researchers can ill afford to ignore the "treble role" of players and must try to understand the factors contributing to dominance of one or the other at any given time.

Who Doesn't Like Games?
Who Doesn't Learn From Them?

Just as there are students who become very excited and involved, so too there are some who recoil from game play, refusing to take part or participating passively. Little attention has been given to trying to understand the characteristics of these students, to knowing the conditions under which games should NOT be employed, or, if they are used, what resistances might be expected.

How Many Games Can Be Used Effectively with a Given Group?

We talk about using gaming-simulations in teaching and mention the good ones we know, but do we *really* mean to recommend unlimited utilization? How many games are "enough"? "Too much?" Aren't there lessons that could be offered about numbers, sequencing, carryover effects, etc., that would help teachers maximize the teaching potential of games?

A friend asked me several years ago whether I thought that the involvement and interest generated in classes by gaming-simulations and the subsequent learning from them derived primarily from their novelty. "What if we used *only* games in teaching" he asked, "and then someone came along with a new innovation called 'the lecture.' Would students show the same improvement in interest, motivation, learning, etc., that they now do when a game is used in lieu of a lecture?" We must seriously confront such questions.

There are, of course, many additional questions that could be raised. In the last selection in the readings section, Charles Elder deals with some of these and alerts us to the dangers of uncritical adoption of the technique.

CONCLUSION

These last sections should not be taken as negative depictions of gaming, but rather as urgings to caution. Gaming *is* serious, and it is too important a pedagogical tool to be treated as a fad or allowed to grow like Topsy, to be unchecked, uncriticized, and unevaluated by its proponents and those interested in maximizing its promise.

The costs of using gaming-simulations in classes and training programs are high: They include not only money, but the time and energy to find a suitable gaming-simulation, obtain it, integrate it into the curriculum, learn to operate it, run it, and lead the post-game discussion. Often additional space and personnel must be found, department chairmen or deans persuaded to endorse if not support the endeavor, etc. These costs must be carefully weighed against a clear set of objectives if wise decisions are to be made. Findings regarding success are often elusive, particularly when the goals entail modification of future behavior in situations such as the one simulated (e.g., consumer credit purchasing, marital decision-making, wise public policy decisions and votes, etc.). Those of us who believe in the power of gaming-simulations to effectuate learning must strive to create better techniques for differentiating between good and bad games and work relentlessly on developing a better understanding of the dynamics of the positive learning experience with a gaming-simulation.

REFERENCES

A. Literature

Allen, Layman E.
    1972 "RAG-PELT: Resource Allocation Games: Planned Environments for Learning and Thinking." Simulation and Games 3 (December): 407-438.

Belch, Jean
    1973 Contemporary Games. Volume 1: Directory. Detroit: Gale.

Boocock, Sarane S. and James S. Coleman
    1966 "Games with Simulated Environments in Learning." Sociology of Education 39 (Summer): 215-236.

Boocock, Sarane S. and Erling O. Schild
    1968 Simulation Games in Learning. Beverly Hills: Sage Publications.

Bruner, Jerome
    1966 Toward a Theory of Instruction. Cambridge, Mass.: Harvard Univ. Press.

Cohen, Kalman and Eric Rhenman
    1961 "The Role of Management Games in Education and Research." Management Science II (January): 131-166.

Coleman, James S.
    1971 "A Decade of Gaming-Simulation." Symposium address to the Tenth Annual Meetings of the National Gaming Council, Ann Arbor, Michigan, 1971.

Coleman, James S. et al.
    1961 The Adolescent Society. New York: Free Press.

    1966 Equality of Educational Opportunity. Washington, D.C.: U. S. Department of Health, Education and Welfare, Office of Education.

    1968 "Social Processes and Social Simulation Games." In Sarane S. Boocock and E. O. Schild, Simulation Games in Learning. Beverly Hills: Sage Publications.

Elder, Charles
　　1973 "Problems in the Structure and Use of Educational Simulation." Sociology of Education 16 (Summer): 335-354.
Gibbs, G. I. (ed.)
　　1974 Handbook of Games and Simulation Exercises. Beverly Hills: Sage Publications.

Fletcher, Jerry L.
　　1971 "The Effectiveness of Simulation Games as Learning Environments: A Proposed Program of Research." Simulation and Games 2 (December): 425-454.

Greenblat, Cathy S.
　　1973 "Teaching with Simulation Games: A Review of Claims and Evidence." Teaching Sociology I (October).

Noel, Robert C.
　　1971 "Inter-University Political Gaming And Simulation Through the Polis Network." Delivered at the annual meetings of the American Political Science Association.

Pool, Ithiel de Sola, Robert P. Abelson, and Samuel L. Popkin
　　1965 Candidates, Issues, and Strategies: A Computer Simulation of the 1960 Presidential Election. Cambridge, Mass.: MIT Press.

Ruben, Brent
　　1973 "The What and Why of Gaming: A Taxonomy of Experience Based Learning Systems." Delivered at the Twelfth Annual Meetings of the National Gaming Council and the Fourth Annual Meetings of the International Simulation and Gaming Association.

Sarason, Seymour
　　1971 The Culture of the School and the Problem of Change. New York: Allyn & Bacon.

Schild, E. O.
　　1971 "Simulation Review: Blacks and Whites." Simulation and Games 2 (March).

Shirts, R. Garry
　　1970 "Games Students Play." Saturday Review 53 (May 16): 81-82.

Taylor, John L. and Rex Walford
　　1972 Simulation Games in the Classroom. Baltimore: Penguin.

Zuckerman, David W. and Robert E. Horn
　　1973 The Guide to Simulation Games for Education and Training. Cambridge, Mass.: Information Resources.

B. Gaming-Simulations

BAFÁ BAFÁ. Designed by R. Garry Shirts. Published by SIMILE II, 1150 Silverado, La Jolla, California.
BLACKS AND WHITES. Designed by Robert Sommer and Judy Tart. Published by Psychology Today Games, Clinton, Iowa.

CLUG (Community Land Use Game). Designed by Allan Feldt. Published by the Free Press, 866 Third Avenue, New York, New York.

DOCTOR'S GAME: (TERMINEX). Designed by Sister Miriam Jude Doogan, Leah Rowbatham, Diane Walker, Judy Foster Karshmer, and Jerry Wallhauser with assistance from Cathy S. Greenblat and Shirley Smoyak. Information available from Dr. Shirley Smoyak, Chairman, Department of Psychiatric Nursing, Rutgers University, Medical School, New Brunswick, New Jersey.

EQUATIONS. Designed by Layman Allen. Published by WFF 'N PROOF, 111 Maple Avenue, Turtle Creek, Pennsylvania 15145.

... ET ALIA ... Designed by R.H.R. Armstrong, Margaret Hobson, and Jim Hunter. Published by Institute for Local Government Studies, University of Birmingham, Birmingham, England.

GHETTO. Designed by Dove Toll. Published by Western Publishing Company, 850 Third Ave., New York, New York.

IMPASSE? Designed by Richard D. Duke and Cathy S. Greenblat. Published by Environmental Simulation Laboratory, University of Michigan, Ann Arbor, Michigan.

METROPOLIS. Designed by Richard D. Duke.

METRO-APEX. Designed by Richard D. Duke. Published by Environmental Simulation Laboratory, University of Michigan, Ann Arbor, Michigan.

ON-WORDS. Designed by Layman Allen, Fred Goodman, Doris Humphrey, and Joan K. Ross. Published by WFF 'N PROOF, 111 Maple Avenue, Turtle Creek, Pennsylvania.

POLICY NEGOTIATIONS. Designed by Fred Goodman. Published by URBEX Affiliates, Inc., 365 Rockingham St., Rochester, New York.

POLIS NETWORK. Designed by Robert Noel. Information available from POLIS Laboratory, University of California, Santa Barbara.

SIMSOC. Designed by William A. Gamson. Published by The Free Press, 866 Third Avenue, New York, New York.

SITTE. Designed by R. Garry Shirts. Published by SIMILE II, 1150 Silverado Avenue, La Jolla, California.

STARPOWER. Designed by R. Garry Shirts. Published by SIMILE II, 1150 Silverado Avenue, La Jolla, California.

THE MARRIAGE GAME: UNDERSTANDING MARITAL DECISION-MAKING. Designed by Cathy S. Greenblat, Peter J. Stein, and Norman F. Washburne. Published by Random House, Inc., 201 E. 50th St., New York, New York.

VALUE GAME. Designed by Thomas Linehan and William S. Irving. Published by Herder and Herder, 232 Madison Avenue, New York, New York.

WFF 'N PROOF. Designed by Layman Allen. Published by WFF 'N PROOF, 111 Maple Avenue, Turtle Creek, Pennsylvania.

RUNNING GAMES:
A Guide for Game Operators

RICHARD D. DUKE and CATHY S. GREENBLAT

T he specific steps to follow to run a game vary, of course, from one game to another. There are, however, four major elements of administration common to all games: (a) preparation; (b) introduction of the game; (c) operation or management of the game; and (d) post-game discussion or critique. Each of these consists of a series of steps that should be followed, and experienced game operators do these quite regularly. In the following pages, we shall attempt to outline the critical elements of game administration and to offer suggestions on how to make your run a smooth and successful one.

PREPARATION

(1) Know what your intentions, aims, or pedagogical purposes are; review the available games; and select one that seems appropriate. This may sound like a very obvious piece of advice, but all too often someone runs a game in a class not because it seems appropriate for what they want to present but because someone spoke enthusiastically about it. A game that works well with one group may not work with a group of different age, background, etc.; what is successful for one purpose may be a disaster for another. Therefore, the first step is to understand whether the game you have chosen is one that is appropriate to the learning aims that you have in mind, or rather, to have a clear understanding of what you want to teach and to select a game in accordance with those goals.

(2) Having selected the game, integrate it with other materials. If the game is to be used in a class, then it should be tied in to the larger perspective of the course outline. If it is to be used in a nonacademic setting, it should be tied in to other materials, topics, and activities of the group. A game that is just "stuck in" as a random event without thought to the ways in which it relates to the rest of the curriculum or endeavor will be less successful than one which is meshed with other tools and topics.

196

(3) Become familiar enough with the game that you can run it well. In some instances, it may be possible to familiarize yourself through prior experience. If you know someone else who is going to run the game, perhaps you can arrange to participate in their run. This provides a familiarity with the mechanics not possible to obtain by simply reading the materials and will give you a player's perspective on the game. Other times it may be feasible to become familiar with the game by running it with a group of friends or a small group of students prior to using it with your "real audience." If it does not require too much time or too many people, this may be a viable and fruitful way to learn the game under conditions in which problems you encounter or slowness of your response is less a problem than in the classroom or with the group that has come together at greater difficulty and expense and may give you greater confidence when you run it later.

If neither of these alternatives seems possible, then the familiarization preparation should take the form of "walking through" a typical round or rounds: go through the material, read the rules from the perspective of a player as well as the perspective of an operator, and be sure that you understand what players are to do at each point. If you know someone who has run the game, ask them for hints to more effective administration.

(4) Be sure you have adequate personnel to run the game. Having gone through the material and become familiar with the game, you should now understand (whether the manual specified it or not) whether you will be able to run the game alone or will require assistance. If the latter, you should now obtain aides in advance and prepare them. It may be that you can use one of the students or other participants to be an assistant for you and that such a person can be selected at the last moment. But don't count on having them capable of performing difficult or complicated operations. Likewise, don't count on preparing an assistant at the same time that you are trying to keep the game running; it simply is too much.

(5) Make up a time schedule for the game. First, this involves being sure that enough time is available for a successful run; a good game squeezed into too little time will become a bad game. Some games can be divided; that is, they can be played one day, stopped, and then continued another day or several additional days. Others require continuous play and even require that initial post-game discussion or critique begin the same day that play takes place. STARPOWER, for example, cannot be started one day, continued into a class period the following day, and then discussed the third day; in order to be used successfully, it must be run in at least an hour and a half of consecutive time. This kind of time block, of course, is not possible in some circumstances and at those times the game *should not be* employed.

If you are running the game in several time periods rather than in one continuous period, you will have to be sure to allow enough time for players to get "recharged" each time. That is, they lose momentum by going out of the game and into their real life world and then coming back, and thus require additional time each round to get back into the swing of things. This sequencing does, however, permit more time for reflection and planning.

If the time schedule you arrange involves one long play period, be sure you recognize the potential fatigue that will take place and allow for breaks if possible or at least for some kind of stretch period during which the players can move around. Coffee and/or other refreshments are also required in this type of time frame. These breaks should fit in with the 'natural rhythm' of the game. The first such break, for example, should not come until players have had a positive experience after initial confusion. If it comes right after an introductory cycle, some will leave if they can. Breaks taken at times of anticipation of something good to come will be easier to end when you want play to resume.

(6) Prepare all the materials. This obvious step always takes longer than anticipated. Whatever the requirement—whether it be to cut out SIMBUCKS and tickets in SIMSOC or to arrange the cards in BAFÁ BAFÁ—the mechanical chores of preparation must be done, and far enough in advance that you don't find yourself hustling through cards and sorting them into piles as the players arrive. Such last-minute preparation always entails a psychic cost even if it doesn't delay the actual commencement of play.

Separation of materials also means careful checking to be sure all things that should be there are in fact there. When games have been run a number of times, little pieces often get lost. A check before you go to the game site will reveal whether things have been moved within your supplies (i. e., from one game to another) or perhaps have been lost or broken in a previous run. It should be obvious that such preparation of materials should be done far enough in advance to obtain new ones in case any are missing. Once everything is there and in usable form, arrange the materials so that they can be rapidly put in place at the game site.

(7) Decide whether to give out materials in advance. If the group does not meet regularly but is coming together for a special occasion to play the game, it would not be easy to give out materials for even a complex game. One might, of course, wish to send a mailing to those participants, but this will involve an expense which must be carefully considered to decide whether it is worthwhile. With a class or other group that meets on a regular basis, the option of giving out materials in advance is, of course, open to the operator. For many games, this would be unnecessary and, in fact, undesirable. The games are simple and can be easily introduced at the time of play. Other games will not operate successfully *unless* players have seen something in advance. For example, the player in SIMSOC should review the basic rules before he start to play. Players of THE MARRIAGE GAME must read a section describing the designers' view of marriage as a social system prior to play or they won't understand the rationale

behind the moves and choices they are asked to make, and thus will not clearly learn about marital decision-making.

With other games, the operator must decide when to distribute materials. With a few it may be advisable to give out some material in advance so the players can begin to think about what they wish to do. This may be most appropriate where the materials contain fairly factual accounts that must be reviewed carefully. For example, we have found it fruitful to give out the players' manuals for METROPOLITICS the day before play. Participants can then review the six options open to them for support and see the pro and con arguments for each of the positions. The material will remain dry, however, until they begin to use it in play of the game. It is as if they were given a straight reading assignment on the materials, and it is difficult to get very excited or enthused about them. Such distribution of materials, however, may make exciting things happen earlier in the play period. In general, prior distribution does not prove as fruitful as one wishes. At the end of the first round, participants in METRO-APEX always urge that they should have been given the materials to review in advance to prevent the confusion they felt at the start. Experience with the game, however, has revealed that the rules don't become clear until players begin to use them; sending out manuals in advance then may do more harm than good, for players may read them and be confused *alone*, thus developing a negative set before they come to the actual play period. It may be better to have them begin together, where at least they can commiserate with one another about the enormity of the task before them! They will soon discover that the confusion is dispelled as they begin to play and to use the instructions they have been given.

If materials are sent out or given out in advance, then there are three main things to keep in mind: (1) keep the amount limited so as not to overwhelm players; (2) send or give out a chronology of events—i.e., the overall sequences and the steps in a given round; (3) don't rely on players' remembering to bring materials with them; have duplicates available for them.

(8) Decide on various dimensions of role assignment. First, you must decide *when* roles are to be assigned or given out. If materials are to be given out in advance, perhaps you wish to also allow players to know what roles they will play. With games such as METRO-APEX or DANGEROUS PARALLEL, the manuals for each player are different; therefore if materials are to be given out in advance, players must be informed of the role they will play when given the appropriate manual. In a game such as METROPOLITICS, these two decisions need not go hand in hand; that is, materials could be given out early but the players need not be informed of their roles until they actually meet for play of the game. In games such as this, where the manuals are common to all players irrespective of role, it may, in fact, be wise to separate the two things—that is, to have players read through the materials in toto, unaware of their game roles. In that way they may pay more attention to elements not critical for them, but central to the model as a whole. Thus, for example, in using THE MARRIAGE

GAME is may be wise not to tell players in advance whether they will be playing males or females; in that way they are more likely to read both sets of charts presented for jobs, marital status, parental status, etc., and to become aware of the disparities built into the game.

A related decision is that of *how* roles should be distributed. Should they be assigned by the game operator? Should they be given at random? Or should players be allowed to choose their roles? The specific game manual may offer some advice on this, and it is hard to advise in general terms.

If the decision is made to assign roles, the next question becomes "how should they be assigned?" Should aggressive students or players be given leadership positions? Alternatively, should quiet people be given leadership positions in hopes that they will emerge from their shells? It is, of course, tempting to give the roles to those you know will play them in an active manner. Sometimes, however, the most potent leaders are those one would not recognize as such on the basis of the ordinary classroom experiences preceding use of the game. Assigning some leadership roles to those who have not displayed outspoken traits may provide those students with the opportunity to show themselves in a very new light.

Finally you must give thought to the *number of* players per role. Perhaps this should have been mentioned earlier, for often a game operator will look at a game and decide that it cannot be used with his group since his group is too large. This decision is reached on the basis of the number of roles vis-á-vis the number of participants. What should be kept in mind, however, is that the best learning experience may emerge if several people play a role together. Our "rule of three" suggests that the best learning takes place when there are three players per role. The mental lethargy which may emerge from one person playing a role is negated by the presence of the second; the two must discuss the strategies they will pursue. The presence of the third person creates an inherent instability and means that more active discussion is likely to take place. In some games, of course, such doubling or trebling on a role may be impossible; in THE MARRIAGE GAME, for example, only one person can play a role. In many others, however, role teams of three can be created.

(9) Prepare the space and the furniture arrangement. This is not simply a physical aspect, for the arrangement of furniture and the arrangement of groups in the room may very strategically affect the interaction that takes place. Of course, the basic problem is to be sure that there is enough space. If you are running SIMSOC, for example, you should try to have four rooms so each region can be in a separate room. For BAFÁ BAFÁ, you must have two rooms. For most other games, the spatial needs are not for multiple rooms but for sufficient space for players to move around as play progresses. Such space, of course, is greater than the space required for an equivalent number of people if they were to be sitting still for a lecture; games with high interaction require more space. Most game manuals will inform you of the amount of space desirable for a run of the game.

Having run games in a number of unfamiliar places, we have discovered the utility of going, *as soon as you arrive on site,* to check out the rooms that have been assigned to you. They may not be suitable and changes may need to be made. Rooms that are too large or rooms with stationary furniture also provide a problem. If you cannot avoid the former, something can be used to set off a part of the room as a "no man's land," and it should be made clear to players that they are to play only in the section marked off for them.

The arrangement of furniture and the location of groups within the room will also prove to be important. Groups or roles that are centrally located are more likely to play central roles than are those located on the periphery. In METRO-APEX, for example, the role of the Air Pollution Control Officers will be quite different if they are placed in the center of the room than if they are placed in a corner; likewise, the role the politicians play will be different if county and city politicians are placed near one another than if they are placed on opposite sides of the room where communication is impeded between them.

Additionally, the nature of the role or the perception of it may depend upon such subtle cues as the degree of crowdedness or the degree of comfort provided by the furniture or furniture arrangement. In running SIMSOC, for example, try to give the largest, most spacious, airiest, sunniest, etc., room with the most plush furniture to the region with major resources of the society as a whole. Correlatively, assign a small, dark, bare, unaesthetic room to the impoverished region. Such physical arrangements often will lend support and drama to those very disparities one hopes will be felt by players. In one run, for example, a player from the Green Region had gone to visit the Red Region. As he emerged from their room, he said in an aside to no one in particular, "Boy, I'm not going back there again—it's too crowded and dirty and unpleasant. I like my room much better." How much that reminded us of the feelings of the rich as they drive through poverty areas, closing their windows and promising themselves to take another route another time! It was good food for discussion in the later critique.

A similar use of space and furniture can be made with STARPOWER. Send the Squares to sit in the most comfortable chairs and put the Triangles in the least comfortable. In THE MARRIAGE GAME, the spatial arrangements recommended to the game operator include the provision that those in the wealthiest neighborhood be given the most space to move around and sit comfortably while those in the poor area have their chairs squashed together, making it more difficult to keep their things separate from one another and more difficult to avoid mixup of papers, forms, etc.

(10) Decide upon a policy concerning visitors and/or observers. This decision will vary from operator to operator and no rule of thumb can be given. We generally refuse to allow visitor-observers; anyone interested in the game must be a participant when we run it. This is for several reasons. First, many people really would like to play but are hesitant to do so. What they require is the additional push given by saying "No, you can't observe but you're welcome to be a

player." Second, the perspective gained by playing the game is often far different from that derived from simply observing and remaining marginal in the process. The third reason for our usually refusing visitors and observers is that it makes some players feel more self-conscious than they otherwise are. We believe this is unfair to them, for they are already struggling to overcome the self-consciousness that comes from playing a role in an ambiguous situation, and the presence of observers may further hinder their smooth transition into the game. Many of those who wish to observe are those in positions of more authority than those who are playing, and such observers may further hinder the players' easy role assumption.

INTRODUCTION TO THE GAME

The way in which the game is introduced to participants by the game operator may well be critical in determining the success of the experience. There are several things that you must present to participants before they are ready to begin play. These will vary in specifics from game to game. The following are a few general suggestions for ways of introducing games to maximize the probability of success.

(1) Your early comments should include references to the following: (a) gaming-simulation as an instructional medium; (b) the purpose of the specific gaming-simulation you are employing that day; (c) the rules of the game in outline form (if players have manuals, you may wish to have them read the rules and simply highlight those of importance; if you are presenting all the rules, then this part of the introduction will, of course, be longer and more complex); and *(d) the roles represented by players in the room.* Instead of your introducing roles you may wish to begin play by having participants introduce themselves in the role. Where this is possible, it often proves a good idea because it begins their interaction.

(2) Do not take too much time for the introduction. Not only will a long explanation of the nature of the game, the rules, and the reasons you are running it, etc., not help create a successful run, but it is likely to kill it. Keep the introduction short! As questions arise later, you will be able to deal with them. Covering all points at the beginning is a poor idea, for players will forget those not seen as relevant because the questions have not yet arisen.

(3) Sound decisive; if you convey the idea that you are sure it will be a good learning experience, you will be more convincing to players. This does not mean that you must sound as though you are an expert on the game and know exactly what will happen. You must, however, explain the rules clearly and decisively so players gain confidence that although they may be confused, you know what is going to happen in general terms!

(4) Explain the expectability of initial confusion. You know from having seen many games that as play begins players are confused. But the novice player may not realize that the confusion he is experiencing and the concern he feels about whether he will be successful in developing a strategy is common to all. Whereas you should not announce that everyone in the room is hopelessly confused and ready to walk out, it will be helpful to them if you say something such as the following: "I know that by now you probably are feeling confused. Don't worry about this. As play begins and you start to use the information you will surely find that it makes more sense than it does now. The rules are complex, so that play of the game can be a rich experience. If they were simple enough for you not to feel confused now, then the game would become quickly boring. So try to relax; take my word that you will feel better as soon as we get into play; and let's get started!"

(5) Acknowledge that you, the game operator, recognize the nervousness and feelings of self-consciousness that some of the players feel. In that way, they will feel that you are less critical of their early attempts to cope with a situation fraught with ambiguity. Likewise, as was suggested in 4 above, you will confirm that the feelings they have are not unique to them and thus dispel many of the doubts they have that they will be able to play successfully.

(6) Sound enthusiastic. This, of course, is related to 3-5 above, but is sufficiently important to warrant separate mention. If you sound enthusiastic you may be able to make the feeling contagious. If you sound hesitant, bored, and as if this is a mechanical exercise and you don't know whether it will work or not, players will pick up that feeling from you. A critical ingredient in the successful run of the game often is the game operator who is able to get players into the right frame of mind as they start.

With your good preparation and a short, enthusiastic introduction, most games will go very well.

OPERATION OF THE GAME

The particulars of administration are highly variable. In one game, the operator may be constantly involved in a variety of management enterprises. In another, the operator may be largely free from such tasks and able to circulate, seeing what is going on and collecting vignettes for use in the post-game discussion-critique. There are a few kinds of activities, however, which are typically engaged in by the game operator. These will be only briefly sketched out here, since they are the elements of game operation most likely to be adequately described in the operator's manual. We conclude the section with some guides to observation during cycles.

(1) Remind players of the rules as situations arise.

(2) Give out necessary resources. In SIMSOC, for example, resources must be distributed at the beginning of each round to all travel agents, subsistence agents, and heads of groups. In METRO-APEX, computer printout reporting changes based on last-cycle decisions must be distributed as each cycle begins. This will be a major element of some games and insignificant in others.

(3) Collect forms which must be submitted to the game operator. Be sure when you get these that you establish a system for keeping track of them, or keeping them in order so that you can later locate them. It is also a good idea to put some kind of mark on forms collected, indicating either the round number or the time at which it was submitted. This is useful for later chronological analysis of what transpired.

(4) Check forms and other materials that are submitted for accuracy. If you can do this while the player is still there, it will save your running around trying to locate him or her or, worse yet, attempting to remember who gave you that incorrect form.

(5) Perform the necessary calculations by hand, with a desk calculator, or on the computer, depending upon the nature of the game you are running and what is available to you. The variations in what is involved in this aspect of game operation are far too great to deal with here. If the calculations are complex, it is usually advisable to have an assistant and to find the time to cross-check one another for accuracy.

(6) Announce the time limits if there are any, and indicate whether there will be any releases from these. That is, if they're not finished, will you add another five minutes or will you collect incomplete forms?

(7) Announce the amount of time left in the round at several intervals. For example, announce there are fifteen more minutes before Form C must be turned in, then give a five-minute warning and a two-minute warning. The level of activity and enthusiasm characteristic of play of many games means the students or players will not be looking at their watches or at a room clock. Your announcement of the time remaining will be the only guide they have and should be done, therefore, with some consistency.

(8) Regulate the rhythm of the game and non-game enterprises. As mentioned earlier some games must be played straight through, in one period (e.g., STARPOWER) or in several consecutive sessions (e.g., SIMSOC). Others (such as METROPOLIS or THE MARRIAGE GAME) can be divided into consecutive rounds, or the rounds can be separated by other course activities. Whichever of these formats has been selected, the game operator's task is to move players back and forth smoothly from one activity to the other or from one round to the

other, reminding them of where they were when they left off and "reintro-ducing" the game as necessary.

(9) Deal with unanticipated consequences. No matter how adequate the preparation, no matter how thorough the instructor's manual provided, there will be unanticipated consequences. It is important to remember that the game operator need not know all the answers or even seem to. Players are always far more tolerant of instructor's hesitation than operators usually believe! A good guide for making decisions about such things is to ask two questions:(a) Will doing what is suggested interfere with game mechanics? (b) Is there a real-world parallel for what is suggested? If the answer to the first is "no" and the second "yes," we usually allow (encourage, in fact) the proposed innovation. To answer these questions, of course, you must have an understanding of the game mechanics and of the system being simulated.

(10) The major activity of the operator during this phase is careful obser-vation and assistance to those who require it. The central point to remember in terms of observation during the game run is that most of the participants are in a situation that is foreign to them both in terms of the subject matter that they are encountering and in terms of the context of game-play. Watch for the "hubbub factor," which indicates that a certain level of involvement has been achieved at the outset and that the players are pretty much involved on their own hooks. At that point you must watch pretty carefully to see which players *haven't* been caught up.

There are a variety of reasons why players will drop out at early stages of the game. These include a general sense of personal unease at being in the person-to-person or role-playing situation and confusion about the forms or mechanics which they have to deal with. The latter may result from a poor presentation of the game, inadequacies of the forms themselves, or just some obtuseness on the part of that particular individual. The problem is best handled by working with him directly until he has mastered the mechanics.

If the problem is the interpersonal kind, it is probably best to assign him to a different role or location which the operator selects thoughtfully and privately. In doing this, you would look for other places in the game room where the individual might fit in better. For example, if the individual is a very passive sort, there may be a table where more passive and thoughtful types of things are going on and the operator can sometimes invent an excuse to tactfully move the person into participating in that role. In some cases, if the individual has a particular perspective that is not represented by the game, a role can be invented and assigned to him and he will be required to generate the momentum to keep the thing going.

Probably the most significant reason people drop out during the course of the game is that their perception of reality differs significantly from that being presented in the game; their dropping out represents an objection to the material being presented. This is the most serious situation of all. It can be dealt with by stopping the game and giving the individual a chance to express his views, but this is usually the worst alternative of all because it disrupts the whole game. A better approach is to take the individual aside until he has made clear the nature of the world as he sees it and why he sees the game as being wrong. If the operator is familiar with the subject matter (and the operator should be or should have some expert on hand), then a private dialogue should go on between the player and the operator until some kind of accord is reached. At this point, the player should be encouraged to set aside or hold in abeyance his views and to continue play from the perspective represented by the game design but with the knowledge that he will be given an opportunity during the critique period to express his own perceptions and to take issue with that presented in the game. If this is done, it is very important to be sure that the player is given the opportunity during the critique period.

(11) It is particularly important to watch the players when they get the results of a particular cycle. If there have been errors in the output or if there is some new material that is confusing them, simple observation of their faces will often reveal that a problem exists. Game operators are often inclined to disappear at the time that the output is presented because if they stay they have to confront the problems which might be involved with mistakes in the game mechanics. Worse yet, they fear they may not be competent to deal with the basic model that is in the game, and the best way to avoid that is not to be around when people are trying to ask questions pertaining to it. A cardinal rule from our standpoint is that the operator should be present and very much alert at the time that the players receive the feedback from their decisions and should confront problems in an organized and helpful way.

(12) Another thing to watch for in player behavior is a general lassitude. This is reflected by people thumbing through newspapers or chatting with one another about stuff that doesn't pertain to the game, or being distracted by such things as radio, television, visitors, or events outside such as the construction of a building next door. This can be the result of poor mechanics—that is the inability of the operator to process things quickly enough to keep people involved—or it can be the result of the distraction itself. For example, if the building next door is being put up with pile drivers and workmen banging things around, the problem was having selected the site in the first place. The introduction of competitive activity, whether it is reading material, a visitor who is not really there to think about or work with the game, or peripheral discussions going on among players can sometimes best be dealt with by either removing the distraction or rearranging the players and putting them with people that they don't know in a personal way.

POST-PLAY DISCUSSION-CRITIQUE

If you think of a "game event" as the whole business from the time you get started until the time it is all over, you should devote between one-fourth and one-third of this to critique. Start-up time should take not more than about ten percent, the initial cycle perhaps another twenty percent, perhaps about 40 percent of the time to actually run the game and then about the last third to go through the critique process to be described. There are three distinct phases of the post-play critique.

(1) The first phase involves letting the players vent their spleens about the things that happened in the game itself. That would be the arguments that have ensued, the oneupsmanship that has gone on, the hurt feelings because of a mix-up in the computer output, or anything that they just want to get off their chests—where the game did them wrong or where they can lord it over some-body else whom they happened to better. At the beginning of the post-game discussion, then, allow an opportunity for catharsis. Many players in the game will have become very emotional and highly involved in what transpired. Before asking them to *analyze* the experience, you must give them an opportunity to vent some of the emotions they have pent up. Some of these feelings will have been obvious to other players at the time they arose; others, however, will not have been manifest. For players to fully understand the experience, they must learn how things felt to others in other parts of the game. Unless you are a trained psychologist, however, you should quickly turn them from the venting of anger and accusations ("you did this," "he did that to me," etc.) to an analysis of why those things were done and what it was about the situation or the role that led to some of the actions taken and interpretations that arose.

(2) The second stage is a systematic examination of the model presented by the game from the perspective of the various roles. This gives everybody a chance to see what happened from the eyes of the other role players.

A good technique is to take some event such as an issue that was presented to them and have the different players speak in turn to that issue from their perspectives.

The world looks different to those in different social situations; likewise, the game looks different to those in different social roles within it. Analysis cannot take place until perceptions of what transpired have been shared between participants. Be prepared for conflicts in views and arguments about "what really happened."

(3) Finally, in the last stage of the critique, urge that the players and the operator should focus on the reality which was represented by the game rather than the game itself. This means "bringing them down" and it involves getting

out of the game situation altogether and addressing thoughtfully and at some length the actual reality that the game simulated.

During this last phase, it is fruitful to have some kind of a conceptual model portrayed visually, with charts prepared in advance, worked up on the blackboard, or transmitted verbally.

It is often helpful to select some issue, event, circumstance, or problem that they *do* understand, either in a real-world context or preferably in both the context of the game and the real world, to use as a theme around which to examine the overall basic model of reality.

This involves "reality-testing" the game against their prior knowledge, refining their previous perceptions, and leaving the game event with sharpened questions and answers about the world in which they live.

THE STUDENT AS RESOURCE:
Uses of the Minimum-Structure
Simulation Game in Teaching

JAMES L. HEAP

University of British Columbia

Since sociological problems are not taken to be problematic by members of society, students in introductory courses often express the feeling that the theorist's problems are hatched in midair. The "leap of faith" between the everyday world and the world of scientific theorizing is difficult to make at first. One way to aid and abet this leap is to try to provide students with a situation wherein their pragmatic purposes at hand coincide, roughly, with those of the theorists. Thus, they begin to confront problems which were never recognized as needing to be solved. One way to involve students with these problems is by use of a minimum-structure simulation game.

To allow the reader to more easily locate and evaluate the minimum-structure type of simulation, it will be contrasted to the better-known type of simulation game in terms of their respective structure, functions, benefits, and limits. In view of the fact that the minimum-structure gaming technique is relatively new and unknown to most instructors, an attempt will be made to analytically delimit and clarify its distinctive structural features.

CLASSIFICATION OF MODELS

The British geographer, J. L. Taylor (1968), has developed a classification system for types of teaching models according to their degree of abstraction

Reprinted from **Simulation & Games**, Volume 1, October 1971. ©1971 by Sage Publication, Inc.

AUTHOR'S NOTE: *An earlier version of this paper was presented at the meetings of the Canadian Sociology and Anthropology Association, St. John's, Newfoundland, June 1971.*

209

from real-life operation. He classifies the *case-study* model as the nearest to reality, where students study and work with detailed descriptions or histories of selected problem situations. At the highest level of abstraction is the machine or *computer simulation,* where human participation is limited to "the construction and manipulation of an *operating* model" (Dawson, 1962: 3). At the next highest level of abstraction is the *gaming simulation,* or what I prefer to call the *sophisticated simulation game,* in which human decision makers are placed in highly structured hypothetical situations and then confronted with a series of predetermined problems. While there are three other model types, for our purposes we need only mention the next most abstract, the role-playing model, in which players take and act out assigned roles within a hypothetical context. Now it is between role-playing and the sophisticated simulation game that I locate the "minimum-structure simulation game." It lacks the structured competitive element of the sophisticated simulation game,[1] as well as the central focus upon roles characteristic of role-playing.

Of all these types of simulations, the sophisticated simulation game has been the pedagogical technique which has received the most attention and praise. This type of game can be used to train students who some day will be confronted with the situation which is simulated, or it can be, and is most often, used to improve the students' understanding of situations with which *others* are confronted. As in role-play, students are provided a hypothetical situation within which they are to act. The difference, however, is the degree to which these situations are prestructured. It is characteristic of this type of game that a set of explicit, formal instructions and rules of how to play are provided. These include the order of events—e.g., companies are formed by players or groups of players; they draw company deed cards and then decide on strategy; dice are thrown to determine who goes first, and so on (Walford, 1969: 88). Also included are the end-game conditions—e.g., "The game continues for a predetermined number of rounds. . . . At the end of that time the auditors are presumed to arrive and the companies' books are laid open for inspection" (Walford, 1969: 89). In that these games are almost always competitive, the end-game conditions usually specify what is required to win—e.g., maximum profits or highest point total. In addition, special iconic or symbolic material must usually be supplied and explained—e.g., maps, identity cards, currency, dice, and the like.

Chadwick Alger (1963: 151-154), among others, has noted the benefits of this type of game. It is claimed that simulation games heighten the interest and motivation of students while offering them an opportunity to apply and test past knowledge. They are reported to give participants greater understanding of the world as seen by the decision maker, by providing a miniature world that is easier to comprehend than the "real world." Students also develop closer relationships with their classmates than is usually the case when other types of teaching techniques have been used, e.g., lectures and discussions. As has been pointed out by Greene and Sisson (1959: 2), the educational benefits of simulation extend to the teacher as well.

There are, however, a number of prohibitive aspects to simulations. The major ones revolve around the time, cost, and labor which are required in order to develop and execute a sophisticated game. Alger's (1963: 165) international relations simulation game cost $55 per session, which pays for graduate, undergraduate, and calculation assistants, plus $10 for supplies. Alger's game operates in weekly three-hour sessions for ten weeks, thereby making it rather expensive. It is also difficult to generalize as to how long it takes to develop an adequate game, but it is safe to say that it is a teaching technique which requires more time and labor than any other technique presently in wide use. One can overcome this hurdle by purchasing a ready-made kit, but these are not generally designed with college-level students in mind and are not always appropriate to a sociological context. Ready-made kits, however, do save the instructor the trouble of having to learn how to devise a game, which is sometimes a challenging task. The snag is that one ends up building his course around the simulation rather than building his simulation around the course.

This last point can become an important one. Unless that which we wish to simulate has the properties of a game which are recognizable to students, the situation can easily arise where the *game* becomes the focus of student interest and involvement. The significant feature, the fact that the game is a *simulation* of phenomena of import to the discipline, is thereby lost in a wave of enthusiasm. This problem closely overlaps with the one of complexity. If the game is too sophisticated, with a multitude of directions, rules, and rounds of play, it becomes too complex for the majority of the students to learn to play adequately, let alone to grasp the simulated nature of the game. The game becomes "unreal," which can have an indelible effect upon the success of the course. If one can afford the time to develop and perfect a sophisticated simulation game, the results would be worth the effort, cost, and time. Unfortunately, time and money are usually in short supply at any university.

THE MINIMUM-STRUCTURE SIMULATION GAME

There is, however, another type of simulation game which generally lacks these shortcomings. Although it is devised to serve a different purpose than are its sophisticated relatives, it provides many of the same benefits. I have elected to label it a "minimum-structured simulation game." It simulates selected cognitive features of a hypothetical situation and operates, insofar as possible, with only one basic, external rule: players are to treat it as a game, as if the situation were real. All other directions and rules are internally and organically embedded in the account as game-furnished conditions or preferential-play possibilities.

This type of game, in the forms I have cast it, has centered around survival situations on islands and planets, but it need not be limited to such settings. For the most recent game-play, students were provided a three-page handout in the form of a letter. This letter contained a diary of the events experienced by them, the survivors, from just prior to the ecological apocalypse, through the escape via "Magnetic Space Coupes" to the landing on "Htrae," the twin planet of Earth in the "Aipotu Galaxy." The final part of the letter consisted of questions concerning what should be done, and how to do it, in order to secure the continuance of the species homo sapiens. Verbal directions were limited to statements that "a game is about to be played" and that "the handout tells what has happened to you so far," which excluded the instructor from the game. After that, I have generally had to goad the students into action[2] by citing certain facts and raising questions regarding the consequences of their inaction— e.g., starvation. Within less than three minutes, discussion usually begins. By the nature of the scenario, players are directed to attend to certain features of the hypothetical situation and grapple with their problematic character. This type of simulation requires that the players mentally transport themselves into the hypothetical situation and treat it, and their presence in it, as real for game purposes.

Some players, however, may not find the account moving. This problem can be overcome if the more vocal in the class, especially the first few to speak up, find the account convincing and begin to take the situation seriously. Thus the potential definition of the situation provided in the account becomes the actual definition when students begin to act in terms of that context—i.e., they become "players." In Garfinkel's terminology (1963: 191) *they* must assign the "constitutive accent." This accenting or actualizing process is important because it generates or reinforces expectations to treat the situation as real. Students who did not find the account itself convincing are thus led to take the game seriously because of group definitions and expectations. The success of this type of simulation is therefore dependent upon the moving nature of the account, the players' imaginations, and the actualization of the context by student-players who take the game seriously. The game is finished when the students run out of energy or ideas or when the instructor calls an end to play. End-game conditions are arbitrary; there are no required winners or losers.

While this type of game might be called role-playing by others, there are two reasons that I have instead called it a "minimum-structure simulation game." The first reason is structural; the second is functional. Players in this type of game do assume roles, but they are not assigned nor are there sets of roles from which they are to choose. The roles they play are the ones they think they would play if the hypothetical situation were real—i.e., they are themselves. This allows the player to focus upon and act in terms of the problem at hand rather than the presentation and maintenance of his/her role.

The second reason for calling this a game revolves around the normative function of this label: it provides the one basic, external rule, that players are to treat the situation as if it were real. By "basic," I mean that it provides an overall frame of reference for players. In the words of Harold Garfinkel (1963: 190): "Basic rules of play serve each player as a scheme for recognizing and interpreting the other players' as well as his own behavioural displays as events of game conduct." "External" is used here in the sense that the rule is not to be found within the account with which members are to deal; rather, it stands logically outside the account, much the same as the title "a short story" stands outside the story to which it is attached. If the technique were labeled role-playing, the consequence would be that the players would concern themselves with their roles, however arrived at, rather than orienting to the problem at hand. By using the label "game," a contextual rule is provided which directs students to treat the hypothetical situation within which they are to act, as if it were real—i.e., what transpires in the account is true, we all experienced it, and so on.[3]

From my experience with high school and college students,[4] it appears that, at least by sixteen years of age, students know to treat this type of exercise as a game, as if it were real. Furthermore, students expect each other, and know that they themselves are expected to treat the situation as real for game purposes. While it might be fruitful to treat students' knowledge of how to play games as a topic, for reasons of time and purpose we shall use this knowledge as an unexplained resource. Also, the problem of what it means to treat a situation as if it were real, will have to be sidestepped by invoking "what everyone knows" about games and how they are played.[5]

While the title "game" provides the external, contextual rule, the attempt is made to embed all other directions and rules internally and organically in the account provided to players as game-furnished conditions or preferential-play possibilities. By having the letter handout appear to be authored by someone other than the instructor—e.g., "Sam Clemens"—a game-furnished condition is provided which does not allow players to ask the instructor questions about the account—i.e., "Sam wrote the letter so you should ask him, if he shows up." Game-furnished conditions are those cognitive features of the account which are to be treated as if they were real.[6] They operate as constraints in environmental, historical, and social senses. The complete list of such conditions in any one game would be long, if not infinite for practical purposes. For example, in the most recent scenario, these were some of the game-furnished conditions: Everyone in the class except the instructor was part of a "we" with the same set of experiences described in the account;[7] we had enough supplies and cybernated equipment to last everyone in the park at least a year; we were on the planet Htrae which is the exact twin of Earth with a level of civilization corresponding to the Earth's in 631 B.C.; we landed in a cleared, flat area, surrounded by forest, that resembles the lower mainland, southeast of Vancouver; we were a group composed of people from all over the United States and the world; and so on.

The preferential-play possibilities refer to the type, range, and degree of choice open to players treating the game-furnished conditions as real for game purposes—i.e., given this situation, what we can do? The list of these possibilities seems also to be infinite for practical purposes. The most important one is presented in the opening of the letter "it is up to us to carry on and begin again." How and if the group should "carry on and begin again" is the central focus of the game and the most important preferential-play possibility. It is with regard to these problems that the preferential-play possibilities, enumerated in the letter handed out to players, were formulated—e.g., "Should we simply try to transplant our old ways?" "Who shall lead, or should we even have leaders?"

Thus, it is characteristic of this type of game that the game-furnished conditions and the preferential-play possibilities are presented in, and as organically a part of, the account provided players. The rationale behind such an arrangement stems from a concern with the basic rule: It is assumed that it will be easier for students to take the game seriously if the account provided them is consistent with the nature of the reality to which they are to transport themselves. By so embedding the conditions and possibilities, it is hoped that students will be led to treat the account as part of the reality and not simply as a preface to it. To provide an external set of directions and rules would be unreal, since it is an essential feature of the documents of everyday life that they do not come with specified directions, but rather the directions are located within the documents by both authors/creators and readers/audience. The directions are socially organized into documents/accounts while at the same time serving to organize that which they are organized into.[8] Thus anyone who is a competent member of the everyday world is, potentially, a competent player of a minimum-structure simulation game. This holds because similar types of interpretive procedures are required to understand the documents within both "finite provinces of meaning," as Alfred Schutz calls them.[9]

PURPOSE AND USES

There is one other aspect of this type of game which differentiates it from sophisticated simulation games. While the latter's purpose is either to train players for some future task or to improve their understanding of a situation, which they probably will never face, the aim of the minimum-structure game is to involve students in generating the order of problems with which the class will deal. While all simulations offer players an opportunity to apply and test the knowledge which is sedimented in their biographies, *this type of simulation explicitly treats the student as a resource.* Rather than the instructor telling the students that this course will deal with the following topics and problems, an appropriate game can be devised which raises questions important to the discipline and allows students to grapple with them. They thereby discover or touch upon the problems with which the course and the discipline, or subfield, are concerned. They come to recognize that these problems are not hatched in midair, but follow logically from the situations of action and the actors' normative orientations (see Parsons, 1937), which are simulated when students attend to game-furnished conditions and preferential-play possibilities within the context of a basic rule.

In order to take full advantage of this type of game, it is advisable to tape record the game play or to take notes. At the next class meeting, an excerpt sheet can then be handed out containing statements and pieces of conversation selected from the total interaction in terms of their relatedness to the issues, topics, and problems with which the course will deal. These excerpts serve as springboards for discussion.

After my most recent use of a game in a political sociology class, the excerpts were used to introduce the ideas of certain theorists, while the game course of play was used to locate their ideas vis-á-vis certain issues. For example, the excerpts "our priorities will be automatic and on our mind," "you're not going to survive as individuals," and "we need a logical system of order and rules if we are to survive as a group" were used to introduce the ideas of Locke, Hobbes, and Rousseau, respectively. As we hoped and planned for, the game-play provided an introduction to the problem of order and was in many ways a "replay" of the historical "dialogue" between the rationalists of the seventeenth and eighteenth centuries and the romantic-conservatives of the eighteenth and nineteenth centuries.

Other possible uses of this type of game would be for the generation of research topics, or at least areas of research, for term papers. Readings could be assigned to be discussed in terms of issues raised during the play of the game. Also, the excerpts which have not been handed out previously could be used as a test requiring students to explain the excerpts or relate them to the works, authors, or ideas discussed in class.

In whatever course such a game is used, to derive full benefit, it should be played early in the term. This suggestion has to do with the class-formation function of gaming: It gets a class going. It is a device to break the ice. While students may have introduced themselves to the class at the first meeting, the active discussion during the game almost requires students to stop and find out names, and address each other appropriately—i.e., pointing, using pronouns, or saying "that guy," has its limits.

By discussing and working together, students begin to become a class, rather than remaining a fragmented group. Student-student rapport is developed. Since the game does not presuppose or require a formal sociological background, their apprehension over starting a new course is assuaged. Their self-confidence is bolstered as they learn that they are capable of discovering and discussing the issues which are central to the course and the issues with which great philosophers have dealt. Of course, student-teacher rapport is also a dividend. While the teacher is not a participant, he or she is looked upon as the benign soul who provided this novel experience, which served to improve their social and self-images. This rapport also develops as a result of the students' involvement in generating the order of problems with which the course will deal. While the instructor has embedded the problems in the game, he or she is not seen as dictating what is to be done about and with them.

ADVANTAGES AND DISADVANTAGES

This type of game, played early in the term, is useful to the instructor in other ways. In the process of taking notes, the instructor allows for sizing up the class. By the nature of the remarks, one can get a notion of student backgrounds, experiences, and present ability to handle ideas. Thus the level at which the course should be aimed can be provisionally determined. By noting who speaks, talkative and quiet students can be located, and by attending to who dominates, class leaders can be identified.

These advantages, to a degree, also result from playing a sophisticated simulation game. However, the minimum-structure gaming technique does have certain advantages relative to its more sophisticated counterpart. First of all, and possibly most important, the minimum-structure game requires considerably less time, cost, and labor to develop and play. No paid assistants are required. Any classroom will do. Materials consist of mimeographed handouts. The time and labor needed to devise such a game are comparatively less because only a minimum of superimposed structure is required. The crucial and time-consuming steps in building a sophisticated game, which Rex Walford (1969: 113-114) calls transforming participants, objectives, and interaction into a game reality and adding constraints and framework to the game reality are just as crucial, but require less time and labor when developing the type of game discussed here. In this type of game, these steps depend upon the instructor's explicated common sense and his/her sociological imagination to develop game-furnished conditions which stimulate the situation of action in the everyday world. Since an enumerated list of basic rules does not have to be developed and made coherent, consistent and workable, time can be spent suggesting different, significant normative orientations for players by formulating appropriate preferential-play possibilities.

By lacking a formalized game structure, this type of game does not fall prey to the snares of complexity, which bog down play and obscure the simulated dimension of the game. In that the minimum-structure game simulates a situation which students can play using common sense, the game cannot be too complex: it cannot introduce totally alien concepts. Even if the formalized game is not too complex, there is the chance that students may become enthralled in "playing the game" rather than in "doing a simulation," which is not a problem with minimum-structured games. Since it does not have this strict game format, the latter type does not create losers, which enhances its class-formation function.

A number of these same advantages, however, can become, or be seen as, disadvantages, or limits. The fact is that in the everyday world the rules of the game do create losers (see Gusfield, 1962), although this is not usually recognized by members. Having to depend upon what members do recognize, or know, keeping the game simple and not introducing new concepts or unfamiliar situations of actions limit the temporal usefulness of this type of game to the first few weeks in the term. While every subfield has its central issues, a number of these are tied into the nature of roles and their interrelationships. In these cases, some form of role-playing might be more suitable than this type of game. Finally, the very fact that this type of game is relatively unstructured means that the instructor cannot be certain as to the type of discussion it will generate. He must either become adept at composing scenarios that channel attention in the direction he wants or else intervene in the games to stimulate or redirect discussion.

The question can also be raised as to what would happen if games came into wide use. Since games have been a novelty until now, would students become bored and lose interest in playing? One can only speculate as to what would actually happen, and it has to be admitted that boredom is a possibility if gaming becomes a common experience. However, I would suggest that it is also possible that the opposite might occur, students might become *more* involved and better players. It seems that if students were exposed to this amount of gaming, then developing their own games, as a class assignment, would be more enlightening than playing them.

CONCLUSION

The minimum-structure simulation game fills a specialized niche, in terms of purposes, uses, and logical location between role-playing and sophisticated gaming techniques. Since it does not require any formal sociological knowledge on the part of players, this type of game can be used at the beginning of a course to immediately involve students in the problems which usually can only be seen from the distant world of scientific theorizing. By simulating a situation which is problematic to the theorist, students begin to discover and generate the order of problems with which the theorist deals. The game-play can thus be an experience which serves to organize and make sense of the course material, while providing the class with a common, and commonsense, base and basis. In the process, student-student and student-teacher rapport develops—i.e., a class is born.

NOTES

1. Rex Walford (1969: 27) has pointed out that, in the American High School Project's "Portsville" and "Metfab" simulation games, the competitive elements are not stressed. Thus the line drawn between minimum-structure and sophisticated simulation games, in terms of superimposed competition, is not hard and fast.

2. To be more precise, I have had to provide the "shock" required to shift the players'

accent of reality from one finite province of meaning (the everyday world) to another (that of games or play). See Alfred Schutz (1962: 231).

3. The "and so on" refers to what Garfinkel (1963: 190-191) has called the " 'et cetera' provision" of any set of rules, and *in this case* includes Garfinkel's "constitutive expectancies."

4. I have used such games over the past three years in sociology with three classes of senior secondary students, two classes of freshmen and sophomore university students, and one class of fourth-year university students in a continuing education program. My sample consists of some 200 people from ages 16 to 50+. The high school classes consisted of three homogeneous groups ranked by the school system as low, medium, and high in "ability" and performance. The most recent playing out of a game was in political sociology taught in the Department of Continuing Education, University of British Columbia in the fall of 1970. It is from this game that my examples are drawn.

5. However, a suggestion for dealing with this problem would be to examine Schutz (1962: 207-259), with special attention to his discussion of "finite provinces of meaning." While the sophisticated simulation game is played within the finite province of meaning of "games," it seems that it might be more precise to suggest that the minimum-structure simulation game is "done" within the finite province of meaning of "play." (For normative purposes, I shall still refer to minimum-structure simulations as "games." See text.) Both of these provinces, however, are "derived" from the "paramount reality" of the "world of working." The differences have to do with the different cognitive styles which define each, and with the actor's shift of "null point," from his Here and Now in the everyday world, to the "game" or "play."

6. This type of game is thus different than the formal games with which Garfinkel has dealt (1963). Whereas in formal games the basic rules circumscribe the "legal" parameters of play and, as a set (constituted by the constitutive expectancies attached to them) define the game, the basic rule of this game is that it is to be treated as if it were real, as a "game." Insofar as the player's set of experiences within this finite province of meaning called "play" are kept consistent with the latter's specific cognitive style (Schutz, 1962: 230), it would seem that we are warranted in assuming that members can mentally transport themselves to, and operate within, that province without the aid of any other specified rules than treat what follows as a game. With this type of game, the "legal" parameters of play are circumscribed by the game-furnished conditions, while the basic rule provides *how* to treat these parameters—i.e., they are real for game purposes. Thus the importance of the game-furnished conditions is greater, and their relation to the basic rule is different than that of formal games which possess a *specified* set of basic rules.

7. Although cultural differences will intervene, the "general thesis of reciprocal perspectives" is assumed to hold as a result of the basic rule. See Schutz (1962: 12).

8. On the social organization of accounts, see Smith (1971).

9. This follows from Schutz's point (1962: 226-229) regarding the everyday world of working as the "paramount reality" upon which all others are based.

REFERENCES

ALGER, C. (1963) "Use of the Inter-Nation Simulation in undergraduate teaching," pp. 150-189 in H. Guetzkow et al. (eds.) Simulation in International Relations: Developments for Research and Teaching. Englewood Cliffs, N.J.: Prentice-Hall.

DAWSON, R. E. (1962) "Simulation in the social sciences," pp. 1-15 in H. Guetzkow (ed.) Simulation in Social Science: Readings. Englewood Cliffs, N.J.: Prentice-Hall.

GARFINKEL, H. (1963) "A conception of, and experiments with, 'trust' as a condition of stable concerted actions," pp. 187-238 in O. J. Harvey (ed.) Motivation and Social Interaction. New York: Ronald Press.

GREENE, J. R. and R. L. SISSON (1959) Dynamic Management Decision Games. New York: John Wiley.

GUSFIELD, J. R. (1962) "Mass society and extremist politics." Amer. Soc. Rev. 27 (February): 19-30.

PARSONS, T. (1937) The Structure of Social Action. New York: McGraw-Hill.

SCHUTZ, A. (1962) Collected Papers I: The Problem of Social Reality. The Hague: Nijhoff.

SMITH, D. E. (1971) "K is mentally ill." University of British Columbia Department of Anthropology and Sociology. (unpublished)

TAYLOR, J. L. (1968) "Some pedagogical aspects of a land use planning approach to urban systems simulation." University of Bristol Department of Geography. (unpublished)

WALFORD, R. (1969) Games in Geography. London: Longmans.

COMPUTER-ASSISTED SIMULATION IN URBAN LEGAL STUDIES

DANIEL A. DEGNAN and CHARLES M. HAAR

Syracuse University　　　　*Harvard University*

This is a report on possible uses of urban gaming simulation by computer in courses relating to urban legal studies. In the spring of 1970, the Environmental Simulation Laboratory of The University of Michigan brought the urban simulation game APEX to the Harvard Law School to be played over a weekend by forty students in the course in land use planning and about ten other persons engaged in urban work, studies or teaching.[1] APEX simulates on a computer the city of Lansing, Michigan, and enables its players to effect changes in the city over a period of years through their decisions as planners, politicians, industrialists, developers, and pollution control officers. Problems in zoning and planning, environmental law, municipal law and administrative law, among others, arise in the simulated environment, as do broader questions of economic and social policy.[2]

The use of urban systems analysis combined with operational gaming which APEX represents has, in the authors' opinion, important possibilities for urban legal studies. This article begins with a brief introduction of urban gaming simulation, with the focus on APEX. The first part then describes what the players experienced and the second part offers an evaluation of urban gaming simulation in terms of its uses and advantages for urban legal studies. While our emphasis in this report is upon urban law, its conclusions seem to apply also to the study of public administration, planning, urban development and related fields, and, generally, to efforts to gain an understanding of the urban process.

APEX uses a full-scale computer simulation or systems analysis of the city of Lansing, Michigan.[3] The twin bases for the computer's simulation of the environment are empirical data from Lansing about population, jobs, households and household types, and so on, and "computer simulation models," which represent urban processes or functions, such as changes in employment, and population distribution, and even voting.[4] Although the level of abstraction from the real Lansing community is high, the empirical data includes the people of the city, in terms of households (which are divided into five socioeconomic types), school population by age groups, and employment. The employment data are built

This article is a reprint of an article appearing in the **Journal of Legal Education**, Volume 23, pages 353-365, 1971.

around two relatively simple concepts, "exogenous" or basic exporting industry and "endogenous" or non-supporting service industry and commerce. The three major employers, for example, the state government, Michigan State University, and the Oldsmobile plant, are exogenous industries; changes in their employment due to outside circumstances are the major link between the simulated city and the larger environment. All data, including land use, land value, zoning, public facilities, tax assessments, are made available by the computer for each of twenty-nine "analysis areas," small geographical units resembling a typical voting ward or district, roughly equivalent to the census tracts.

The computer simulation in APEX, however, rests on the simulation models which represent urban processes and interactions. The importance of these models can be realized from a brief description. GROW is a computer model representing employment and population growth generated by exogenous or basic industries; SPREAD, a model of general employment and population growth which is also capable of distributing that growth in different parts of the metropolitan area; SELL represents the residential real estate market; VOTE represents the voter responses to bond issues; ELECT represents the voter responses to candidates; and AIR offers a model of changing air characteristics and contaminant levels. The most difficult aspect of the simulation is the linking of these models so that they affect each other realistically. GROW, for example, provides the overall metropolitan growth and SPREAD then generates from this information not only the additional employment, households, housing, commercial areas, and public services and facilities that accompany this growth, but also their geographical distribution in the metropolitan area. SELL reports the demand for land, homes and businesses generated by GROW and distributed by SPREAD.

The purposes of gaming simulation and of APEX in particular are divided between research or planning, and teaching. Operational gaming,[5] through its use of role-playing and games theory, enables its users to introduce alternative policies and choices into the urban model. Game operators can examine and test a variety of processes and assumptions as the models reveal these. The organizational and psychological processes through which the actors or players come to decisions can also be tested. These purposes cannot be separated from teaching, however. Duke and others speak of "heuristic" gaming, since the gaming simulation provides a way of discovering the principles and patterns of urban processes. The game's designers and others can test and evaluate their understanding of the processes which the models represent, planners can test the consequences of their decisions, and teachers and students can use gaming as a mode of understanding.[6]

Richard Duke, the principal developer of APEX, offered the following summary of the immediate benefits of gaming simulation for the law student: The games should enable the student: (1) to see the metropolitan area in a broad perspective; (2) to achieve this perspective in a variety of roles; (3) to familiarize himself with metropolitan problems, their scope and importance, and with the variables—people, jobs, land, air, votes—which are involved in these problems and processes; (4) to see how decision-assisting tools can be used, especially analytic tools such as projection techniques and synthetic tools such as programming; and (5) the most important, to uncover and evaluate the goals, objectives, and decison-making criteria in the problems with which he is most concerned. (The simulated environment is itself designed to force the student to test and re-test the assumptions built into the simulation.) In addition, in gaming simulation the player will act out, in a real time context, legal problems raised in the simulated city. This differs from a moot court in that all players are participants in the world in which the dispute arises, and the effects of the decision are felt in that world. For example, an industry may be forced to close because of pollution control enforcement.

Our own evaluation of gaming simulation is based in part upon these goals, but more particularly upon what we think it can contribute to legal studies. In the first section of this report, we shall concentrate on what the players experienced; in the second section we shall relate this experience to the goals of urban legal studies.

I

APEX is designed to enable eighteen decision makers, representing city and country politicians, city, county and regional planners, land developers, industrialists, and air pollution control officers, to interact with each other and with the simulated city. At Harvard, approximately forty-five players shared the eighteen roles, usually two to a role such as manager of an electric power plant, or city planner. At the start, each player received a computer printout, the data describing what has happened in his role in the last few years—the production of his cement plant or power company, its costs and profits; or the rules administered by the air pollution agency, its budget, and current emissions and air pollution level; or county and city tax rates and tax base, operating budget, capital improvements, bond issues.[7]

The first cycle plunged the actors immediately into decision-making, forcing them to absorb information as quickly as possible with the aid of manuals for each role and advice from the game's operators. Decisions, recorded on forms, were fed to the computer, which returned the results for that year to each player, telling what had happened in that year to his business, the city, or his political fortunes as the result of his own decisions, the decisions of other players, and changes in the city simulated by the computer. Decisions had to be made in interaction with the players in other roles. Developers, for instance, immediately began negotiating with each other, with industrialists, city and county planners, and city and county politicians. The physical arrangements in the room where the game was played facilitated easy communication among the players. County politicians and planners were at tables opposite city politicians and planners, some distance apart. Developers, industrialists, regional planners and air pollution control officials were grouped about in between, the whole forming a rough circle, with space for moving about.[8]

APEX can be played in up to ten cycles, each cycle a year in the city's life. The game played at Harvard was a short, intensive one. One cycle was played Friday afternoon, one on Saturday morning and one Saturday afternoon. A final session on Sunday was devoted to a general critique of the experience. The three cycles played with the first computer printout, amounted to four years in the city's life.[9]

We requested ten of the players, in different roles, to submit a one or two page statement of their experience of the game and their evaluation of it. There was also a general critique by all of the players and operators at the end of the game. The account of the players' experience which follows is based upon these reports, the general critique and our own observations. We shall start with a section from the report of Michael Heaney, one of the two air pollution control officers, continuing with brief impressions of each role.[10]

Regulation Changing. This was at once one of the hardest and most creative aspects of the APCO role. Apex is programmed so that at the game's beginning, the APCO finds himself with very weak county ordinances ("Laws") and agency regulations ("Rules"), including low fines and tax pollution standards; no rule-making authority or subpoena power; little data on the polluting industries; and no record of administrative or judicial action. During the first cycle, we attempted to discover reasonable emission and other pollution standards, in order to request changes from the County Board. This proved particularly difficult for law students because of our virtual ignorance of the technical aspects of the role, and of our inability to make sense of the myriad statistics presented to us initially in our APCO print-out and manual.

By the end of the second cycle, however, we had progressed to the point where we sought successfully a comprehensive new pollution control 'act' from the Board. The act included a substantial increase in fines; stricter emission and other standards, all of which were raised to the level of Laws rather than Rules; reinstatement of our rule-making authority; and a pollution tax scheme under which qualifying industries were permitted to pay an incremental tax on their pollution, in lieu of being fined, and under which industries demonstrating bona fide efforts to abate their pollution might be exempted from all or part of the tax. This act, passed with little opposition during the second cycle, was strenuously attacked during the third (after a re-election had placed two pro-industry Commissioners on the County Board), and barely escaped repeal.

Industry Regulation. This function, difficult or impossible at the outset of the game, was well underway toward the end. Early in the second cycle (and before the County Board had adopted our new act), we decided to bring suit against one of the gamed industries, Peoples Pulp, under the existing Rules. The case was first continued pending administrative efforts at a settlement; then later in the cycle brought to trial where the industry was found in violation on one of two counts. Although sentence was stayed pending efforts on the part of People Pulp to comply with APCO regulations, the suit served to give industries notice of APCO's intentions and the local court's backing.[11] During the third cycle, all of the gamed industries came forward with pollution control proposals for their plants, and agreements and tax exemptions were worked out with APCO for each. Additionally, many of the simulated industries undertook similar actions, presumably on the basis of our inputs during previous cycles. The APCO print-out for the fourth cycle indicated continued progress in the reduction or elimination of pollution in most of the studied industries.

The "regional planner," confronted by an enormous amount of data about the city and its suburbs, not only embarked upon a two or three year study in the manner of planners, but, more unusually, as Professor Duke remarked, began to take action. The planners persuaded the county commissioners to repeal the present zoning ordinance and substitute "planned unit development," under which geographical or area zoning was scrapped and performance standards were substituted. If an industrialist, for example, met the standards for buffer zones, adequate access, minimum pollution, etc., his application would be approved to locate in any site he had chosen.

In the major court case of the game, the developers brought suit to have the planned unit development ordinance declared invalid, as beyond the powers granted by the zoning enabling act. Oral argument was heard, and, as best could be done in a limited context, cases and arguments based on the land use planning course, and expert testimony from one of the assembled professorial experts, were presented by the parties. The court delivered itself from the bench of an opinion ruling the ordinance to be illegal, and the ensuing dilemma is described by a regional planner:

> Between the second and third cycles we tried to develop some general and preliminary standards since the county had already passed the ordinance accepting the pud (planned unit development) idea. Further pressure for a delineation of the standards came from the institution of a suit by the Developers. Cycle three started with the trial and we lost, perhaps on the merits, but certainly because there was no time to reprogram the computer.

> Feeling defeated and confused with only one cycle left to do long-range planning, we were given all the powers of developers, industrialists, and regional planners (the urban development corporation).[12] We were also given a large HUD grant to use as we saw fit. Given the news of the new plant looking to locate in the area, we decided to work on putting it in the best place for the people who need it, the city, and the county. Since the county got the same benefit wherever in the county it was located, we concentrated on getting it close to the unemployed in the city core while displacing as few people as possible. We acted as liaison between the city and county, finally settling on dividing the plant up by putting small parts assembly in the city core, within walking distance of many of the unemployed, and placing the large assembly line in the county just down the highway and rail line from the plants in the city core.

The "county commissioners" quickly organized into committees in order to handle their heavy workload, caused in part by having "people after them" constantly (as could be seen in the room where the game was played). This development enabled them to handle budgeting, capital improvements, planning, zoning and development, requests for zoning variances, and air pollution problems, not to mention seeking votes for the next election. When three of the five county commissioners were turned out of office at the end of the second cycle, the two remaining commissioners found that their past experience enabled them to have a great influence on policy. The commissioner who handled the budget, for instance, reported that he had entrenched himself through past favors. The industrialists, however, were disappointed by the performance in the office of the new commissioners elected by their efforts, although as the game closed, the commissioners had been persuaded to consider more relaxed air pollution controls. One of the county commissioners reported:

> The chief value of playing a County Commissioner was his ability to interact with all participants in the game. In at least one instance this interaction was intensive, for the Air Pollution Control Board had to come to the County Board every time it wished to adopt some procedure or rule. The discussion of the air pollution statute, as drafted by Steve Barrett, provided not only many insights into air pollution control, but also into statutory drafting and interpretation. When the statute came up for review in the last cycle, the issues raised of fairness and procedural due process changed the Board into a court reviewing the constitutionality of the statute.

> As a County Commissioner I was also deeply involved in land-use development. In individual cases I was called upon by land developers to grant zoning changes; the difficulty of weighing all the factors in deciding on the most appropriate use for the land raised many of the problems of the course material. In addition, the game provided the unacademic factors of bribes, political contributions (I received several thousand dollars in contributions for granting one change) and persuasion.

> On another level I was involved in the discussion of the proposal to enact a planned unit zoning ordinance for the county While the plan was declared unconstitutional, the debates in the County Board raised all the problems of various types of regional planning, zoning, etc. The discussion was especially aided by the participation of 'Commissioner' Frank Logue, director of the National Urban Fellows, who provided many interesting ideas for change. The other area in which I spent a good deal of my time was in the hearings on rate increases by electrical companies. This task coincided with much of the research I had done this summer while working in Washington, so it provided an excellent means of trying to apply what I had learned about electrical utilities.

The "developers" began the game as owners of land in different parts of the city and county and with varying amounts of cash and indebtedness. Since the developers were free to invest anywhere, much of their effort was focused on deciding where to invest and on the types of development—residential, commercial or industrial—to undertake. Several developers combined to create an industrial park. The developers were especially dependent upon directions of the city's growth, and of course upon zoning restrictions. The game's principal law suit, the action to have the new zoning ordinance declared invalid, was brought by the developers. At the trial, the developers argued that planned unit development made the value of all of the land in APEX dependent upon the ad hoc decisions of a planning board, since the use of land in each case would depend upon board approval. Even when a use was approved for one parcel, the developers contended, one could not know the potential uses or value of adjoining land. Although the planners, in defense, insisted that the ordinance applied general performance standards to which a developer need only conform, the ordinance, as has been noted, was held to be *ultra vires* the enabling act.

Several developers complained about lack of cooperation from the city. Land was easily available in the suburbs, but in the city it was necessary to approach officials for joint planning of projects. And when the city was approached, its response was "come in with a project" or "see the planner." The city was also charged with failure to seek outside industry. The "city politicians" replied that they had been too busy with time-consuming work on operating budgets and capital budgets to do the developers' work for them, that the city had made zoning changes at the developers' requests, and had brought in four or five new industries.

One of the major decisions of the city politicians, made in conjunction with the urban development corporation, was to sponsor intensive development in a central city area occupied chiefly by low income people. In an effort to attract investment and to improve employment and other conditions, the city also became more and more expansive in initiating capital improvements. One of the city politicians reported, however, that it was discouraging to have the computer report a continuing downturn in the central city despite these efforts. To halt this trend, local expenditures were greatly increased in the last cycle, and the city was also seeking federal grants.

The problem for the "industrialists," besides prices, production levels, new capital expenditures, and land purchases, was air pollution abatement: The industrialists had to determine whether to switch from burning low grade fuel to burning a higher grade, the cost of fuel and of the equipment for burning it, the cost of abatement devices, the threat, in at least one instance (the pulp company), of being put out of business because of an inability to meet the standards for sulphur dioxide emissions and odor, even with new equipment. One industrialist complained that the game did not allow for elements such as labor problems and mergers with other companies, but agreed that his firm, at least, had a major problem in meeting air pollution control standards.

At the end of the first cycle, the pulp company was brought to court for violation of the old pollution regulations, and although the case was continued, this brought heavy pressure against all of the gamed industries. As the industries invested in studies by consultants, and began switching to higher grade fuels, new pollution standards were enacted. These carried heavy penalties, but also provided for a "variance," under which a graduated tax would be assessed in lieu of compliance. In a second court action, one company was found to be in violation of the new ordinance, and as the game ended an industrialist had sued to restrain enforcement of the pollution ordinance. The issue centered on the control officers' holding out the promise of a "variance" to extract agreements from the industrialist to install new equipment, fuel, and abatement devices.

Although these individual reports were submitted some days after the game had ended, their universal conclusion was that the gaming simulation had been a valuable experience. One developer reported that he had "lost his shirt" in the first round due to inaccurate advice from a game operator, rendering the game thoroughly unenjoyable "qua game" after that. Nevertheless, he returned for each session "because the educational experience was intense and valuable." An industrialist said that he had himself worked five years as an industrialist and the role had not been as simple as APEX had portrayed it. The real-world industrialist was not as single-mindedly devoted to profits as his APEX counterpart, knowing that profits depended upon community well-being and, in addition, having a personal interest in the community. His report concluded, nevertheless, that "Apex is excellent and most complaints are trivial." Its most valuable function, this player thought, was to reveal the different interests at stake in matter such as air pollution abatement, and to induce solutions which would take these interests into account. Common notes in these reports were an initial sense of confusion caused by having so much data to handle, and a sense that the game had ended too soon. While detailing extraordinary efforts to control and manage their complex roles, the players evinced a sense of frustration at not being able to do more. A summary of a questionnaire answered by many of the players appears in the margin.[13]

II

The comment by the player who had worked in industry and who then played an industrialist's role raises an important question preliminary to any evaluation of gaming simulation: Does the simulation represent reality or does it merely create a fascinating and intensely exciting game? The authors, three other law professors, two directors of urban institutes, and several professors and graduate students in urban planning participated in the game or helped to conduct it. Our experience, besides teaching, combined public administration, politics, the practice of law, including municipal law, planning, and business. The response of this group, from our conversations with them, seemed to be that although the simulation compressed time and space, and restricted or limited the number of relationships possible, the problems which arose in the simulation showed a remarkable similarity to "real life." Certainly the authors found this to be so.

The several lawsuits relating to pollution control, for example, created strong pressures for settlement or compromise, while the developers' successful effort to have planned unit zoning declared invalid was devastating in its effects on planning. Most important, in the intense interaction between the players—politicians, planners, developers, industrialists—realistic economic, social and personal forces seemed to be at work. Although it is harder to judge the simulation model as a whole,[14] few false notes were detected. Among things indicating that the model was realistic were the empirical data from which the model was built; the realism of the players' roles, since these were integral to the overall model; and paradoxically, the model's limitations. Since the model did not pretend to everything and since the areas of simulation or of abstraction were readily identifiable, one could take a critical approach to the simulation while at the same time benefitting from the intense experience it offered.

This brings us to the question with which this report began, the uses of urban gaming simulation for urban legal studies. A program in urban law is bound to depend upon a series of courses such as land use planning, local government law, federal urban legislation and administration, housing and urban development, environmental law, federal and local taxation, yet any such program exists against the background of the urban process, a process which is extremely complex and difficult to grasp. In a course in land use planning, for example, it is hardly enough to present the relatively coherent set of appellate decisions on zoning, even when supplemented by textual material. A major goal must be to understand the relationship of planning and zoning law to the urban phenomenon, the influence that zoning has upon the city. The authors see urban simulation as a promising method of relating specific areas of urban law, such as land use planning on environmental law or public housing law, to broader questions of the urban process.

In the weekend's gaming, legal problems relating to land use planning arose on a series of levels. On the level of "primary activity," where law touches the citizen before any question arises of remedy or litigation,[15] developers, for example, conformed their investments to the constraints of zoning and public development policies. The developers had to take into account zoning, taxation, capital expenditures for roads, utilities, schools, and so on. They extended their activities to other levels, by obtaining zoning variances, supporting expenditures through political pressure, and by bringing an action to attack the validity of the new zoning ordinance. Among public officials, the planners, starting out with data concerning employment, income, quality of housing, directions of urban growth and evaluation of needed public facilities and amenities, found themselves engaged not only in making policy, but in legislative drafting, informal procedures and formal hearings, lobbying, redevelopment decisions under federal grants-in-aid, and full-scale litigation.

Issues arose in the simulated city as they would arise for lawyers in practice, but with differences important to the study of law, notably the compression of time and space. The lawsuit over the zoning ordinance was instituted after only two cycles of playing time, in the midst of other planning and zoning questions. The organization and administration of air pollution control moved from initial planning to legislative drafting to administrative procedures in the same time period. Thus, the game was not primarily a clinical experience or a moot court, but a teaching method in which the simulation unfolded a series of urban legal problems in their natural environment. The players experienced not only specific urban functions such as planning, budgeting, private development, but through the compression of time and space, saw the city as a whole as a combination of interrelated urban processes.

The functions of computer simulation were also appraised in actual use. After the initial shock of receiving data-heavy computer print-outs, the student soon became familiar with both the data and the models simulating relationships between population and employment growth, land use, housing uses and densities, transportation and other capital improvement. Role-playing, in which particular decisions were made on the basis of computer data and in interaction with other players, quickly established the contrast between individual decisions and the computer simulation, which measured large-scale and long-term effects on the city. A sense developed of what systems analysis could and could not do, an expectancy, for instance, that more sophisticated zoning and land use models could be created, or that the action of pressure groups could be simulated.

The game also combined intense involvement in the players' role with a demand for an objective critique or appraisal of what was taking place. This effect seemed to be caused by the nature of simulation, which purports to present a realistic model of urban processes and a realistic role for the player. Precisely because he was intensely engaged in playing the game, we think the player felt a need to appraise and evaluate the simulation. In an extended game, this evaluation should range from an examination of the usefulness of the simulation and the degree of its correspondence with reality, to a critique of the institutions and policies around which the game has revolved, and of the larger social and economic issues, such as the extent to which the city's development or lack of it has depended upon private investment policies and speculation in land.

Of course there were limitations to the simulation, some of them already apparent in the reactions of the players and ourselves. The processes of the city which lent themselves to quantification were economic and physical ones primarily—although role-playing and even some modelling was able to simulate other forces. The simulation model might have been lacking in validity—although we do not believe that it was—and so far it has proved difficult to test models except by a kind of natural reaction to them. The simulation was limited to the data and models which were incorporated in it, and the ingenuity of the players could at times push the game past these limits. (The game at Harvard was not programmed for floating zones or planned unit development, a fact which may have influenced the court's decision in the zoning case.) Above all, the simulation was an abstraction, a model of some aspects of urban reality.[16]

Gaming simulation as we experienced it, however, functioned in a way which was complementary to the use of legal materials. The simulation presented a series of relationships and demanded judgments and decisions about them both with respect to urban processes and legal institutions which were a part of these processes or which interacted with them. Legal issues were not built up through a series of appellate cases but through instances of law's functioning on many levels. In the appellate cases studied in law school, the student realizes vicariously some of the experience of hundreds of lawyers and judges. Urban gaming, however, placed the student in a role situation based upon the processes of an actual city, so that he became in effect the planner or politician or developer weighing plans for the future against the resources available today.

The student participating in the simulation was also confronted with the difference between the law in books and the law in action. He saw that one cannot litigate every case, that private arrangements and negotiations were the bases of action. To some extent the student also saw the consequences of his decisions for the urban environment: What happened, for example, when air pollution standards were revised and enforced, or when different zoning policies were adopted?

As the students themselves recognized in their brief experience, simulation could be used in a variety of ways. The game could be presented for intensive play or played a cycle at a time, as a supplement to courses, interspersed with discussion, legal research, background study in systems analysis, urban problems, and so on. Within the gaming simulation, a variety of techniques offered itself. It was possible, for instance, to create an urban development corporation as was done in the third cycle. Simulation could have been utilized by a number of courses, land use planning, local government law, environmental law, housing and urban development, as an introduction to urban legal studies, or as an intensive experience in systems analysis and urban processes.

From our encounter with APEX, two things stand out to make us interested in further experimentation with urban gaming simulation. First, gaming simulation promises a way of relating specific legal questions to the entire urban process—a way not dependent upon appellate cases but closely resembling the law's actual functioning on a number of levels. Second, there is the quality of gaming itself. It engages the players in decision-making and action, in what Aristotle called the exercise of practical reason. Heuristic gaming—especially, it seems to us, as it involves legal problems—demands that we give critical attention to our own decisions and to those of others with whom we interact. It is neither clinical experience nor pure academics, yet urban gaming simulation has possibilities for placing law studies in the matrix of some of the more difficult problems of the city.

NOTES

1. The game at Harvard was made possible and was conducted by Professor Richard D. Duke, its principal author and the Director of the Environmental Simulation Laboratory, with the help of the IBM Corporation, for whose 1130 computer the game is programmed. Important assistance also came from the staff of the Laboratory and from Professor Alan Schmidt of the Graduate School of Design of Harvard University and Mr. Francis Ventre of the Department of City Planning, Massachusetts Institute of Technology. The financial support, so indispensable, came from the efforts of Derek C. Bok, Dean of the Harvard Law School.

2. Apex is probably the most sophisticated and most computer-oriented of a relatively brief line of urban games designed to reproduce for their players elements of urban interaction and growth. See Mark Nagelberg and Dennis L. Little, Selected Urban Simulations and Games (Middletown, Connecticut: Institute for the Future Working Paper WP–4, 1970); Allan G. Feldt, Operational Gaming in Planning Education, 32 Journal of the American Institute of Planners 17 (1966); Richard L. Meier and Richard D. Duke, Gaming Simulation for Urban Planning, 32 Journal of the American Institute of Planners 3 (1966); Peter House and Philip D. Patterson, Jr., An Environmental Simulation Laboratory, 35

Journal of the American Institute of Planners 383 (1969); Paul H. Ray and Richard D. Duke, "The Environment of Decision Makers in Urban Gaming Simulations," in Coplin (ed.), Simulation in the Study of Politics (Chicago, 1968), p. 149; Richard D. Duke, Gaming Simulation in Urban Research (East Lansing, Michigan, 1964); Manual for Region, An Urban Development Model (Washington, D.C.: Washington Center for Metropolitan Studies, 1968).

3. The systems analysis of cities entails a mathematical model, usually designed and programmed for computers, of some of the interacting elements of a particular city or of cities generally: elements such as population changes, housing, jobs, industry, transportation, capital improvements. Systems models, like models in economics, necessarily abstract from the details of the real, yet they offer a method for controlling vast amounts of empirical data, for understanding the relationships and interactions of a city, and—assuming the continued applicability of the empirical data and the correctness of the models of urban functions and relationships—for predicting future trends. See Ira S. Lowry, A Short Course in Model Design, 31 Journal of the American Institute of Planners 158 (1965); Britton Harris, New Tools for Planning, Ibid., p. 90; William D. Coplin (ed.), Simulation in the Study of Politics (Chicago, 1968), p. 1; Britton Harris, "Quantitative Models of Urban Development: Their Role in Metropolitan Policy-Making," in Harvey S. Perloff and Lowdon Wingo, Jr. (eds.), Issues in Urban Economics (Baltimore: John Hopkins Press, 1968), p. 367; Dennis L. Little, Models and Simulations—Some Definitions (Middletown, Connecticut: Institute for the Future, Working Paper WP–6, 1970).

4. The information which follows is taken principally from M.E.T.R.O. (Metro Project Technical Report 5, Tri-County Regional Planning Commission, Lansing, Michigan, January, 1966); APEX: A Gaming Simulation for Air Pollution Experience in a Simulated Environment (a draft report on APEX by the Environmental Simulation Laboratory, School of Natural Resources, The University of Michigan, December 1968); and Ann Cochran, Introduction to APEX—A Gaming Simulation Exercise (Comex Research Project, a joint research program by the School of Public Administration, University of Southern California and the Environmental Simulation Laboratory, University of Michigan, October 1969).

5. Operational games contrast with mathematical or pure simulation models in which policy design and choice cannot be mathematically simulated. Gaming simulation, in which human players interact with the simulated environment, adds, through role-playing, human policy design and choice, made in interaction with the simulated city. Gaming simulation arises out of the need, simply stated, to include human actors. Its premise is that many dynamic, complex situations such as one encounters in the city cannot be modeled in straightforward mathematical formulations or in computer programs. See Ray and Duke, op. cit., pp. 153-158, 160; Meier and Duke, Gaming Simulation for Urban Planning, 32 Journal of the American Institute of Planners 3 (1966), p. 5. For a description of pure simulation see Wilbur A. Steger, The Pittsburgh Urban Renewal Simulation Model, 31 Journal of the American Institute of Planners 144 (1965); and, generally, see note 3.

6. METRO, APEX's parent game, was designed for research and experimentation in urban systems analysis, as a teaching device, and as a way in which planners and administrators could measure the results of various policies. APEX was designed for teaching in the program in air pollution control at the University of Southern California and for the study of urban processes generally and their specific relationships to air pollution control.

7. Although most of the players were students in the course in land use planning at the Harvard Law School, they were joined by students in urban planning from Harvard and MIT, by the directors of two urban institutes, and by some law professors, including one of the authors, who played the role of "city politician." The other author acted as judge for the legal actions brought during the game.

8. It seems useful to mention some of the practical details of operating the game. An 1130 IBM computer, for which the game is programmed and which is available at most universities, is needed nearby, preferably not more than ten minutes away, since the computer's print-outs are prepared in the interval between cycles.

A computer operator and a team of game operators conduct the game and advise the players. Except for the first game, these operators can be drawn from the university community, since the game is designed so that persons in the area will be trained to operate it. The major initial cost is for training, and the principal operating costs are the time of the computer operator and of the game operators. Our idea, worked out in conjunction with some of the faculty at MIT and the Harvard Graduate School of Design, was to interest several institutions in installing the game, so that a pool of skilled operators could be made available and the expenses of training shared. Persons from these schools assisted us, although Professor Duke's own staff conducted the game.

9. The game at Harvard was conducted in something of a pressure cooker. APEX is designed for more extended use, in which the play of one cycle at a time can be geared to courses, allowing for further development of legal and other procedures arising in the course of play and for class discussion based upon game problems. Both APEX and METRO, however, are also played in intensive sessions.

10. The authors wish to express their gratitude to the following students who have given us permission to quote from their reports: Michael K. Heaney (air pollution control officer) of the class of 1970 and Mark A. Willis (regional planner) and Allan R. Abravanel (county politican) of the class of 1971 of the Harvard Law School.

11. To the "judge" in this action, who was one of the authors, its realism stood out: while pollution control laws had to be enforced, the ability of People's Pulp and possibly of others to continue to do business was in question. The parties fortunately saw that air pollution reduction and not punishment was the goal.

12. The role of regional planner was a new one, improvised for the game at Harvard. Professor Duke added the urban development function.

13. A questionnaire designed to elicit their evaluation of APEX was completed by twenty-six of the players between the second and third cycles of play. They were asked whether they felt "that Apex would contribute significantly to a senior or graduate level course in some field other than air pollution control?" Twenty-five said yes, one said no. Asked to suggest which field or fields, the players named planning 19 times, law 11, government 8, political science 7, and other fields 6. Twenty-three players found the game to be realistic, 25 found themselves more aware of interrelationships as the game progressed, 24 thought that the interdependencies shown in the game had reflected reality, and 25 thought the game to be an effective teaching device. Principal criticisms were the amount of information the players were forced to absorb, the limited time for play, and the belief of some of the developers and industrialists that the county politician's or county planner's role would have been more rewarding. (In an extended game, the roles would have been shifted.)

Commenting on possible use of the game for first year law students, many felt that a background in property law, zoning and planning law, or administrative law would be needed, although others thought that with more preparation and playing time, the game would be useful for first year students. Comments on use of the game for second and third year law students stressed more concentration on planning techniques and decisions, more time for study of legal and technical aspects of both planning and pollution control, and greater use of legal processes and administrative practices in decision-making.

14. See Henry M. Hart, Jr. and Albert Sacks, The Legal Process: Some Basic Problems in the Making and Application of Law (Cambridge, Tentative Edition, 1958), p. 207. et seq.

15. Whether a model's assumptions are valid seems so far to be a question answered only in terms of the design and utilization of individual models. See Paul H. Ray and Richard D. Duke, "The Environment of Decision Makers in Urban Gaming Simulations," in Coplin, Simulation in the Study of Politics, pp. 149-152; William D. Coplin, "Introduction," in Ibid., pp. 1, 2.

16. See note 3. Coplin, Simulation in the Study of Politics, contains several critiques of both gaming and pure simulation, e.g., Matthew Holden, Jr., Comments on Ray and Duke's paper, p. 176. Holden finds in METRO an overly simple "public goods" economic concept and a disengagement from the political problem of disruption and violence.

THE ROLE OF MANAGEMENT GAMES IN EDUCATION AND RESEARCH

KALMAN J. COHEN and ERIC RHENMAN

Carnegie Institute of Technology Business Research Institute of the Stockholm School of Economics

HISTORY OF MANAGEMENT GAMES

During the past four years, management decision games have become increasingly popular as a means of entertaining executives, as a training device for management, and as a teaching aid in university business schools. Before trying to assess the role of management games in business education and their potential for both teaching and research, it is desirable to view them in historical perspective. Direct outgrowths of military war games, business games were made possible by recent developments in operations research and electronic computers.

Development of War Games

When we try to trace the historical development of war games, we find that their origins are unclear. It is possible that such games are direct outgrowths of chess and similar board games. One author[1] boldly declares: "The game of chess is the oldest form of war game, and modern map maneuvers have grown out of the game of chess by a long process of evolution." A more cautious statement would be to point at their obvious similarity and common historical background. This does not mean that early war games were nothing but parlor games, played for pleasure. As is well recognized by the students of the history of chess, this and other similar board games were at a very early stage used as symbolic equivalents to warfare.[2] From this beginning, it was probably a very short step to attempt to use these games for planning and training purposes.

"Although it is not possible to be definite about the origin of war games, it seems that the games very early took on a formal and abstract character. By formal is meant that there were definite rules covering what the players could do and what the immediate outcomes of each action would be. By abstract is meant

Reprinted from **Management Science**, Volume 7, January 1961, pages 131-166, with the permission of the Institute of Management Sciences.

that the rules, the playing board, the pieces, etc., were not specific representations of real life phenomena."[3] Very interesting surveys of these early games
with many detailed examples are given by Andersson,[4] Thomas[5] and Young.[6]

The most significant improvement over the early board war games was the
NEW KRIEGSPIEL introduced at Schleswig in 1798.[7] This was the first game in
which actual maps were used to replace the older kinds of game boards. The map
was divided into a grid of 3,600 squares on which pieces were moved in a way
which resembled the ordinary marches of troops, due account being taken of
the topographical features of the ground. The very complications of this game
led to some criticism of it, however: "This war game is a bad product of the
refined military education of the period, which has piled up so many difficulties
that it was incapable of taking a step in advance."[8]

In the latter part of the nineteenth century, the development of war games
branched out along two directions, RIGID KRIEGSPIEL and FREE KRIEG-
SPIEL, corresponding to the opposing demands of realistic games and playable
games.[9] The introduction of new formal rules designed more accurately to
reflect the changing nature of war characterized the development of the rigid
variety. Random effects were incorporated by the use of dice. Extensive charts,
tables, and calculations were used to incorporate the exact details of troop
movements, effects of fire, etc. In contrast, the experienced judgment of human
referees played a major role in the free variety of war game.

As a result of the increasing popularity of both free and rigid varieties of war
gaming in Prussia, such games rapidly spread into other countries during the
latter part of the nineteenth century. Military games were introduced into the
British Army in 1872.[10] The British games were shortly afterwards copied in the
United States, "notably at West Point, where the game was played a great
deal."[11] Although extensive original work was done in the United States to
further the rigid variety of KRIEGSPIEL by the introduction of new technical
apparatus and extensive tables based on data from the American Civil War and
Franco-Prussian War,[12] the greater ease of administration of the freer version of
war games led to the increasing popularity of the latter not only in the United
States but in most armies of the world.

A summary of the present evaluation of rigid versus free versions of war
games is provided by Weiner:

> At the present time the two major forms of war games, the free play and the rigid
> play, still exist. Both have been employed as techniques for analyzing and evaluating
> military tactics, equipment, procedures, etc. The free play game has received support
> because of its versatility in dealing with complex problems of tactics and strategy and
> because of the ease with which it can be adapted to various training, planning and
> evaluation ends. The rigid play game has received support because of the consistency
> and detail of its rule structure and its computational rigor. In addition, the develop-
> ment of large capacity computing machines has made it possible to carry out detailed
> computations with great rapidity and made it possible to go through many different

plays of a game. With these developments the number and types of war games have increased.[13]

Although by the beginning of the twentieth century knowledge of war games was widespread throughout the world, it was Germany and Japan that made the most extensive use of such games during the present century. As part of their preparations for World War II, the Japanese engaged in some extremely ambitious war games at the Total War Research Institute and the Naval War College of Japan.[14]

Here military services and the government joined in gaming Japan's future actions: internal and external, military and diplomatic. In August 1941 a game was written up in which the two year period from mid-August 1941 through the middle of 1943 was gamed, was "lived through" in advance and, of course, at an accelerated pace. Players represented the Italo-German Axis, Russia, United States, England, Thailand, Netherlands, East Indies, China, Korea, Manchuria, and French Indochina. Japan was played, not as a single force, but as an uneasy coalition of Army, Navy, and Cabinet, with the military and the government disagreeing constantly—on the decision to go to war, on X-day, on civilian demands versus those of heavy industry, and so on. Disagreements arose and were settled—in the course of an afternoon, at the pace of this game—with the military group, by the way, as the more aggressive one, winning arguments.

Measures to be taken within Japan were gamed in detail and included economic, educational, financial, and psychological factors. The game even included plans for the control of consumer-goods plans, incidentally, which were identical with those actually put into effect on December 8, 1941.[15]

During the three centuries in which war games were extensively developed and used, they have become a highly appreciated tool both for instruction and for analysis. However, their limitations for the latter purpose have often been emphasized. Since the contents of war games were formulated from the knowledge already possessed by experienced officers, these games should, according to this opinion, never be regarded as a research technique which could add to the fundamental store of human knowledge. However, this view is at variance with an enthusiastic claim like the following:

War gaming is the traditional final step after the preparation of the war plan; it is universally regarded as the best peacetime test of a plan. Recently, the technique of war gaming has been modified to make it a method for solving problems previously thought to be beyond analysis and answerable only by appeal to the judgement of experts.[16]

In our discussion of management games below, we shall consider at greater length both the use of gaming as a technique of analysis and the relative merits of rigid and free games. The possibility of transferring experience gained with war games in these two areas to similar problems in connection with management games is the justification for this brief review of the historical development of war games.

Other Factors Influencing the Development of Management Games

Even if there is no doubt that war games are in many respects direct ancestors of today's management games, it is also clear that the development of the theory of games and the availability of electronic computing machinery have also had their influence. The mathematical theory of games characterizes general conflict (i.e., game) situations, and it provides definitions for a set of rather illuminating concepts, such as strategy, coalition, game value, and game solution. Although game theory offers very little for the analysis and nothing for the solution of the very complex situations involved in many business games, it is evident that it has greatly stimulated interest in problems connected with the general notion of a game.

The impact of modern computing machinery on the development of management games is more direct. It is true that some of the management games that have been developed do not require an electronic computer. It is also probable that the use of computers in games has not been advantageous in all respects, a question which will be considered more fully below. On the other hand, the use of computers has provided an opportunity for the designers of games to incorporate in them a great deal of realistic complexity while still keeping their administration relatively simple. An electronic computer also adds considerably to the drama of game play. Therefore, it is our assessment that the availability of electronic computers has been the most important factor determining the speed with which management games have developed during the last four years.

AMA Top Management Decision Simulation

Development of the first widely known business game was undertaken by the American Management Association in 1956. The AMA has clearly stated that its first management game was a direct outgrowth of military war games.[17]

> In the war games conducted by the Armed Forces, command officers of the Army, Navy, and Air Force have an opportunity to practice decision making creatively in a myriad of hypothetical yet true-to-life competitive situations. Moreover, they are forced to make decisions in areas outside their own specialty; a naval communications officer, for example, may play the role of a task force commander.
>
> Why then, shouldn't businessmen have the same opportunity? Why shouldn't a vice president, say, in charge of advertising have a chance to play the role of company president for fun and for practice? Why not a business "war game", in which teams of executives would make basic decisions of the kind that face every top management—and would see the results immediately?
>
> From these questions grew AMA's Top Management Decision Simulation. After an exploratory visit to the Naval War College, a research group was formed and work

began on a game which would eventually become part of an AMA course in decision making. This in turn, it was hoped, might lead to a sort of "war college" for business executives.[18]

Although this AMA game grew out of experience with military war games, its development also was dependent upon recent advances in the field of operations research and electronic computing machines.

The AMA TOP MANAGEMENT DECISION SIMULATION provides an environment in which teams of players, representing the officers of firms, make business decisions. The game consists of five teams of three to five persons each. These companies produce a single product which they sell in competition with each other in a common market. The basic decision period in the game represents a quarter of a year. At the American Management Association's seminars the participants generally simulate five to ten years of company operation by playing the game from twenty to forty quarters. There are six types of decisions which each team must make every quarter. They must choose a selling price for their product, decide how much to spend for marketing activities, determine their research and development expenditures, select a rate of production, consider whether or not to change plant capacity, and decide whether marketing research information about competitors' behavior should be purchased.

The financial picture is extremely simplified in the AMA game. All expenditures are treated as current cash flows. Sales made during one quarter are converted into cash at the end of that quarter, and this cash is available for covering expenditures during the following quarter. A basic budgetary restriction prevents borrowing, so all budgetary allocations must be covered by cash on hand.

In order to reduce the computational burden for the participants (as well as the key-punching effort for the operators), the AMA game was constructed to allow each company only a discrete number of decision alternatives during any quarter, these ranging from three to nine per decision variable. This was rationalized on the basis of a "concept of marginal change," meaning that a firm is allowed to make only small incremental changes in the values of any variable from one quarter to the next. At each decision stage, the computer automatically prints out a complete table of admissible values for the firm's entire set of decision variables.

After all firms make their decisions for a quarter, these are key-punched and fed into an IBM 650. The computer determines in just a few minutes all the complex interactions which result from the players' decisions. It is only the relatively severe time pressures under which the AMA game is usually played that makes the use of an electronic computer essential.

The mathematical model itself is comparatively simple and the amount of information to be processed is relatively slight, so that all of the computations necessary for a single quarter of play can be performed on desk calculators in about 45 minutes (compared to a total processing time of about 5 minutes for key-punching, 650 processing, and printing).[19] Using desk calculators would have two main disadvantages, however. First, the likelihood of making arithmetical or transcription errors would be increased, and the second, the rate at which play could be carried on would have to be substantially reduced.

General Management Games

As a result of the very enthusiastic reception accorded to the AMA Top Management Decision Simulation by business executives and educators, the notion of management gaming became fairly widespread. The AMA game was available only to participants in executive decision-making programs held at the AMA Academy in Saranac Lake, New York. As a result, several similar games were developed independently, both to improve the game and to enable others to use it more conveniently and more cheaply. The two most important offshoots of the first AMA game were the IBM MANAGEMENT DECISION-MAKING LABORATORY[20] and the UCLA EXECUTIVE GAME No. 2.[21] Both of these games utilize IBM 650 computers.

The IBM game differs from the AMA game mainly in added complexity on the marketing side. The total market in the IBM game is divided into four different areas. Three firms compete against each other, each of these being based in a different region. The fourth region, in which no firm is located, is a common marketing area where no firm has any natural advantage. Each firm may choose to pursue different pricing and advertising strategies in the various markets. Transportation charges are incurred when sales are made outside a company's home area, these being less for goods sold in the common competitive area than for sales made in the home area of a competitor. The principle of marginal change is absent in the IBM game, so each firm is allowed a great deal more choice in its decisions than in the AMA game. Aside from these changes, the IBM game is very much like the AMA game, and the general pattern of play and organization of teams are very similar.

In the UCLA EXECUTIVE DECISION GAME No. 2 (which quickly obsoleted their short-lived Game No. 1), anywhere from two to nine firms consisting of three to six players apiece compete in a single-product industry. The financial constraints have been relaxed to allow net cash to fall below zero, which means that the firms may in effect borrow money. When this happens, costs are incurred representing loan negotiation charges, interest, factoring expense, etc. The size of these costs prevents firms from completely ignoring cash limitations, although they have more flexibility in this regard than in either the IBM or AMA games. The concept of marginal change is lacking. In other respects, the gross appearance of the UCLA EXECUTIVE DECISION GAME No. 2 is very much like that of the AMA game, although the underlying model is slightly different.

Many other general management games have also been developed. Most of these games are designed to be used with an electronic computer. Some of the recent top management decision games have been developed by UCLA,[22]

Carnegie Institute of Technology,[23] Westinghouse Electric Corporation,[24] University of Washington,[25] Pillsbury Mills,[26] University of Oklahoma,[27] and Indiana University.[28] In addition, a few top level business management games have been developed which are feasible to operate without the use of an electronic computer. The most famous of these was developed by G. R. Andlinger of McKinsey and Company.[29] Some other representative non-computer games are the Greene and Sisson TOP OPERATING MANAGEMENT GAME,[30] Boeing Airplane Company's OPERATION INTERLOCK,[31] Esso Standard Oil Company's PETROLEUM INDUSTRY SIMULATION,[32] Herron's EXECUTIVE ACTION SIMULATION,[33] and Vance's MANAGEMENT DECISION SIMULATION.[34]

Both computer and non-computer games have also been developed abroad. Most of those known to the authors are general management games similar in structure to the early AMA and IBM games.[35] A partial exception is the game of the Compagnie des Machines Bull which approaches the Carnegie Tech game in complexity.

Some of these newer general management games differ considerably both among each other and from the earlier games. Stochastic or random elements have been explicitly introduced into some of them. A few of them are intended as quasi-realistic models of particular industries such as the Oklahoma and the Esso games (the petroleum industy), the Carnegie Tech game (the packaged detergent industy), the Westinghouse Business Simulator (electric generators), the McGuinness game (property and liability insurance),[36] and the McKinsey-IBM BANK MANAGEMENT SIMULATION (commercial banks).[37] While some of the newer games are still relatively simple, several of them are considerably more complex than the original top management games, the prize for complexity so far going to the Carnegie Tech Management Game.

The Carnegie Tech Management Game

The Carnegie Tech Management Game has been developed to mirror more realistically than earlier games the problems of running a company. In most business games, for example, if players choose a specific production level; they get that much output. To produce a desired amount of finished goods, players in the Carnegie game must not only request it, but they must also coordinate maintenance, overtime policies, hiring policies, raw materials purchasing, and other variables that affect the extent to which scheduled production and actual output coincide.

Instead of making a dozen decisions every "quarter," players in the Carnegie game have to record between 100 and 300 decisions every "month" of simulated play. To make these decisions, they get on a regular basis or through special purchase several hundred pieces of information about their own performance and about their relations with competitors, suppliers, customers, and financial institutions.

Three teams of players form the "detergent industry" in the Carnegie game, and they make decisions to cover a month of operating time. Each firm can market up to three products at one time in four regional markets.

Each firm has one factory, located in the Central Region, which has a raw materials warehouse and a warehouse for finished goods. Factory facilities can be used interchangeably to produce different mixes of product. In each of the four marketing regions, a team leases a district warehouse for finished goods, from which deliveries of detergent are made to the customers in the game.

All products can be manufactured from a basic set of seven raw materials which must be ordered from suppliers one to three months before they are delivered to the factory. Raw material prices fluctuate, and discounts are available for prompt payment of bills to suppliers.

The players schedule production. However, as noted above, to get what they schedule, they must make sensible decisions about raw materials, maintenance, changes in plant capacity, overtime, hiring, and firing. Finished goods can be consigned to any of the five warehouses and can be moved from one warehouse to another. Inventory run-outs at a warehouse carry penalties for future sales.

Products are not available for delivery to customers until the month following their manufacture. Sales in any one month depend on the total market for detergents; on consumers' reaction to product characteristics such as "sudsiness"; and on the teams' decisions about selling price, advertising expenditures, and distribution outlays. Consumer behavior may not be the same in the four market regions.

By spending money for product research, firms can generate new products. Not all new products will be worth marketing, though, and laboratory reports on their characteristics will be only partially reliable. A team may buy market surveys to get better, but still not perfect, estimates of consumer reactions to new products. A firm may also buy market research information about what consumers think of their current products and estimates of competitors' prices and expenditures by product and region for advertising and distribution.

Firms can expand production capacity and company-owned warehouses by building new facilities, but new construction takes six months to complete. Additional space in the district warehouses can be leased as needed. Expenditures for maintenance must be large enough to cover repair and renovation of existing facilities.

Firms must plan to have enough cash to meet their financial commitments. If they need additional funds, they can defer the payment of bills and negotiate short-term bank loans. By making application four to six months in advance, they can obtain capital by issuing debentures or common stock. Funds from outside sources are available only if the company's financial position satisfies a number of realistic constraints.

Players in the Carnegie game must plan and budget carefully to perform well. For some sequences of action, plans must be made 12 to 15 months ahead to achieve desired results. In addition to the recurrent obligations for raw materials, production, marketing, and research, firms must deal with other financial variables such as depreciation, income taxes, dividends, and investments in new construction.

The Carnegie Tech game has been played at the rate of one move a week by graduate students during the regular school year, using teams of five to ten players to represent each company. Periodically, players report on their performance and their plans to a faculty board of directors.

Despite the complexity of the Carnegie game, it was possible to formalize it sufficiently to program it for an electronic computer. However, this game is slow and expensive to play. Making a set of decisions takes at least two to three hours, not fifteen minutes; and teams need to play 20 or 30 moves to experience the long-run consequences of their decisions. The computation of results for each move ties up an IBM 650 computer (with a RAMAC memory unit) for about 45 minutes. Faculty guidance and supervision take a great deal of time.

In the Carnegie Tech game, as in most top management games, the ostensible object of play is fairly vague. The participants are not told that they are supposed to maximize profits, maximize sales, maximize total assets, or maximize anything at all. Each company must decide for itself what goals, both short-run and long-run, are appropriate to them in the light of what they know of the business world. Furthermore, no simple standards are established by which the administrators of such a game can determine which firm has done best. Some of the most fruitful discussions in the feedback sessions following play of these games revolve about the choice of criteria for "winning the game."

Functional Business Games

In contrast to the general management games, which are designed to give people experience in making business decisions at a top executive level and in which decisions from one functional area interact with those made in other areas of the firm, a number of management games have been developed which focus specifically on problems of decision-making as seen in one particular functional area. Functional area games are frequently much simpler than top management games. Indeed, in some functional business games optimal solutions or best strategies can readily be ascertained once the mathematical models are known. In contrast, the possibility of obtaining optimal strategies in even the simpler top management games has never yet been realized. Because of their relative simplicity, it is quite feasible to operate many functional area games without the use of an electronic computer.

The AMA has devised a second game, a functional area game known as the "Materials Management Simulation."[38] The participants in this exercise manage for a firm the overall material flow, including raw materials purchasing, inventory control and production scheduling, and finished goods distribution to the field warehouse. Each team has a well-defined objective, to minimize the total annual cost incurred in overall company operations subject to the constraint that sales requirements must be met. The model built into this game simulates flow of materials in a manufacturing firm. There are two plants, each manufacturing two products, and the products are both marketed in two regions. Sales demand will vary from month to month, and selling prices are fixed. In this game, as in virtually all functional area games except those in the marketing area, there is no competition among firms. Although several teams may play the game simultaneously, there is no interaction among the firms.[39]

Some other well known functional area management games which have been developed include Tulane University's PRODUCTION-MANPOWER DECISION GAME,[40] General Electric's DISPATCH GAME[41] and MARKETING STRATEGY SIMULATION EXERCISE,[42] University of Pennsylvania's SMART[43] and INVENTROL,[44] Kroger's SUPERMARKET DECISION SIMULATOR,[45] Boeing Airplane Company's OPERATION FEEDBACK ON MANAGEMENT CONTROLS,[46] several non-computer functional business games published by Greene and Sisson,[47] and Burrough's STEPS.[48] At the Stockholm School of Economics, an Inventory Game has been designed to be a part of a short course in Inventory Control.[49]

In functional area games, players have to worry about making detailed decisions in only a single area. Other aspects of the business firm, if present at all, are treated in a very sketchy fashion. In those functional area games which are simple enough for an optimal solution easily to be derived, the object is usually to teach the participants a simple standard point. For example, Greene and Sisson's MATERIALS INVENTORY MANAGEMENT GAME essentially involves the use of the economic order quantity formula, i.e., the ordinary square root lot size formula used in inventory control. "This game has been designed to illustrate the use of the Economic Order Quantity formula in inventory control and some very simple demand forecasting. The game is useful for introducing the problem of reordering and EOQ to inexperienced people."[50] Greene and Sisson's PERSONNEL ASSIGNMENT MANAGEMENT GAME is designed to illustrate and teach the value of linear programming. "In this game a very simple linear programming problem is illustrated; the problem is to assign a set of resources (accounting teams, as an example) to jobs to be performed. This game has been designed to be solvable by the 'assignment' method of linear programming."[51] Tulane University's PRODUCTION-MANPOWER DECISION GAME is designed to indicate the use of linear decision rules with quadratic cost criteria. The model consists entirely of the Holt, Modigliani, Muth, Simon model developed in connection with a paint factory, and the optimal strategy consists in using the decision rules these authors have expounded in two articles in *Management Science.*[52]

General Features of Business Games

Although the number of management games is already large and increasing steadily,[53] and although the differences among them are considerable, it is possible to abstract some general features found in most of the games that have been designed to date.

The simulation of the environment to make possible feedback of the results of their actions to the players is the fundamental "game" idea, and this feature is found in all the games described above. The characteristics of the environment have always been expressed in logical or mathematical relations. Some of these relations—the rules of the game—are always completely made known to the players, while the remaining relations—which describe the detailed intrinsic characteristics of the environment—are usually made known to the players in only a vague qualitative manner. The representation of the environmental characteristics by a set of formal relations makes it necessary for the environment to be represented in the game by a computer. Whether the computer used is electronic or human is fundamentally of little importance, since its role has been the same in all cases: passive interpretation of the environment's response to the actions of the players as dictated by the logical and mathematical formulas describing the environment.

The interaction between the players and the environment is thus the core of all these games. Already in the first AMA game, and later in all the various games where players compete against each other, the interaction between players is another essential idea. To some extent in the early games, but particularly in a game like the Carnegie Tech game, another interaction that has been of prime interest in the design of the game is the interaction between functional groups within a "company."

A final common characteristic of all these games is their simplicity. Even a very elaborate game like the Carnegie Tech game is still a radical simplification of the reality it purports to simulate. These simplifications result both from our lack of knowledge of how the world really is and from our desires to keep the games "playable."

THE USE OF MANAGEMENT GAMES AS A TEACHING DEVICE

Types of Current Uses

Despite the fact that they have existed for only four years, management games are now being widely used as a teaching device. Three kinds of uses can be distinguished: in a graduate or undergraduate university business curriculum; in an internal company training program; and in an executive development program run externally either by a university business school or by a group such as the American Management Association.

Management games in one form or another are now being used in several major business schools as a part of their overall program for training students. At Carnegie Institute of Technology, the Carnegie Tech Management Game is played throughout the academic year by second-year students in the Masters' and Ph.D. programs. Business games have been incorporated into regular courses as part of the work in business policy, operations research, or elementary business administration at such business schools as the University of California at Los Angeles, Michigan State University, the University of Pennsylvania, Indiana University, the Stockholm School of Economics, and the Copenhagen School of Economics and Business Administration.[54]

A great many large and well-known firms such as IBM, Westinghouse, General Electric, Procter and Gamble, Kroger, Pillsbury, Esso, and Remington Rand either already are or soon will be using business games as part of their internal executive training programs.[55] The Boeing Airplane Company has run these games for more than 2,000 of its management and pre-management trainees.[56] Most companies that use management games in their internal training programs have either developed their own games or else modified existing games for this particular purpose.

The first use of business games for training purposes was in connection with executive development programs run by the American Management Association. The AMA has continued to pioneer in this phase of the training use of business games. Some universities (e.g., UCLA and Carnegie Tech) in their executive development programs have also introduced business games as a subsidiary part of the training program.

Reasons for Use

There are at least two commonly mentioned reasons why management games are used as a training device in business teaching. To begin with, it is generally conceded that playing these games can be a lot of fun. A very high degree of personal involvement and competitive spirit is usually engendered by these games. Most participants seem to enjoy them enough to welcome the opportunity to repeat the experience. Much more important, however, is the fact that some developers and users of business games feel that important aspects of an educational program can be taught much more effectively through the use of management games than in any other way.

Critics, however, point out that it sometimes seems doubtful that the proponents of games have been clear about what they were trying to accomplish. To quote a speaker at the first national conference on management games:

> I wonder if the speakers who have spoken so far have actually had any specific purpose in mind. Have you gentlemen had one or two general hypotheses? Do you have any specific purpose other than that everybody has a whale of a time in playing these games?[57]

Since the question, "Why should we build business games?" is related both to the question, "How should we build them?" and to the question, "What can be accomplished through using them?", a certain systematic experimentation in the design of the early games is probably very valuable and should not be criticized. However, as our knowledge of games is increased through continued experience in designing and playing them, this knowledge should be formalized in several ways. First, it is important to know what advantages games of various types might have when compared to other teaching devices. Second, but equally essential, it is necessary to understand how to design a game which will possess certain desired properties in use.

Some Hypotheses About the Relations Between the Design and Administrative Characteristics of Business Games and Their Educational Properties

A first step toward establishing a systematic body of knowledge about the relations between the design and administrative characteristics of business games and their educational properties is presenting an organized set of hypotheses. On the basis of published experience of others and our own observations from the play of various games (particularly the Carnegie Tech Management Game), the hypotheses summarized in Table 1 have been formulated.

We do not claim that these hypotheses have been tested and proved correct. In their present formulation, they are generally nothing more than reasonable assumptions. However, we hope that empirical data will be presently gathered by ourselves and others that might be used, if not for rigorous scientific verification or falsification, at least for additional arguments for or against these statements. [58] In that way it should be possible ultimately to come to a reasonably good understanding of both the relative educational advantages and disadvantages of various games and of the way in which a game possessing certain desired educational properties should be designed. Such knowledge will obviously be of considerable value when formulating, buying, renting, or administering a business game.

The hypotheses contained in Table 1 are arranged in three groups:

(1) The center of the matrix shows the relations between design and administrative characteristics and intervening variables. The + sign in row a, column 1 thus correspond to the hypotheses: "As the complexity of the game increases, the subjective realism will also increase, ceteris paribus."

(2) Relations between intervening variables are all gathered in the bottom line: "+2" in column 4 stands for the hypothesis: "As the dynamics of the game situation increase, the time pressures on the players will also increase, ceteris paribus."

(3) Possible relations between design and intervening variables and the properties of the game as a teaching device are indicated by the second column from the right. "+g, +2" in the fifth line should thus be read: "As the importance of random factors in the game and the dynamics of the game situation increase, the game will provide more experience in decision-making under conditions of uncertainty, ceteris paribus."

Discussion of Table 1

All of the educational properties of a business game listed in the righthand column of Table 1 (except for "excitement and enjoyment of play") will be discussed below. In so doing, we shall spell out the set of hypotheses contained

TABLE 1. Some Hypotheses Concerning the Relations Between the Design and Administrative Characteristics of a Business Game and Its Properties as a Teaching Device

Intervening Variables:

Design and Administrative Characteristics	1 Subjective realism	2 Dynamics of the game situation	3 Players' subjective understanding of the gross relationships of the game	4 Time pressures on the players	5 Need for information exchange within a team	6 Sensitivity of game results to quality of play	7 Feasibility of experimentation by players	8 Degree of competition
a Complexity of the game (the number of decision variables and the degree to which they are interrelated)	+		−	+	+	+	−	
b Number of players per team					+			
c Number of moves per real time unit				−	+			+
d Amount of automatic information feed-back provided players			+	+	+			
e Opportunities to buy information			+	+	+			
f Dynamic nature of the constraints limiting action		+				+	−	
g Importance of random factors in the game		+				−	−	
h Long-range influence of decisions		−				+		
k Equality (initial position, rules, team size, opportunities) of teams								+
m Validity of institutional information contained in the game	+							
n Multidimensionality of goals				+		−	+	
o Use of human environment	+			+			+	
p Degree of quantification of the game structure	+						+	
q Discovery of the rules an important part of play			−	+				
Relations between intervening variables			−2	+2	+2		−2, −4	

Educational Properties of the Games as a Teaching Device:

Factors influencing these properties	Description
+1, +2, +6, +8	A Excitement and enjoyment of play
+a, +f, +5	B Importance of the interrelations between functional areas
−a, −c, +0	C Possibility of manipulating the game to create a desired problem situation
+1, +3, +6	D Credibility of results
+g, +2	E Experience in decision-making under conditions of uncertainty
−d, +e, +q, +7	F Importance of systematic collection of information
+F, +6, −4, −8	G Importance of techniques for systematic analysis
+b, +4, +a	H Importance of efficient solutions to organizational problems
+m, +1	K Teaching institutional facts
−4, +h	L Importance of long-range planning and policy-making
+c, −p, +q	M Importance of imaginative and creative behavior in a realistic situation

(+ indicates a direct relation, − indicates an inverse relation. Further explanation is given in the text.)

in Table 1 regarding the factors which have a direct influence on these properties. Cross-reference between the text and Table 1 is facilitated by the incorporation of key symbols in the text. A single capital letter enclosed in parentheses at the end of a sentence, e.g., "(B)," means that the preceding sentence gives an extended definition of the educational property identified by this letter in Table 1. A single lower-case letter or numeral enclosed in parentheses is similarly inserted in the text after an extended definition of one of the design and administrative characteristics or intervening variables of Table 1; an algebraic sign (+ or −) will precede the lower-case letter or numeral to indicate that this particular variable has a direct (+) or an inverse (−) relation to the educational property being discussed in that paragraph. The algebraic sign will generally correspond to such expressions in the text as "this will increase the attainment of (the educational property)" (+) or "this will have an adverse effect on the attainment of (the educational property)" (−).

The Interrelations Between Functional Specialties. In general management games, the participants become vividly aware that the various functional specialties of a business are closely interrelated (B). What one does in the marketing area may have serious ramifications on finance and production. While this fact is frequently stated in most business training programs, it does not seem to be driven home so forcefully to students as it is through participation in management games.

We believe that one of the factors that determines the usefulness of a business game for this purpose is its inherent complexity (+a). Also of importance in this regard is the perception by the players that the constraints which limit action are changing, e.g., that during the play of the game first the production facilities, then the financial situation, and finally the limits of the market are the "bottle necks" (+f). To the extent that the need for exchanging information within the team (+5) can be increased by other means, this also improves the value of the game for teaching the interrelations between functional areas.

Two Advantages over the Conventional Business Case. Business games place students in an essentially dynamic situation. Their decisions at any point of time are affected by what they and their competitors have done in the past and affect the alternatives available to them in the future. The dynamic nature of business games most significantly differentiates them from the case book method of instruction. This is of value for several reasons.

Business cases, as well as business games, can present to students the materials from which business managers have to make decisions, and the students can be asked to formulate the managers' decisions. There are two drawbacks of cases, however. [59] First, there is seldom a chance to try a problem more than once. It is very difficult to give students, through cases, repetitive experiences of moderately increasing difficulty. Second, it is usually impossible for students to get any direct feedback on the consequences of the decisions that they have made. The only "evaluations" of these decisions are the opinions of fellow students or instructors. These judgments are, of course, only tentative or informed guesses. Although such evaluations may in fact be correct, it is sometimes difficult to convince the student of their validity.

Management games can overcome these two limitations of the case method. First, in a dynamic situation the student is exposed to similar kinds of problems over and over again, and it is possible for the administrators to manipulate the game to create a desired problem situation (C). Second, the consequences of his decisions are automatically fed back to him from the computer in a way which, if the game is a reasonable simulation of the world, usually convinces him of the correctness of the results (D).

According to our observations, however, the possibilities for the administrators to manipulate the game decrease rapidly with the complexity of the game (−a), and this can be counterbalanced only to a limited extent by fewer moves per real time unit, that is, by having a longer time for the administrators to plan their intervention between each move (−c). Generally, the use of a human environment (e.g., a board of directors, a union leader, or a bank representative) will facilitate certain types of manipulations (+o).

The credibility of the results (D) is often an important property for which business game designers strive. This will be enhanced the more realistic the game appears to the players (+1). If the players feel that they understand the gross relationships of the game, i.e., if they can perceive a certain logic in the relation between their actions and the game's outcome (+3), then the credibility of the game will also be increased. The same effect will follow if the game is very sensitive to the quality of the decisions made by the players (+6).

Risk and Uncertainty. Uncertainty is an important characteristic of the environment in which business firms actually operate which frequently gets neglected in business school courses, but which can be very convincingly illustrated in a management game (E). The industrial organization operates in a world of uncertainty, and unless business managers learn to adapt to and cope with an uncertain environment, their decisions will suffer. In some business games, uncertainty is explicitly introduced by means of random or stochastic terms (+g). While other business games are in fact strictly deterministic, they still present an apparently uncertain situation to the players, since the consequences of identical decisions by one firm may differ from period to period because of rapid changes in the general game situation which might depend on a business trend, random factors, or perhaps the competitors' actions (+2). In a framework of either true or quasi-uncertainty, participants must try to predict relevant aspects of the uncertain environment, and to adopt decision rules which will enable them to maintain a flexible position, since the world may not turn out exactly as they expect.[60]

Systematic Collection of Information. In most business games, a major problem for the participants is to infer the characteristics of an unknown environmental structure. Most games do not tell the players the exact rules whereby their decisions and the decisions of other firms are translated into economic consequences. Quite obviously, however, the more the participants are able to estimate relevant characteristics of the environment, the more able will they be to make sound business decisions. In the real world, business firms are continually devoting large amounts of money and time trying to learn the nature of the industrial and economic environment in which they are operating. One object in playing management games is to get students used to exploring their environment (F). By playing business games, students can learn that it is rational sometimes to take particular actions solely for the sake of information gathering, since the kinds of decisions that should be made when information gathering is essential may be very different from the kinds of actions that are optimal in a situation where the nature of the environment is known with certainty.

The problem of inferring the structure of an unknown environment which management games pose to participants is especially relevant to the real business world because this occurs in a framework of dynamic and competitive uncertainty. This property of the game might easily be strengthened by decreasing the amount of information that is automatically provided to the players (−d), by increasing the opportunities to buy information (+e) and by making the discovery of the rules of the game one of the objects of play instead of informing the players about everything in the beginning of the game session (+q). It is also of great value from this point of view if the game is designed or administrated in a way that makes experimentation by the players feasible (+7).

A Field for Systematic Analysis. The total task situation which a business game imposes upon students provides an opportunity for learning and reinforcing a variety of analytical tools (G). In the process of trying to infer the characteristics of the environment in which they are operating, participants have to use (whether implicitly or explicitly) statistical inference and statistical estimation. Statistics and probability theory may also be useful in connection with other parts of management games, such as purchasing marketing information, attempting to forecast both endogenous and exogenous variables in the game, etc. In playing business games, it is sometimes possible to derive analytical decision rules which lead, if not to optimal, at least to good or improved decisions. Some games have been designed and are administered with the intention of encouraging and facilitating the use of analytical techniques. This is generally achieved by any measure that facilitates and encourages the systematic collection of information (+f), and probably also by making the game results very sensitive to the quality of the players' decisions (+6). Other games, because of the severe time pressures they place upon participants, generally stress an intuitive, "top of the head" approach to decision making (−4). We also have observed in some cases that when the competitive aspects of the game are too dominating (−8), the players' interests in systematic analysis are reduced.

Organizational Problems Are Illustrated. While most management games do not explicitly have in their object matter psychological and organizational variables, nevertheless psychological and organizational interactions become very important in the play of the games (H), since the operating decisions are usually made by teams of several players. Depending upon the number of participants in a firm (+b), the amount of time pressures which are exerted (+4), and the perceived complexity of the decision-making problem (+a), a considerable degree of organizational specialization may develop. Problems of decision-making within an organizational context are usually present to some extent in the play of business games. By participating in such games, students can learn a great deal about the dynamic interactions of subgroups within an organization, the processes of goal formation in an organization, and the distortions which may be caused by subgoal identifications which are different for various members of the firm.[61]

Teaching Institutional Facts Many institutional facts can be taught through the use of business games (K). For example in developing its Business Simulator, Westinghouse designed the output forms which come to the participants from the computer to be as close as possible to many of the standard operating forms used in the company. One of the somewhat unexpected by-products which was found during the plays of this game is that some Westinghouse executives and management trainees have learned a great deal about the meanings of these report forms, even though this should already have been familiar to the participants. Another example is a statement made by Greene and Sisson about one of their games: "The game will illustrate what facts are important and may give some idea of the approximate quantities involved. For example, the student would learn in the Retailing Inventory Management Game that markups in retailing range from 20 to 60 percent and not from 5 to 95 percent."[62]

In order to achieve the goal of using a business game as a device for teaching institutional facts, it is not only important that the game contain correct institutional information (+m), but it is also essential that the players generally perceive the game as being meaningful and realistic (+1).

The Value of Planning and Policy-Making. Playing a management game may focus the players' attention on the importance of establishing policies and on making long-range plans (L). To make this possible it is essential that time pressures are moderate (−4). In contrast to on-the-job experience which frequently tends to emphasize "putting out fires," dynamic business games may be structured to have their entire emphasis on long-run strategies (+h).

Assessment of the Educational Value of Management Games—Some Cautions. Although we have reviewed a great many kinds of things which we think can be learned by participation in business games, we must again caution the reader that no objective empirical evidence has been amassed which proves either that these concepts can actually be taught by the use of management games or that they can be taught more effectively by games than in some other ways. It is always well to remember that the use of business games is not free; in fact, it can be quite expensive. There are three different types of costs which can be distinguished in connection with the use of games. First, there are the capital costs (costs of people's time, of computer time, and of materials and supplies) which are involved in the development of any management game. Second, there are the out-of-pocket costs of running the game (the cost of computer time, materials' cost, the clerical cost, and the cost of the administrators' time). Third, and possibly most important, there is the opportunity cost to the participants in playing the game. By opportunity cost, we refer to the fact that nobody's time is worthless, and that while playing a business game the participants are thereby prevented from using that time in any other way. It is here that the really critical work has to be done in connection with the training use of management games. People who have been intimately connected with them very definitely feel that the types of concepts enumerated in the preceding sections *can* be learned through the play of business games. But the real question certainly is, can they be *better* learned in this way than in any one of a number of alternative uses of the participants' time? It is here that an economic question enters, for the relevant problem is to teach particular attitudes and skills in the *lowest total cost manner possible.*

Several other cautions should be noted. First, playing business games will not produce a complete manager. No game yet developed includes all of the kinds of problems that managers have to face; in particular, personnel, psychological, and organizational problems have not yet been introduced in these games to anything like the degree to which the more analytical kinds of problems in marketing, finance, and production have been incorporated.

The problems of information gathering are not adequately handled in many games. There is a danger that the participants in these games will get used to having information come freely and easily from computers, and that they will fail to realize how difficult it might in fact be to obtain some of this information in the real world. How this deficiency might be overcome by designing a business game in which great stress is placed on imaginative and creative behavior in a realistic situation (M) will be discussed in the next part of this paper.

There is also some danger that the participants will feel too strongly that they really know how to run a business as a result of their experience in playing management games. While those of us in the business of management education do feel that we have some useful things that can be taught to students in one way or another, there always is the danger of a bright young graduate from any kind of training program feeling that now he knows how to solve all the world's problems. "Now instead of 'thinking' they know how to run a company, they may really 'believe' from their experience with a game that they can run a company better than its present managers."[63]

To the extent that business games are used to teach institutional facts to the participants, it is very important that the institutional facts taught are true. The more that management games are used to demonstrate techniques, to sell ideas, and to teach facts about the world, the more important it is that these games indeed be realistically valid. There can be a danger from using the wrong kinds of games as an indoctrination tool, because students may be less skeptical toward "facts" that they think they are learning from the experience of playing a business game than they are from information gleaned through books or lectures. This particular danger can be mitigated by not using games as a sole training device, but by using them judiciously in combination with older teaching tools and in programs having definite teaching objectives.

SOME SUGGESTIONS FOR IMPROVING BUSINESS GAMES

Our preceding discussion has indicated that no single ideal business game exists. Formulated in positive terms, we have suggested that there should be considerable flexibility in the use of this new teaching instrument. The historical developments surveyed earlier should not be viewed as ultimately leading to a *few* good games. Rather, we have reason to expect that the future will see the development of a growing number of different games, each designed to fulfill a particular educational purpose. An important problem is to use our accumulated experience with games to increase our understanding of the relation between the ways in which a game is designed and used and its pedagogical value. We have already suggested how this empirical knowledge might be organized.

Another important task is the development of new types of games, or rather new features in games, for the purpose of achieving particular educational goals, e.g., using analytical techniques, encouraging long-range planning, improving skills in solving organizational problems, etc. Our purpose in this section is to suggest a few new design characteristics for business games and to discuss the educational properties that might thereby be achieved.

The Dynamic Business Case—Warps Industries, Inc.

As a background to some of our suggestions for new design features in business games, we shall first review another type of educational tool, which could be called the "dynamic business case." One of the most interesting cases of this type has been developed by a Swedish research institute. [64] This case, called "Warps Industries, Inc.," deals with a medium-sized company confronted with the problem of expansion. Newly discovered copper ore deposits offer new business opportunities to the company, but considerations of financing, labor, internal organization, community relations, etc., present a number of inter-related problems. These are complicated by the fact that Warps Industries is located in a predominantly agricultural area and already, before the proposed expansion, plays a very important role in the community.

The case is hypothetical, but nonetheless very realistic. When it was first used (in March 1959), it was simultaneously handled by four groups of about ten persons per group, although there was no overt interaction between the groups.

The referee group consisted of five persons. It took several days of continuous work to "solve" the case.

From the very beginning, the general background is outlined in great detail. However, when the case is presented to the group, the situation is dynamic. Certain, but not all necessary, information about the discovered ore deposits, financial requirements, the labor situation, etc., are given. The participants are also told that a few specific studies have been undertaken which later will provide them with additional information. On the whole, however, the group is on its own. It has to work through the case, define what it regards as problem areas, analyze these, and make the necessary decisions. As the group works on the case, events are occurring. New information comes in as a result of studies, some of which the group might have ordered itself (which means that not the same information is given to all groups). New problems also might arise, partly depending on each group's action or inaction. For example, one group decides to keep its plans secret from the public for awhile. "Next morning" it finds the local newspaper headlining an exaggerated story that Warps Industries plans to shut down its present factories and go into the mining business. Other realistic and dynamic features characterize this case. When and if the group wants, it may see its bank, and depending on its earlier behavior and upon how it presents its plans, it might receive help from the bank in solving its financial problems. When and if the group wants to see some other authorities, they were generally made available. This case has been favorably accepted when played by a great number of businessmen and educators in Sweden.

Let us summarize the characteristics of the Warps Industries case. It is similar to a business case in certain respects, since the group is confronted with a realistic problem described in text. Another similarity is that the group itself has to define what might be problem areas and decision areas (even though the structure of the case does give some clues). However, this case also has some properties that generally have characterized business games in contrast to the conventional Harvard Business School type of case. The situation is dynamic. The amount of information given to the group depends on its own actions. The environment changes as the group works with the case, to some extent reflecting a feedback of the group's actions. This case also includes some role playing elements, for the participants might be called upon to conduct face-to-face negotiations with a union leader, a newspaper man, or a banker who might be presented to the group. Another definite characteristic of a dynamic business case is that its development requires a great number of man-hours of work by various specialists, plus a talented case author. Such a case, like a business game, needs to be "debugged" before it can be publicly played.

A Comparison Between Business Games and Dynamic Business Cases

In using the title "dynamic business case" for the Warps Industries type of case, we are emphasizing one of the characteristics that makes it similar to a business game. As defined earlier, *dynamic* in this context means that the behavior of the team at one stage of the game will influence the situation that it will meet later. However, this similarity is not the only reason why comparisons are interesting. The dynamic business case undoubtedly contains some features which would be very useful if incorporated in a business game. Let us therefore extend the comparison.

In the games that have been described earlier, all rules have been presented to the players at the beginning of the game, and the whole model of the environment has been put into mathematical form. We have already indicated that this has several advantages. A computer can handle all situations which arise during the whole course of the game, and a feeling of realism is gained through the "objectivity" of the machine.

On the other hand, this rigid approach has put very severe limits on the design of business games, in comparison with dynamic business cases. One of these is overemphasis on the quantitative aspects of the game. Particular features of a business are more easily and more completely translatable into a mathematical and logical language than are others. Cash flows, prices, advertising, and investment budgets are all important in real life, but in a business game they assume lives of their own separated from the physical world of expenditure control, product design, campaign planning, and technological development that is not so easily represented by a mathematical model.

Another important restriction on a game of the traditional type is that the task of the players is restricted to choosing from among a limited number of prescribed alternatives. In real life, a number of problems exist which have a quite different character. The discovery of a problem, the invention of a solution, and the elaboration of all details of a plan are all parts of what can be called "creativity," and which go far beyond the task confronting the participants in even the most complex of the presently available management games.

The Business Game Case

The suggestions for improving business games that we shall present here can be summarized as an attempt to combine the best features of management games with some valuable ideas gleaned from dynamic business cases. To this amalgam, we shall apply the name "business game case." It essentially involves a relaxation of the requirements that all the rules of the game and the whole environment be specified in a complete computer program and that all the rules of the game be presented to the players at the start of the game. While this might require the introduction of a group of human referees, some of these features can, at least to some extent, be obtained in other ways. Before presenting some specific suggestions as examples of what might be accomplished, let us point out that our suggestions for the future development of management games essentially consists in introducing some "free" or "non-structured" elements into what to date have all been "rigid" or "highly structured" games. The direction that we propose for future work with business games is entirely analogous to the historical outgrowth of FREE KRIEGSPIEL from RIGID KRIEGSPIEL which occurred for military games during the latter part of the nineteenth century.

In most top management decision games, the financial aspects are extremely simplified. In some games, all expenditures and receipts are treated as current cash flows, and no external sources of funds are available. Those games which make some provisions for external financing prescribe rigid criteria which must be met to determine the firms' eligibility for securing additional funds, and these criteria are told to the teams at the start of play. What we would propose as an alternative which could present a greater challenge to the players' creativity is that they be informed at the start of play merely that all financing possibilities which generally exist for a firm are also open to them in the game, without the details being fully spelled out. When a firm wishes to obtain new capital, it will be asked to prepare a report stating how it proposes to obtain the funds. It will then be confronted with a human "representative" of a "bank" or "brokerage firm" with whom its officers discuss their proposal. The banker's or broker's decision is then given to the firm. Several rounds of negotiations, involving offers and counter-offers, can be accommodated in this scheme.[65]

Labor costs and labor productivity, if present at all in most business games, are usually very passive factors which do not interact with the players' decisions. To introduce a more prominent role for these factors in management games we suggest that periodically, for example once every simulated year, the firms be confronted with the necessity for negotiating a new contract with their labor unions, subject to the threat of a work stoppage. The roles of union leaders can be played by actual labor representatives brought in from the outside, by faculty members, or by another team of players in the game. The results of the negotiations will directly affect the wage rate and labor productivity. If the union is represented by another team, "fairness" can still be retained if the game contains a formula for estimating the minimum wage offer which the union should accept, a floor which could depend upon average wages in the industry, changes in wage rates last year, stability of employment, the firm's past profits, etc.

Public relations are an important aspect of business life which is completely absent in the typical management game. During the play of a game strikes, rapid price changes, excessive profits, and other circumstances might be deemed by the administrators or referees to cause the type of adverse publicity which might affect the market for the firms' products and the selling price of their stock. The exact type of behavior which could cause such adverse publicity would not be explicitly told the players in advance, however. Part of the administrators' equipment for running the game would include a set of descriptions of such bad press campaigns. A company which is hit by such adverse publicity would be required to prepare a plan of counteraction. The suitability of this program would be judged by the referees, and the adverse effects of the bad publicity accordingly reduced.

Many business games do incorporate cyclical patterns in the market demand function, and they usually automatically provide the teams with periodic reports on the state of the economy, estimates to total demand, etc. We propose that when a certain stage of the game has been reached, the players are merely warned about the risk of an impending change in the business cycle. The firms would then be allowed to purchase whatever information they want, and this would be given to them. If the cyclical pattern used is taken from a real period, at least in principle there would be no difficulty in answering the players' requests. The players' feelings that the game is realistic as well as fair could still be maintained, and yet they would be induced to use their abilities to forecast the general business trend.

Another element of extreme artificiality that is present in most business games is that the players must at fixed intervals overtly reach specific decisions relating to the entire gamut of company operations which is included in the game model. [66] In real firms, the length of time elapsing between decision dates is variable, depending upon information feedbacks coming from the environment. Furthermore, when new decisions are reached, they usually pertain only to selected aspects of the company's activity, not to the whole dimension of their possible behavior. An important function of the successful business manager is to recognize when the behavior of the environment departs sufficiently from previously held anticipations to make revisions of plans worthwhile, and to select those aspects of overall company behavior for which new plans should be formulated. One way of providing a game environment in which these skills could be practiced is to reformulate the game in terms of continuous play, so that information feedbacks from the environment would occur much more rapidly than the teams could possibly reach completely new decisions, with the possibility that firms could make new decisions at any time, rather than only at fixed, predetermined intervals. The mechanics of recording decisions would also be simplified so that when previously made plans are to be unchanged, the players need take no action specifically to indicate this. Thus, only essentially new decisions or policy changes would be required from the teams. Along with the introduction of continuous play, it would be worthwile to restructure the decision variables in the game to allow the players to arrive at decisions consisting of policies (these being essentially equivalent to strategies in the formal game theory sense)rather than specific actions.

Since existing management games usually have teams of several players representing companies, they provide some experience within teams for practicing human relations and organizational skills. However, greater scope for human relations problems to appear, including some new dimensions of negotiating behavior, would result from allowing the teams to interact with each other outside of, as well as within, the structure of the computer simulation of the game environment. We would propose that the teams in a business game be allowed to conduct, to whatever extent they desire, face-to-face negotiations regarding the possibilities of selling raw materials from one company to another, cross-licensing of new products, sharing warehouse space, borrowing funds, merger, etc. It would be extremely easy to accommodate any changes which were mutually agreed upon by the teams into the computer simulation of the game environment, provided that the game is programmed initially with such possibilities in mind.

Research and development activites have been introduced in various ways in different management games, generally amounting simply to the specification of an expenditure of some number of dollars. In the IBM game, the effect of R&D expenditures is qualitatively indistinguishable from the effect of advertising expenditures. In the Carnegie Tech game, R&D expenditures buy a certain number of chances in a random drawing from a universe of potential new products. We would propose that the players' manual for a game should state only that research and development activities can be undertaken. Any team wanting to do so would then have to state not only how much it wanted to spend, but also for what type of activity. In coded form, this information then can be entered into the computer, which, probably through some basically stochastic mechanism, would determine the results of the research program.

The foregoing are all suggestions of ways in which greater freedom can be introduced into management games. The broader scope of actions which business game cases would present to the participants should greatly stimulate imaginative and creative behavior by the players. A serious defect of existing business games from the pedagogic viewpoint is that any creative urges which are aroused during their play generally are aimed only at finding loopholes in the rules or model of the game. Relaxing the restriction that all the mechanisms and variables incorporated in a management game be representable by a set of mathematical and logical relations can make a business game case a very useful interpretive medium for illustrating all aspects of business activity.

In conclusion, let us once more indicate that to achieve the advantages which a business game case offers, we do not discard completely the representation of the economic and industrial environment in which the firms interact by means of a computer program. However, we do, at critical points, supplement the mechanisms contained in the computer program with the experienced judgment of human referees. This avenue for the future development of management games is entirely analogous to the branching out of FREE KRIEGSPIEL from the RIGID KRIEGSPIEL trunk in the military game tree.

THE BUSINESS GAME IN THE SOCIAL SCIENCE LABORATORY

The Confusion of Concepts

Management games can be useful not only as a training device in a business education program but also as a research tool. Many of the developers of such games have been as interested in using them for research as for training purposes. After an introductory definition of a few essential concepts, we shall examine some of the possible research areas where management games might be a valuable laboratory tool. The emphasis will be both on the type of use that is most attractive and on the methodological problems that the scientist will meet.

In a discussion of the use of management games for research purposes, it is illuminating to compare this research tool with other parallel concepts and developments. The success of natural sciences like physics, chemistry, and biology in the use of laboratory experimentation has always been a challenge to social scientists. But the equipment of the latter for performing laboratory experiments has been meager, and this has often been thought to be one of the major reasons preventing more rapid progress in the social sciences.

Recent developments in electronic computing machinery and in mathematical model building have led to widespread interest in applying computer simulation techniques to study social science problems. It is beyond the scope of this paper to evaluate the potential of simulation techniques in the social sciences.[67] However, since many people seem to make an incorrect identification between fields of computer simulation and business games,[68] let us merely observe that these are quite different concepts, although they have a twofold interrelationship. As we have seen, a computer simulation is a part of many business games; in particular, the computer is used to simulate the economic and/or industrial environment in which the competing firms interact. However, we can also look upon business games as an example of the development beyond the type of computer simulation in which the entire relevant system is simulated by a computer program, to the more complex types of simulations that are possible when men and machines (preferably electronic computers) are allowed to interact in large-scale laboratory experiments. Although this type of man-machine simulation has to date mainly been used in the development of military systems,[69] the broader concept of simulation on which it rests should be kept in mind when we later discuss the uses of business games for the development of efficient business procedures.

Operational gaming is another concept which is also often, and easily, confused with business games. Like Thomas and Deemer,[70] we want to use the term operational gaming to emphasize the use of gaming to find an optimal solution to a game. Both operational gaming and business games have important roots in the development of war games. However, in contrast to business games, operational gaming has another direct root in the theory of games, in that operational gaming provides a method for attempting to solve games (in the strict sense in which solution is defined in the theory of games). We shall consider in the following section the extent to which business games might be used in an operational gaming context, that is, for finding an optimal solution to a competitive business situation.[71]

Business Games for Solving Management Problems

To our knowledge, nobody has as yet attempted to use an existing management game to discover optimal patterns of business behavior.[72] In principle, such uses are possible, if care is exercised to make the structure simulated by the management game sufficiently realistic and if the participants making the decisions are sufficiently well aware of good business practice to behave in a reasonably intelligent manner. Military war games have been used in a similar way.[73]

The idea that management games can be used for discovering optimal business strategy rests upon an implicit assumption that formal solutions to games can be discovered by repeated plays of the game. The modern mathematical theory of games provides us with precise notions for the meaning of such concepts as "strategy" and "solution to a game." In only special cases, however, have easy computational rules been provided which will always lead to the solution of a game. It has usually been impossible to solve most business games in formal terms to discover optimal strategies, at least for games sufficiently complex to be any kind of reasonable approximation of reality. In trying to use repeated plays to solve complex business games, two different kinds of problems arise, probabilistic difficulties and strategic difficulties.[74]

The probabilistic difficulties are easiest to get at, for the theory of Monte Carlo is directly applicable to these. The problems essentially arise from the fact that where random elements are explicit factors in a game, the actual sets of numbers drawn from the statistical distributions will differ from one play to another, and the kind of strategy that works well with one set of random numbers may be a very bad strategy with another particular set of random numbers. What is actually required are a number of repeated plays in order to attain estimates both of average outcomes and some notion of the dispersion or variance of these outcomes. Standard Monte Carlo techniques can frequently be applied to problems of this kind, so that it is sometimes possible to determine the number of repeated plays that would be necessary to attain any particular degree of accuracy.

The conceptual difficulties regarding the strategic structure of games are frequently, but erroneously, lumped together with those arising from probabilistic difficulties. Strategic uncertainty arises because each team in a competitive game is free to choose its actual strategy from a broad range of permissible strategies. The kind of strategy which works well when pursued by one team in a particular game may work very badly in a different play of the same game against other opponents who are pursuing different strategies. The kinds of theorems derived from the Monte Carlo theory that are applicable to the true probabilistic difficulties really say nothing at all about what would be required to handle the strategic difficulties. These strategic uncertainties arise in competitive games even where no explicit random or stochastic elements are introduced. Even in the simpler games, where there are no stochastic elements, a very large number of game plays is frequently needed in order to appreciate the variations in the results that can occur because of differences of strategies adopted by the competing teams.

The difficulties that are inherent in the use of games to try to find optimal strategies have been exceedingly well summarized as follows:

It should be emphasized that there is no body of theory that sanctions the common use of operational gaming to seek a solution of a game through repeated plays. This is not to say that solutions cannot be found by gaming. It does mean, however, that no theoretical guarantees, such as Monte Carlo has, extend to the use of gaming to tackle probabilistic and strategic difficulties simultaneously. One could conceivably, for a zero-sum two-person game, separate the two classes of difficulties by first using Monte Carlo methods to estimate a pay-off matrix in effect, and then using one of the well-known iterative methods to solve the corresponding game. Such a scheme shows conceptually how gaming could be used in a rigorously justified way. It should be emphasized that no such scheme corresponds to what is now done in practice, for no existing computing machine approaches the enormous size that would be required to handle the monstrous pay-off matrices that would result and the appalling number of plays that would be required.[75]

There is one valuable research by-product in the simulation of business operations which may arise from working with business games. To the extent that the formulators of such games try to have realistic simulations of the industrial or economic environment programmed into the game, it is frequently necessary to extend our empirical knowledge. If one is trying to construct a quasi-realistic management game for the petroleum industry, for example, then one has to find out in fact how the different promotional and pricing policies adopted by firms in this industry will affect both the total market and their share of the market. While business games themselves certainly do not provide any answers to this question, the attempt to construct realistic games may lead to new areas of empirical investigation. In this regard, it should be noted that games per se have no advantage over outright computer simulations of business operations.

Economic Research

Since management games are generally structured so that firms are competing with each other in an economic environment, it is quite natural to think of using business games for some areas of economic research relating to the theory of the firm. The particular area of micro-economic theory about which we know least is oligopoly market structures. Business games are frequently formulated in terms of a relatively small number of competing firms, and the participants are generally acutely aware of the influence that their actions have in calling forth reactions by their competitors. Hence, general management games represent typical oligopoly situations, and they might conceivably be used to explore some of the unsolved problems of oligopoly behavior.[76]

The most interesting research that could be conducted with business games would, from a professional economist's viewpoint, be research which would tell us something about the way firms behave in reality, and not simply research telling us something about how students or management trainees behave in a laboratory situation. Therefore, in order for management games to be most useful to economists, it is necessary to ascertain that business games are valid representations of the environment in which firms operate and that game players make their decisions in much the same way as real executives in actual business firms. These are fairly stringent requirements, and certainly nobody has yet shown that any existing business games satisfy them. However, we might someday be able to devise management games which will meet these criteria.

It is evident, however, that as we succeed in designing more and more realistic games and perhaps approach the requirements stated here, we will simultaneously meet many of the problems of complexity that we are trying to avoid when going from the field to the laboratory. The game will, e.g., be complex enough to make measurements almost as formidable as in field research. The network of relations in the mechanism under study will be so large that it will be very difficult to understand and explain what is happening in an experiment. The design and administration of the game will therefore require real ingenuity. In the final analysis, we shall have to admit that a compromise between satisfactory realism and tolerable complexity will always be necessary.

When and if these methodological problems are overcome, games could be exceedingly useful in exploring various aspects of oligopoly behavior. For example, the effect on individual firms' price and output behavior of such variables as number of competing firms, relative size of competing firms, price elasticity of the market, geographical locations of firms, ability to introduce new products, extent of product differentiation, etc., can all be explored much more easily through laboratory play of such a game than they in fact seem to be explorable through observations, interviews, and statistical studies of actual business firms.

Another class of economic problems which could be explored through research with the type of business game just described would be problems in the area of the "behavioral theory of the firm."[77] This newly developing area of micro-economics focuses on the impact of organizational and institutional factors on business decision-making within firms. By suitably manipulating the organizational structure of firms competing in the business game, it would be possible to try to determine the effects of different organizational structures on price and output policy. Again, in principle, these effects are determinable through analysis, observation, and interviewing of actual firms in industry, but as we know, these latter are all exceedingly time-consuming and costly operations. Hence, it is certainly worth exploring the possibility of being able to investigate such phenomena within the confines of a laboratory.

One final example of the type of economic research that might be pursuable with a suitable management game is the basing point system. If the economic environment in which the firms competed were similar to that of such industries as steel and cement, which for long periods had adopted basing point pricing, one could explore the possibilities of such a basing point system emerging through normal competitive tactics on the part of the game participants without any kinds of explicit formal collusion. In principle, several other aspects of oligopoly behavior which are relevant to anti-trust issues could be explored in laboratory situations through the use of properly structured business games.

Organization Theory Research

Since World War II, great advances in the study of organizational behavior have been made. Not only has a fairly well-defined theory of organizations begun to emerge but a number of interesting laboratory experiments for analyzing particular organizational questions have been devised. [78] One problem which is usually raised in connection with laboratory experiments designed to explore the effects of changing organizational structure or communication patterns on behavior is the question of how transferable are the results determined in such a laboratory situation to the world of business, military, or governmental organizations in which researchers are more fundamentally interested. In this regard, management games may have a significant advantage over previous laboratory experiments. It is quite obvious that the structure of even a simple business game, such as the AMA, UCLA, or IBM game, resembles much more closely the objective task environment found in business organizations than do the simple experimental situations which have hitherto been used to study organizational phenomena. [79] A more complex game, such as the Carnegie Tech Management Game, goes even further in the direction of providing a close approximation to the business world. The type of management game described in the preceding section as being potentially most suitable for economic research would provide a laboratory situation endowed with a large degree of face validity with respect to the actual business organizations in which we are interested.

Again it will obviously be necessary to find a compromise between realism—which may facilitate the transfer of results from laboratory to real life—and tolerable complexity—which makes it possible to draw any conclusions at all from the experiment in the laboratory. It is therefore likely that the simpler laboratory experiment and the more complex game will supplement each other in the same way as, e.g., do the chemist's laboratory test and his pilot plant study. When collecting empirical data to test his fundamental theories, the scientist accepts the artifical test tube experiment. But when he wants to test, e.g., a complex production process, these simple laboratory experiments are not regarded as reliable. He knows that what works in the laboratory might cause him considerable trouble in the full-scale plant. This is why he wants to test the process in a pilot plant designed to make experimentation possible. Considering its size, cost, and purpose, a laboratory for experimental games like the Rand Logistics Systems laboratory [80] really is a pilot plant test station. But even with this limitation, organizational "pilot plant tests" should be very valuable. A simple test which shows that an organization works in a tolerable way provides valuable knowledge.

More refined experiments could also be carried out. [81] By varying the procedures according to which teams of players are selected and organized in playing a management game, it is possible to explore a great many features of organizational behavior. It is possible, for example, to study the effects of variations in team size on performance, morale, and adaptability to change. By changing the hierarchical nature of different firms' organization structures, we can explore the effects of such hierarchies both on external performance in the market and on such internal features as goal formation, organizational identification, goal conflict, influence etc.[82]

We can alter the amount of time allowed for players to make decisions and try to get some notion of the effects of time pressure. As time pressures are increased, will we get a shift from long-range planning to "putting out the fire" behavior? To what extent can problems of time pressure be traded off by establishing more complex organizations with a greater amount of work specialization? Are there differential effects of time pressures which come early, in the middle of, or late in game play?

The pattern of information flows within the firms can also be manipulated experimentally. By restricting the amounts of information that flow to particular team members, it is possible to study the effects of such information flows on team performance, goal formation, etc. Finally, another obvious area of organizational research would be the effects of stability of team membership on performance and morale.

Psychological Research

Since business games present complex problem-solving tasks to individual team members as well as to the teams as a whole, it is possible to use the environment generated by such a game as a laboratory situation in which to explore various kinds of psychological problem-solving and learning behavior. The type of management game which we have described as suitable for economic and organizational research would also provide a laboratory situation for individual psychological research, the results of which might be readily transferable to the industrial situations in which industrial psychologists are fundamentally interested.

It is possible that the performance of individuals in business games may someday be used for executive testing purposes, and that game performance will be used as one kind of psychological test score, along with Rorschach test scores, TAT scores, intelligence test scores, etc., as one of the criteria upon which the hiring or the promotion of management personnel may depend. At present, these are only suggestions, and no scientific evidence has been amassed to prove or disprove the potential value of business games in this type of application.

Although no formal studies have been made, experience to date with simulation exercises [i.e., management games] does not suggest any correlation between success in a particular simulation and success in the real business world—nor have any specific discriminating factors emerged. This is undoubtedly one of the most controversial uses of simulation. On the other hand, while there are many pitfalls standing in the way of this avenue of development, people are, in fact, being evaluated today by techniques equally crude: for example, the personal interview. Since simulation puts the participants under stress, careful psychological observation during an exercise might uncover some information about a man's reactions and ability to perform under pressure.[83]

For the present, however, success in playing a business game must be considered to prove nothing about a person other than that he is a skillful player of a particular business game. Since existing games represent only a portion of the decison-making activities of the manager, we should be exceedingly cautious in trying to judge management potential on the basis of game performance. This is, however, an area where much future research undoubtedly will be done.

SUMMARY—THE FUTURE OF MANAGEMENT GAMES

In the last four years there has been a growing interest in management games. It is possible to incorporate a strikingly high degree of realism in these games, particularly those played with the help of an electronic computer. The participants can become exceedingly absorbed by the game.

Everyone who attended that first [AMA Business Game] course took the simulated exercises very seriously. They put in long hours, frequently ignoring the scheduled break periods. . . . Two members of one team almost came to blows over the company's strategy and had to be stopped by their "president."[84]

Because of the high degree of emotional and psychological involvement of the players in such a training game, they can become highly receptive to learning new ideas.

Many of the techniques and concepts which we feel can be learned by playing business games can also be learned in other ways. The real question still to be answered is what kinds of materials are management games more efficient in getting across to participants than are other types of learning experiences? Perhaps the question should be put another way: What are the characteristics of participants which make them better able to grasp certain techniques and concepts through the use of games than from other kinds of learning activities? We have presented a tentative set of hypotheses which might prove useful in formulating business games which will have particular types of training values.

In summing up, it must be emphasized that business games are not yet a panacea for business education. No existing management game, nor even the whole lot of them taken together, provides for a complete business education. Many areas which are of extreme importance to the manager are not adequately incorporated in either the object or the context of existing games. In particular, important lacks are those human relations, psychological, and organizational factors which so far have been least successfully quantified by operations researchers and management scientists. In an attempt to remedy some of these deficiencies of traditional games, and to introduce greater scope for creative and imaginative behavior on the part of the participants, we have suggested that the traditional pattern of highly structured, rigid business games be modified by the introduction of some non-structured, free elements, producing a new type of training instrument which we have named the "business game case." While business games, whether of the conventional or the game case variety, may be valuable as part of a business school curriculum or as part of an executive training program, we have a long way to go until the day (if it is ever reached) that these games can provide for a complete education.

We have an ever longer way to go in the actual research uses of business games. We are now in the position where everybody is talking about the research uses of management games, but very few people are really doing any research with them.

"Come then! Let the play begin!"[85]

NOTES

1. Farrand Sayre, *Map Maneuvers and Tactical Rides,* Springfield Printing and Binding Company, Springfield, Massachusetts, 1908, as quoted by Clayton J. Thomas, "The Genesis and Practice of Operational Gaming," *Proceedings of the First International Conference on Operational Research,* Operations Research Society of America, Baltimore, 1957, p. 65.

2. H. J. R. Murray, *A History of Chess,* Oxford University Press, Oxford, England, 1913, pp. 46-50.

3. M. G. Weiner, "An Introduction to War Games," P-1771, The RAND Corporation, August 17, 1959.

4. Kingsley S. Andersson, "The Game of War," Technical Operations, Inc., Burlington, Massachusetts, 1960.

5. Thomas, op. cit., pp. 64-81.

6. J. P. Young, "A Brief History of War Gaming," ORO-S-689, Operations Research Office, The Johns Hopkins University, Washington 15, D. C., October 23, 1956.

7. By Georg Venturini. See Thoman, op. cit., p. 66.

8. Von der Goltz, as quoted by Thomas, op. cit., p. 66.

9. See Thoman, ibid., p. 67, and Weiner, op. cit., pp. 8-10.

10. By Captain Baring of the Royal Artillery. See Thomas, op. cit., p. 68.

11. Sayre, op. cit., as quoted by Thomas, op. cit., p. 68.

12. By W. R. Livermore, then a Captain in the U. S. Corps of Engineers. See Thomas, op. cit., p. 68 and Robert D. Specht, "War Games," P-1041, The RAND Corporation, March 18, 1957, pp. 7-10.

13. Weiner, op. cit.

14. James R. Jackson, "Learning from Experience in Business Decision Games," *California Management Review,* Vol. 1, No. 2 (Winter, 1959), pp. 92-107, and Specht, op. cit., pp. 1-4.

15. Specht, op. cit.

16. Alexander M. Mood, "War Gaming as a Technique of Analysis" P-899. The RAND Corporation, September 3, 1954.

17. Another outgrowth of war games, which it is beyond the scope of this paper to discuss, is the development of political games. See Harold Guetzkow, "A Use of Simulation in the Study of Inter-Nation Relations," *Behavioral Science,* Vol. 4, No. 3 (July, 1959), pp. 183-191.

18. Franc M. Ricciardi et al., TOP MANAGEMENT DECISION SIMULATION: THE AMA APPROACH, Elizabeth Marting, editor, American Management Association, New York, 1957, p. 59. This book contains an extensive description of the original AMA game.

19. Ibid., p. 94.

20. The IBM game is described in MANAGEMENT DECISION-MAKING LABORA-TORY, *Instructions for Participants,* IBM, no date given.

21. The UCLA game is discussed in James R. Jackson, "UCLA Executive Decision Games," Management Sciences Research Project Research Report No. 58, University of California, Los Angeles, December 9, 1958 (reprinted in *Proceedings of the National Symposium on Management Games,* Center for Research in Business, The University of Kansas, Lawrence, Kansas, May, 1959, pp. VI-9 through VI-15) and also in James R. Jackson, "Learning from Experience in Business Decision Games," op. cit.

22. Executive Decision Game No. 3, briefly described in Proceedings of the National Symposium on Management Games, op. cit., pp. I-7 through I-9 and VI-14.

23. Carnegie Tech Management Game, discussed in K. J. Cohen et al., "The Carnegie Tech Management Game," to be published in *The Journal of Business,* Vol. 33, No. 4 (October, 1960); in William R. Dill, "The Business Game," *Carnegie Alumnus,* Vol. 45, No. 2 (November, 1959), pp. 4-9; and in *Proceedings of the National Symposium on Management Games,* op. cit., pp. I-3 through I-7.

24. East Pittsburgh Business Simulator, presented in Robert H. Davis, "The Business Simulator," Operations Research, East Pittsburgh Division, Westinghouse Electric Corporation, mimeographed, December 12, 1958.

25. Total Enterprise Business Game, described in Albert N. Schrieber, "Gaming—A New Way to Teach Business Decision Making," *University of Washington Business Review,* April, 1958, pp. 18-29.

26. Pillsbury Management Game, cited in *Proceedings of the National Symposium on Management Games,* op. cit., pp. IV-5 through IV-9.

27. William Viavant, "A Management Game for the Petroleum Industry," *Proceedings of the National Symposium on Management Games,* op. cit., pp. II-1 through II-3.

28. Executive Decision Game, discussed in E. W. Martin, Jr., "Teaching Executives Via Simulation," *Business Horizons,* Vol. 2, No. 2 (Summer, 1959), pp. 100-109.

29. Business Management Game, described in G. R. Andlinger, "Business Games—Play One", *Harvard Business Review,* Vol. 36, No. 2 (March-April 1958), pp. 115-125.

30. Jay R. Greene and Roger L. Sisson, *Dynamic Management Decision Games,* New York, John Wiley and Sons, 1959, pp. 58-69.

31. Cited in *Proceedings of the National Symposium on Management Games,* op. cit., p. VI-20.

32. Discussed in several mimeographed notes issued by the Baton Rouge Refinery, Esso Standard Oil Company, no dates given.

33. Lowell W. Herron, *Executive Action Simulation,* Englewood Cliffs, N. J., Prentice-Hall Inc., 1960.

34. Stanley Vance, *Management Decision Simulation,* New York, McGraw-Hill Book Company, Inc., 1960.

35. Reports on some European games may be found in three papers presented at the 6th Annual International Meeting of the Institute of Management Sciences, Paris, France, September 1959: (a) Torben Agersnap and Erik Johnsen, "A Decision Game of Managerial Strategy as a Research Tool"; (b) MM. Aubert, Bourges, Minthe, Anstett, and J. B. Tricaud, "Contributions et Experiences en Matiere de 'Management Games' "; (c) R. Renard, "Remarques sur les Experiences de Gestion". These papers are reprinted in MANAGEMENT SCIENCES *Models and Techniques,* C. West Churchman and M. Verhulst, eds. Pergamon Press, Inc., New York, Oxford, London, and Paris, 1960.

The Danish game has also been described more extensively in Danish: Torben Agersnap and Erik Johnsen, "Oekonomispil—et nyt redskap for forskning og undervisning i virksomhetsledelse," *Ehrvervsoekonomisk Tidskrift,* 1958, No. 1.

While other games have not yet been described in the literature, some information about them is available from the designers—generally the research department of a computer manufacturer. The Compagnie des Machines Bull has developed other games besides the one described by Aubert et al., op. cit. The Sewdish computer manufacturer Facit Electronics has developed a "Sales Campaign Game."

36. John S. McGuinness, "A Managerial Game for an Insurance Company," *Operations Research,* Vol. 8, No. 2 (March-April, 1960), pp. 196-209.

37. BANK MANAGEMENT SIMULATION, (mimeographed), IBM, July 7, 1960.

38. A summary of this game is contained in Clifford J. Craft and Lois A. Stewart, "Competitive Management Simulation," *The Journal of Industrial Engineering,* Vol. 10, No. 5 (September-October 1959), pp. 362-363.

39. Although most general management games are interactive games, in which several teams play simultaneously in direct competition, a few of the more recent top-level management games have been developed as non-interactive games. Perhaps the most notable of these are the AMA's GENERAL MANAGEMENT SIMULATION (the third business game developed by the AMA) and the McKinsey-IBM BANK MANAGEMENT SIMULATION.

40. "Production-Manpower Decision Game: General Description," Tulane University Computer Center, mimeographed, no date given.

41. Developed by Alan J. Rowe, and described in *Proceedings of the National Symposium on Management Games,* op. cit., pp. II-3f, and in A. J. Rowe and R. R. Smith, "Now Training for Production Is a Play-It-To-Win Game," *Factory Management and Maintenance,* March, 1958, pp. 146 ff.

42. George J. Feeney, "Marketing Strategy Simulation Exercise III," Marketing Services Research Service, General Electric Co., multilithed, February 3, 1960.

43. Developed by John F. Lubin, and described in *Proceedings of the National Symposium on Management Games,* op. cit., pp. II-4.

44. Ibid., pp. II-4f.

45. Cited in *Proceedings of the National Symposium of Management Games,* op. cit., p. VI-20.

46. Ibid.

47. MATERIALS INVENTORY MANAGEMENT GAME, PERSONNEL ASSIGNMENT MANAGEMENT GAME, RETAILING DEPARTMENT MANAGEMENT GAME, PRODUCTION SCHEDULING MANAGEMENT GAME, and MARKET NEGOTIATION MANAGEMENT GAME, presented in Greene and Sisson, *Dynamic Management Decision Games,* op. cit., Chs. 4-8 and 10.

48. Robert Boguslaw and Warren Pelton, "STEPS, a Management Game for Programming Supervisors," *Datamation,* Vol. 5, No. 6, November-December, 1959, p. 13.

49. Bertil Hallsten, "Lagerekonomispel," Stockholm School of Economics, mimeographed, 1960.

50. Greene and Sisson, op. cit., p. 11.

51. Ibid., p. 19.

52. Charles C. Holt, Franco Modigliani, and Herbert A. Simon, "A Linear Decision Rule for Production and Employment Scheduling," *Management Science,* Vol. 2, No. 1 (October 1955), pp. 1-30; and Charles C. Holt, Franco Modigliani, and John F. Muth, "Derivation of a Linear Decision Rule for Production and Employment," *Management Science,* Vol. 2, No. 2 (January 1956), pp. 159-177.

53. As of the middle of 1960, there were over 100 different business games in use, according to an estimate in Max D. Richards and Fred W. Kniffin, "Business Decision Games—A New Management Tool," *Pennsylvania Business Survey,* Bureau of Business Research, The Pennsylvania State University, June and July, 1960, Part I, footnote 4. In surveying the history of management games, we have been able to cite only a representative sample of these games.

54. "In Business Education, the Game's the Thing," *Business Week,* July 25, 1959, p. 56 and E. W. Martin, Jr., op. cit., p. 104.

55. "In Business Education, the Games's the Thing," op. cit., p. 56.

56. Ibid.

57. Martin Shubik, quoted in *Proceedings of the National Symposium on Management Games,* op. cit., p. I-12.

58. Discussions of the experiences of various users of business games are now beginning to appear in the literature. We feel that the value of such discussions could be considerably increased if empirical observations were systematized along the lines we are suggesting here. Compare, e.g., Paul S. Greenlaw and Stanford S. Kight, "The Human Factor in Business Games," *Business Horizons,* Vol. 3, No. 3 (Fall, 1960), pp. 55-61.

59. See William R. Dill, "The Business Game," op. cit., p. 7.

60. In evaluating the performance of players in any business game which incorporates elements of risk or uncertainty, it is important to distinguish between *good analysis* and

good results. For example, random factors may result in a "good" decision having unfortunate consequences on a particular occasion, and yet this same decision, if adhered to, may result in successful long run performance.

61. One particular type of organizational problem which is almost universally absent in existing business games is the problem of follow-up to insure that the decisions which have been made are in fact implemented or else suitably modified. This area of management control is receiving considerable emphasis in the Mark II version of the Carnegie Tech Management Game currently being developed.

62. Greene and Sisson, *Dynamic Management Decision Games,* op. cit., p. 3.

63. William R. Dill, "The Business Game," op. cit., p. 9.

64. Studieforbundet Naringsliv och Samhalle (SNS, = The Swedish Council for Social and Economic Studies), Stockholm. For a report (in Swedish) about the case, see "SNS—orientering," 1959, available free from SNS.

65. The only game known to us that to some extent incorporates this idea is the FRENCH BULL-game, cited in footnote 35.

66. The most noteworthy exception is the third AMA game, GENERAL MANAGEMENT SIMULATION. In this game, there are two classes of decision variables, "policy decisions" and "action decisions." Policy decisions, once made, remain in effect until specifically altered by the players. Action decisions are for only a single period at a time. Players overtly make either type of decision whenever they themselves choose, rather than at rigidly fixed decision dates.

67. A review of attempted applications of computer simulation techniques in economics and a discussion of the methodological issues involved may be found in Kalman J. Cohen and Richard M. Cyert, "Computer Models in Dynamic Economics," to be published in *The Quarterly Journal of Economics,* Vol. 75, No. 1 (February, 1961).

68. In fact, the American Management Association attaches the name "simulations" to their three business games. Authors associated with the AMA, in writing about business games, frequently do so under some such heading as "Competitive Management Simulation." See, for example, Craft and Stewart, "Competitive Management Simulation," op. cit.

69. E. G., See William W. Haythorn, "Simulation in RAND's Logistics Systems Laboratory," *Report of the System Simulation Symposium,* American Institute of Industrial Engineers, New York, 1958, pp. 77-82; Robert L. Chapman, John L. Kennedy, Allen Newell, and William C. Biel, "The Systems Research Laboratory's Air Defense Experiments," *Management Science,* Vol. 5, No. 3 (April, 1959), pp. 250-269; and Murray A. Geisler, "The Simulation of a Large-Scale Military Activity," Ibid., Vol. 5, No. 4 (July, 1959), pp. 359-368.

70. Clayton J. Thomas and Walter L. Deemer, Jr., "The Role of Operational Gaming in Operations Research," *Operations Research,* Vol. 5, No. 1 (February, 1957), pp. 1-27; on p. 6, Thomas and Deemer "define *operational gaming* as the serious use of *playing* as a primary device to formulate a *game* to solve a *game* or to impart something of the solution of a *game.*" (Italics theirs.)

71. For a discussion of the relationship between operational gaming, Monte Carlo, and simulation techniques, see Thomas and Deemer, ibid., pp. 4-6.

72. The use of other methods of simulation for this purpose appears to be fairly common by now. Compare D. G. Malcolm, "Bibliography on the Use of Simulation in Management Analysis," *Operations Research,* Vol. 8, No. 2 (March-April, 1960) pp. 169-177.

73. The Germans in the present century used military games as a means for rehearsing their 1918 Spring offensive in World War I, and in World War II preliminary studies based on war games were used in planning the invasions of France in 1940, of the Ukraine in 1941, and of the potential invasion of England which never did take place. See Thomas, "The Genesis and Practice of Operational Gaming," op. cit., p. 68.

74. See Clayton J. Thomas, "The Genesis and Practice of Operational Gaming," op. cit., pp. 76-79.

75. Ibid., pp. 77-78.

76. At least one publication in this area has already appeared. See Austin C. Hoggatt, "An Experimental Business Game," *Behavioral Science,* Vol. 4, No. 3 (July 1959), pp. 192-203. There have been two papers presented at National Meetings of the Operations

Research Society of American reporting on some further interesting developments in the use of business games for oligopoly research. For abstracts of these talks, see George J. Feeney, "Experiments with Man-Machine Decision Systems," *Bulletin of the Operations Research Society of America,* Vol. 8 (1960), p. B-25 and Austin Hoggatt, "Business Games as Tools for Research," Ibid., p. B-96.

77. See R. M. Cyert, E. A. Feigenbaum, and J. G. March, "Models in a Behavioral Theory of the Firm," *Behavioral Science,* Vol. 4, No. 2 (April, 1959) pp. 81-95.

78. See James G. March and Herbert A. Simon, *Organizations,* New York, John Wiley and Sons, 1958, and the references cited therein.

79. See e.g., Anatol Rapoport, "A Logical Task as a Research Tool in Organization Theory," in *Modern Organization Theory,* Mason Haire, ed., New York, John Wiley and Sons, 1959, pp. 91-114; and Donald F. Clark and Russell L. Ackoff, "A Report on Some Organizational Experiments," *Operations Research,* Vol. 7, No. 3 (1959), pp. 279-293.

80. Geisler, "The Simulation of a Large-Scale Military Activity," op. cit.

81. A description of both the physical instrumentation and the business game models currently being used for organization theory research at Princeton University is found in John L. Kennedy, James E. Durkin, and Frederick R. Kling, "Growing Synthetic Organisms in Synthetic Environments," Department of Psychology, Princeton University, multilithed, presented at the 1960 Meeting of the Eastern Psychological Association, New York, April 16, 1960.

82. The effects of organizational hierarchies within teams playing the Carnegie Tech Management Game on influence is discussed in W. R. Dill, William Hoffman, H. J. Leavitt, and Thomas O'Mara, "Some Educational and Research Results of a Complex Management Game," Graduate School of Industrial Administration, Carnegie Institute of Technology, dittoed, October, 1960.

83. Craft and Stewart, "Competitive Management Simulation," op. cit., p. 361.

84. Ibid., p. 358.

85. This is an English translation of the final sentence in the Prologue of Leoncavallo's *I Pagliacci:* "Andiam! Incominciate!"

TEACHING WITH SIMULATION GAMES:
A Review of Claims and Evidence

CATHY S. GREENBLAT

T eachers of social science are constantly seeking better ways of conveying an understanding of systems concepts and of presenting existing theory and empirical data to students. In recent years, simulation games have gained considerable popularity as potential vehicles for such improvement in pedagogy (compare Greenblat, 1971). These games are currently utilized in schools and colleges for a variety of purposes: heightening interest and motivation; presenting information and principles; putting students into situations in which they must articulate positions, ideas, arguments, or facts they have previously learned; or training students in skills they will later need.

This paper stems from two interrelated problems concerning the claims and evidence for teaching with simulation games. The first is a problem in the *teaching of sociology* and can be summarized by the statement, "Those who have used games tend to be highly enthusiastic and to report very favorable outcomes, but the empirical evidence to systematically test their claims is still limited." The field is new, and much remains to be done. Many claims are as yet untested; others are only partially tested; and methodological difficulties are just beginning to be successfully overcome. Educators, game designers, and researchers have not yet fully explored the nature and range of appropriate applications of simulations to education, nor have they identified the sorts of simulations most effective for various types of learning and learners.

The second is a problem of the *sociology of teaching,* and can be phrased in terms that are somewhat similar to the above, yet critically different: "The empirical evidence to systematically test the claims concerning the consequences of teaching with simulations is still limited, yet utilization is spreading rapidly." Perhaps it is presumptive to think that teachers—even teachers of social science who seek "hard data" to support their research hypotheses and the statements

AUTHOR'S NOTE: *I am very grateful to Marvin Bressler of Princeton University for his criticisms and suggestions on an earlier version of this paper, and to Richard Stephenson of Douglass College for his suggestions regarding this version.*

they make in classes—are often concerned about *evidence* of the effectiveness of their teaching techniques. Perhaps most of us teach more by "faith" in our methods than we would like to admit, for how easy it is to make cavalier assumptions of cause and effect relationships between our teaching and students' good examination or term paper performance. Perhaps when we purport to "see" that traditional methods do not work, we accept our innovations on similar faith. In the absence of evidence for older or newer methods, perhaps personal preference would seem the best criterion for decision-making, and the papers reviewed in this article would have interest to researchers, but not to teachers. Departmental and institutional politics, however, often seem to operate to allow the traditionalists to be considered innocent until proven guilty, while the innovators are more frequently pressured to prove their contentions of superior (or at least equal) results with their new methods. Hence, perhaps, we can account for why the latter (and this author) occasionally go through the scientific process of looking for evidence in support of our personal observations. Although they may not constitute the *basis* of the decision to use a new teaching technique, positive findings help sate the skeptics and critics. And the process of inquiry and investigation leads to improvement of our understanding of the learning process.

These two problems, then, provide the raison d'etre of this paper: it seems fruitful, even at this early stage, to summarize the existing claims and the available evidence, so those concerned can see where we are and where we have to go.

THE POPULARITY AND PROMISE OF SIMULATION GAMES

Beliefs in the potency of games as pedagogical devices stem from several sources, including: (a) the view that the mind is an instrument to be developed rather than a receptacle to be filled; (b) the consequent position that modes of teaching are needed which will help to develop people who are excited about learning and know *how* to learn, rather than people with vast funds of information, much of which will soon be obsolete (Sprague and Shirts, 1966: 15-16); (c) the desire to develop modes of promoting engagement and curiosity, ways of looking at events and processes, and awareness of resources for finding answers; (d) the idea that students learn not because learning is a goal in and of itself, but because learning leads to goal achievement, and consequently information transmission must be seen as facilitating if it is to be effective (Coleman, 1967a: 69-70); (e) the belief that learners learn to act by acting and, hence, should be made to interact with material in an active rather than a passive way (Coleman, 1967a, 1967b; Abt, 1970; Bruner, 1961: 81); and (f) the view that, particularly in the social sciences, students must learn to examine the social world, picking out relevant variables and examining their nature and consequences; hence, modes of teaching about social systems are essential.

Simulation games relate directly to many of these notions. They represent modes of getting students to learn by provoking inquiry rather than by "feeding" information. When students cannot participate in "real" situations because direct experience is too expensive, too time-consuming, or otherwise infeasible, simulation provides an opportunity for vicariously experiencing at least some of the elements of that situation. Students experience the results of their decisions as they encounter basic processes such as decision-making, resource allocation, communication, and negotiation:

> The participants quite literally cannot avoid making decisions. They face the reality that "no decision" has as much consequence as any other decision. They must usually decide, moreover, on the basis of incomplete information, constricted by time limits, and with only partial understanding of why their previous decisions produced the consequences they did. This seems to fit much of what we know about real-life decision-making [Sprague, n.d.: 12].

In these ways students may be surrounded with environments similar to those they might not face until much later in life or might never directly experience. Finally, the classroom environment and atmosphere during the operation of a simulation differs notably from that which usually prevails. The instructor running a simulation and post-game discussion plays a role very different from the instructor teaching in a more traditional manner. There seems to be spillover to later classes, allowing more learning to take place as structural impediments are mitigated or removed.

CLAIMS ABOUT GAMES

In order to review the claims and evidence concerning the use of simulation for teaching, books, published and unpublished articles and monographs, newsletters, and advertisements from games' publishers were reviewed.[1] The numerous propositions (phrased explicitly or implicitly) found in these materials were organized into six general categories, and are listed here in detail in order to present a relatively complete inventory prior to reviewing the available research and offering suggestions for further investigations. In general, the sources are not identified in the list that follows, as the same claims are made repeatedly, and thus multiple sources would have to be cited for each. The claims presented below are phrased in simple descriptive form (for example, "participation in simulation games generates greater interest in the topics simulated"); almost all, however, have explicit or implicit counterparts which are *comparative* (for example, "participation in simulation games generates greater interest in the topics simulated; this increment in interest is larger than the increment with other modes of teaching"). In the interest of economy of space, only the descriptive form is given here; the reader should add the comparative form to each entry.

(1)   Motivation and interest

   (a) Participation in simulation games is itself interesting and involving.

   (b) Participation in simulation games increases interest in the *topics* simulated.

   (c) Participation in simulation games increases interest in the *course* in which the simulation is employed.

   (d) Participation in simulation games increases interest, enthusiasm, and commitment to *learning in general.*

(2)   Cognitive learning

   (a) Participants in simulation games gain *factual information.*

   (b) Participants in simulation games acquire *explicit referents for concepts* used to describe human behavior; abstract concepts such as "organization," "power," "stratification," and "negotiation" take on concrete meaning.

   (c) Participants in simulation games learn *procedural sequences.* "The actors must, of course, learn the rules, comprehend the essential features of the environment, understand the implications of the alternatives open to them, and develop increasingly elaborate strategies. They must be taught to operate the simulated system, in this instance in the hope that they will acquire a better concept of the larger system through a highly concentrated experience" (Meier, 1967: 157).

   (d) Participants in simulation games learn *general principles* of the subject matter simulated (e.g., the need for social control, good communications, and long-range planning).

   (e) Simulation games provide simplified worlds from which students can stand back and understand the *structure* of the everyday, "real" world. "Games seem to display in a simple way the structure of real-life situations. They cut us off from serious life by immersing us in a demonstration of its possibilities. We return to the world as gamesmen, preparing to see what is structural about reality and ready to reduce life to its liveliest elements" (Goffman, 1961: 34).

   (f) Participants in simulation games gain in *explicitness:* "The capacity to identify consciously elements of a problem in an analytic or technical sense."

   (g) Participants in simulation games learn a *systematic analytical approach.*

   (h) Participants in simulation games learn better *decision-making skills.*

   (i) Participants in simulation games learn *"winning strategies"* in those situations simulated.

(3)   Changes in the character of later course work.

   (a) Participation in simulation games makes *later work* (e.g., lectures, reading) *more meaningful.*

   (b) Participation in simulation games leads students to *more sophisticated and relevant inquiry,* for discussion of the simulation leads to questions about real-world analogies.

   (c) Class discussions following a simulation will involve *greater participation* by class members, as they will have had a shared experience.

(4)   Affective learning re subject matter

   (a) Participation in simulation games leads to *changed perspectives and orientations* (e.g., attitudes toward various public and world issues, attitudes toward the importance of collective versus individual action, attitudes toward deviant life styles).

(b) Participation in simulation games leads to *increased empathy* for others (e.g., national decision makers, ghetto residents) and increased insight into the way the world is seen by them.

(c) Participation in simulation games leads to *increased insight into the predicaments, pressures, uncertainties, and moral and intellectual difficulties of others* (e.g., decision makers, ghetto residents).

(5) General affective learning

(a) Participants in simulation games gain increased *self-awareness.*

(b) Participants in simulation games gain a greater *sense of personal efficacy and potency.*

(6) Changes in classroom structure and relations

(a) Use of simulation games promotes better *student-teacher relations.*

(b) Use of simulation games leads students to *perceive greater freedom to explore ideas.*

(c) Use of simulation games leads to *students' becoming more autonomous,* thus changing teacher-student relationships.

(d) Use of simulation games leads to *students perceiving teachers more positively.*

(e) Use of simulation games produces *more relaxed, natural exchange between students and teachers.*

(f) Use of simulation games leads to increased *knowledge of other students* (by students) and greater *peer acceptance* (Abt, 1970: 121).

(g) Use of simulation games involves a *diminishing of the teacher's role of judge and jury.*

(h) Use of simulation games leads to *teachers perceiving students more positively.*

EMPIRICAL STATUS OF THE PROPOSITIONS

Research dealing with various aspects of teaching with simulations varies in both volume and quality. Hence, the level of verification of the specific propositions is very uneven. Many are generally supported by anecdotal reports, but few have received careful attention through systematic testing.

No evidence, for example, was found for any of the propositions relating to changes in the character of later course work (section 3 of the list), though Feldt (1966: 20) reports as follows concerning the Community Land Use Game (CLUG):

No systematic attempt has been made as yet to determine the overall effectiveness of the game. The volunteered opinions of persons who have played it, however, are generally highly favorable. Many students have reported that course work in municipal finance, decision theory, zoning law, urban design, and economics, as well as courses in urban ecology and geography, have been much more meaningful and important to them as a result of having played it. Others have reported that they could better understand changes in urban growth patterns they had observed after having seen similar patterns of development in the course of the game. On the whole, it seems reasonable to argue that the game provides a certain amount of field experience to the participants which makes further course work in planning and related areas more significant to them.

There is a similar paucity of data concerning the propositions about changes in classroom climate and relations (section 6 of the list). Discussions of the reasoning behind these claims were found; Abt (1970: 31), for example, notes that the teacher role is to decide what concepts can be taught most effectively, by what method they can be communicated most memorably, and at what point review and evaluation are needed for "closure." The teacher also runs the "debriefing" or postgame analysis; "this should be a structured, directed discussion of the limitations and insights offered by the game and of the performance of the players in both representing and solving their problems effectively. Here the players will consider the teacher as a vital part of the game's operation and resolver of meaning, rather than as a person who interferes with classroom activities." Boocock and Coleman (1966: 219) speak similarly in suggesting that games are self-disciplining and self-judging, thereby changing the teacher's role. No evidence , however, was located concerning this set of propositions. Some could easily be tested using student or teacher questionnaires or sociometric techniques, but if this has been done, it has not been reported in the literature surveyed.

Other propositions have received more scrutiny by researchers. Some work has been done concerning affective learning re subject matter (section 4 of the list) resulting from simulation participation. Studies with the LIFE CAREER game suggest that participants gain an appreciation of the complexity of "real-life" planning. This is illustrated by students' statements and by increases in the number who after playing believe that "it is hard to plan your life in advance" (Boocock and Coleman, 1966: 229-231). Players were also found to gain a deeper understanding or empathy for the roles they played in the game.

For example . . . boys who worked on teams assigned to a profile of a girl student tended, by comparison with boys who planned for a boy in the game, to move toward a more "liberal" attitude toward the appropriate role for women. This was indicated by their greater willingness after the game to agree that women who wished to do so should have jobs and other interests outside of the home and family [Boocock, 1967: 332].

Responses to a checklist of twelve personal characteristics that subjects thought described what a typical dropout was like were utilized to measure changes in empathy for dropouts. Boys who played the potential dropout role developed more positive images if they already had some feeling of identification; girls playing this role became more negative toward dropouts. Thus, the researcher points to the need for caution in using this kind of teaching device: "If simulation games do indeed change attitudes as effectively as they appear to do, one must be clear as to the direction and desirability of changes" (Boocock, 1967: 332).

Other evidence concerning the development of empathy is largely anecdotal. The following are examples that convey the nature of such reports:

> Finally, one student said: "Another important thing I now understand better is communism. I realized to my amazement when we made out the decision sheet for Ingo that even those of us who claimed to be staunch conservatives were advocating pushing economic development as hard as we could, disregarding as far as we dared the validators. Our 'motto' was to bring the people as close as possible to starvation without there being a revolution. Now I can see why communism has such appeal among the underprivileged nations of the world. The temptation to move swiftly ahead at all costs is a strong one indeed, and this is especially true if you have nothing to conserve (thus why be a conservative?) [Carlson, 1969: 70].

Goldhamer and Speier (1959: 79) report that RAND gamers claim players acquire "new insights into the pressures, the uncertainties, and the moral and intellectual difficulties under which foreign policy decisions are made." The instrument used is not described in their report.

Robinson (1966: 118) found some attitudinal change in Inter-Nation Simulation participants, and discussed the possible significance of evidence of even minimal attitude change:

> The challenge to educators is to find ways to penetrate the firm 'set' that college students bring to courses on politics and international relations. Students' perspectives and orientations toward public and world affairs are ordinarily developed by the time they reach college and are largely beyond amendment, except for some unlikely confrontation with an irreversible learning experience. It would be too much to claim, at this point, that simulation constitutes such an experience, but our data indicate small and important differences whereas other methods yield virtually no differences. We cannot yet boldly recommend revision of instructional policies, but we have grounds for confidently recommending further research and replication to help decide whether our vision of an effective technique is a reality or a mirage.

Less evidence is available concerning general affective learning (section 5 of the list). Earlier it was reported that male participants in LIFE CAREER changed their attitudes toward women pursuing careers. The researchers also report a before-after change in the percentage of female participants who agreed that "an intelligent girl with a good education really should do more with it than just get married and raise a family." This, Boocock and Coleman (1966: 232-233) suggest, is evidence not only of attitude change, but also of learning to feel greater control of one's environment. In another report of LIFE CAREER participants, Boocock (1967: 333) reports that "at the same time that they become aware of the amount of information they must assimilate and the amount of planning necessary to make intelligent career choices, many students also gain confidence in their ability to do so." The evidence is not given, however, and there is no discussion of persistence of the increase in confidence.

Researchers investigating the outcomes of play in a legislative game report more change in the experimental group than in the control group away from the belief that "people like me don't have any say about what the government does." Impressionistic data were also gathered concerning an increased sense of interconnections between various aspects of a situation and interdependence in the environment (Boocock and Coleman, 1966: 234). Boocock and Coleman (1966: 233) and Inbar (1970: 241) report gains in feelings of personal potency by participants in COMMUNITY DISASTER. Some players in trial runs of THE MARRIAGE GAME have reported to the designers about considerable growth in self-awareness and insight into others as a result of the simulation experience. One girl who participated with her real-life partner reported in a personal note to this author, "It's sad and hard to admit, but I think I learned more about someone and our relationship through 8 hours of participation in the game than through 3 years of living together." Again, however, the evidence is anecdotal.

More evidence is available concerning the propositions relating to cognitive learning (section 2 of the list), but the empirical status of these claims is still weak. Specific learning from playing LIFE CAREER and a legislative game was measured by means of "test-like" items (Boocock, 1967, 1966). Participants in both games scored higher in before-after score differences than did those in control groups. Since players in each game served as the control group for those in the other game—and, therefore, the control groups were not taught the same material by an alternative method—evidence is available for the descriptive claim about learning but not for the comparative one. Participants in the legislative game were also found to change in their perception of the best strategy in the game. There is, however, no evidence that the learning of game strategy was paralleled by learning of "real-life" strategy.

Anderson (1970: 49-51) attempted to test learning with a consumer game by a paper-and-pencil test and also by an exercise in which students had to *apply* what they had learned by reading contracts and selecting the best. The control group in this case was taught the same material by conventional methods. He reports that there were no differences in factual learning for those taught by simulation and those taught by conventional methods; behavioral learning was better with simulation for some subpopulations.

Other researchers used student questionnaires to examine the extent of learning (e.g., Attig, 1967). Alger's (1963: 179) students reported six kinds of learning with the INTER-NATION SIMULATION: (1) vividness and understanding beyond what one got from textbooks; (2) realization of the complexities of conflicts between rival nations; (3) importance of having reliable knowledge and the importance of communication in international relations; (4) better understanding of the problems and goals of nations not like the United States; (5) better understanding of the problems of decision-making; and (6) difficulties of balancing the requirements of internal and external affairs.

Many researchers (e.g., Boocock and Coleman, 1966; Boocock, 1966; Feldt, 1966; Inbar, 1970) report that learning is in general terms. They suggest that students gain in awareness or that they are sure that they learn but find it difficult to specify just what the content of the learning is. Participants often are able to articulate general dimensions of their learning, but not specific facts.

In summary, researchers express a conviction that cognitive learning takes place, but measurements provide limited data. Self-reports of participants suggest learning, but information about how many or what proportion of participants report this is generally lacking. Learning that takes place seems to be difficult to specify or articulate, but it appears that it is learning of principles and procedural sequences and the acquisition of referents for concepts, rather than learning of factual information.

Finally, the greatest amount of discussion in the literature centers on claims about the heightening of motivation and interest (section 1 of the list). There is a great deal of anecdotal information concerning student involvement by simulations. Researchers report written and oral statements by participants testifying to the "motivating and self-sustaining quality of the activity" (Boocock and Coleman, 1966: 224). Some describe the degree to which students have evidenced emotional involvement (anger, disappointment, frustration) as evidence of interest (Feldt, 1966: 20; Sprague, n.d.: 16). Other behavioral indicators of interest are implied in the following statements: "After the first day students agreed to bring their lunches for the rest of the week so they could have an extra 40 minutes play each day" (Boocock, 1967: 330); "An indication of student enthusiasm for the game was the number of students who desired to attend class for purposes of conferences and negotiations on days when the other section was meeting" (Attig, 1967: 26); "It is not at all uncommon for players scheduled to play for four hours to insist that they be allowed to continue to play for an additional four hours, often without stopping to eat" (Feldt, 1966: 20).

A study by Robinson et al. (1966: 57) is the only one found which utilized several indicators and which investigated the extent to which interest generated by the simulations was greater than interest generated by other modes of teaching. One course was taught entirely by simulation and another by case study. The measures of interest employed were:

(1) comparisons of students' interest in the course with interest in other courses taken simultaneously;

(2) perceptions of interest in this course;

(3) perceptions of interest in political science;

(4) preference for case or simulation;

(5) descriptive evaluation of the course at the end of the term;

(6) reports of the amount of students' reading on the subject of the course;

(7) use of special reading materials placed on reserve in the college library;

(8) visits to the professor's office;

(9) attendance in class; and

(10) rate of participation in "lab" section.

As the reader will note, differences in many of these measures—particularly the behavioral indicators (6-10)—may reflect differences in variables other than interest: e.g., clarity of presentations, understanding of materials, and so on. In addition, due to the unusually extensive use of simulation and case study, the measures are measures of interest in the course as well as interest in the method of teaching (i.e., simulation). Robinson's findings indicated that "case method succeeds more than simulation in eliciting student interest as measured by students' perceptions, but measures of student behavior indicate that simulation succeeds more than case in affecting student interest and involvement."

Little attention has been focused on the question of whether the interest manifested by students participating in a simulation games is due to the novelty of the approach and represents a Hawthorne effect. If such were found to be the case, it would suggest use of simulations only until these "novelty effects" wore off or diminished beyond some point. Furthermore, as Cohen (1962: 371) suggests in his critique of simulations, "no one has ever systematically investigated the possibility, for example, that those whose interest and enthusiasm are visibly enlisted by games are those who are most interested and enthusiastic about the subject in the regular class, and hence that there is little net increase in interest in the subject." This is a problem of study design, and will be discussed in the next section.

In summary, then, it appears that there is a considerable amount of anecdotal material about affective involvement of students at the time of participation, and some evidence that student interest in simulations is very high. There is, at the moment, little hard data to show that such participation leads to greater interest in the subject matter, the course, or learning in general.

METHODOLOGICAL PROBLEMS IN GAMES RESEARCH

The foregoing discussion has pointed to the fact that many of the claims made about consequences of teaching with simulation games are supported by anecdotal data rather than empirical evidence. An additional problem arises in discussing even the little evidence that has been compiled; this problem stems from the methodological shortcomings of many of the research studies. A great many do not meet the canons of scientific inquiry; the deficiencies include both conventional ones and specific problems with this type of research.

The first category includes problems of research design, sampling, operationalizing of concepts, and lack of control for relevant sample characterisitcs in the analysis.

Many of the studies suffer from poor research design: "after-only" tests which preclude measurement of change; lack of control groups even where the intention is to draw conclusions about the value of simulations compared to other techniques; failure to consider Hawthorne effects; and poor criteria for accepting or rejecting hypotheses.

A related set of problems stems from the sampling techniques. In many instances, the subjects were students available to the researcher, and were known to be unrepresentative of students at that academic level. In other cases, the participants were of the age and ability level one would ordinarily teach with the game being investigated, but were sampled while out of their student roles—e.g., high school students at an organizational conference for a weekend. Their definitions of the situations and thus of the activity might well be different from the definitions by the same students if they were playing the game in a class. Surely teaching and learning take place outside classrooms, but, if generalizations are to be drawn about the use of simulations in schools, the relevant sample would seem to be students in classes.

An additional sampling problem is the very heterogeneous nature of some of the samples. Several games were run with multiple groups of people of differing ages and academic backgrounds. These subsamples were then merged to create a large enough n for subsequent analysis. It is difficult to define the relevant population for generalization from such findings, and subsample differences are not reported.

Another problem in many studies relates to poor operationalizing of concepts. Frequently the multidimensionality of such concepts as "interest" is not recognized; "learning" is often treated as a "yes" or "no" thing; and little attention is paid to the tenacity of the learning. The measures employed often bear little relationship to the learning objectives specified by the game director or by the administrator. Thus Shirts (1970: 82) remarks,

> It seems to me that there are several possible reasons for the discrepancy between the research results and the expectations of knowledgeable persons. For one thing, it appears that the design of most of the research has been guided more by what is convenient, tidy, and available than by an honest attempt to determine the impact of the use of simulations. For what other reason would a person compare simulations with traditional didactic methods on their effectiveness in teaching students numerous facts and ideas as measured by an objective test? Books and lectures present carefully processed ideas and facts to the students in grade-A, enriched, homogenized form. The students, in turn, have been trained by many years of conditioning to accept this rich diet and to return it to the lecturer on demand. In simulations on the other hand, facts are frequently hidden in scenarios, messages from other participants, decision-making forms, statements by people in hot debate, and announcements from the directors or the simulated mass media. Other facts are specific to the game being played and have to be translated before they have any meaning in the real world.

Raser (1969: 133), too, attributes the problem of poor evidence to inadequate measuring devices; "emphasis has been on developing measures to tap the acquisition of punctiform data—the retention of factual material. Only a few isolated efforts have been made to measure learning of concepts and relationships, or even to discover how much learning is facilitated, and none have [sic] been directly applied to gaming."

Finally, a general methodological shortcoming in the studies reviewed is the frequent lack of control for relevant student characteristics. In most instances, only scores for the total sample are reported. In many studies, such factors as sex of players and sex balance of groups, social class background of participants, intelligence, school records, and pregame motivations of participants are not employed as independent variables to examine the differential effects of participation on students varying in these characteristics, although some researchers (compare Stoll, 1969) have found them to be important. Robinson (1966) controlled for cognitive style, need for achievement, need for affiliation, and need for dominance in analyzing his data; such factors are suggestive of the nature of further differentiations that might be made in attempts to ascertain who learns what from simulations.

In addition to these general problems, there are also problems specific to "games" research. Many are difficult to overcome, but the researcher contemplating investigations or the teacher reviewing studies should be aware of these dilemmas.

There are a number of game variables that may be important in explaining differential outcomes, including length of play, size of playing group, and amount and quality of pregame preparation (compare Inbar, 1966). A special instance is the conditions of administration: how the simulation is employed. In some instances, research has been done where simulations were used in courses with correlate readings, lectures, and discussions. Testing for the consequences of use under these (recommended) conditions of administration is quite different from testing the teaching potential of the same simulation when it is run in an afternoon as part of a conference. In addition, characteristics of the game administrator have been found to be of considerable importance in explaining the outcomes of participation (Inbar, 1966); yet discussion of such factors as administrator competence, preparation, and familiarity to students is often lacking. A final set of game variables infrequently controlled is the roles played by the participants. Those playing different parts often have different experiences, but this has not generally been examined.

A large problem arises because, in many studies, "games" and "simulations" are at least implicitly treated as homogeneous. Despite great variations in games, conclusions from studies with one game are generalized to games in general. In point of fact, even the best-designed study could not provide a full test of any of the propositions outlined above; it could reasonably yield only generalizations about the particular game investigated or possibly that type of game. This idea of the heterogeneity of games also presents difficulties of interpretation of reports in which the findings from several studies are put together to test hypotheses about games (Cherryholmes, 1966).

Finally, there is the problem of definition of the proper "end" of a simulation experience. An integral part of all instructional simulations is the postplay discussion and analysis. Students are often unable to observe or to know of the overall course of events and thus to draw relevant conclusions without a discussion of what happened. Debriefings also focus upon the "fit" between the simulation model and the real world and thus allow students to learn from their play activites. The studies reported, however, generally fail to note whether post-play questionnaires and other measuring instruments were administered before or after such debriefings. In this author's opinion, the appropriate measurement time is *after* discussion, as the discussion is *part* of the simulation, not an appendage to it. This, of course, contributes to greater problems of administrator effects, but differential administrator competence *is* an important factor in accounting for the learning outcomes of the use of simulation games.

CONCLUSION: FUTURE DIRECTIONS

The foregoing review points to the lack of sufficient data to prove that games meet their pedagogical promise. For a number of reasons, this should not be taken as an argument for abandonment of their use in teaching.

First of all, it is difficult to tell at this point whether the lack of evidence in support of the propositions stems from poor outcomes or poor measurements. Some of the difficulties plaguing the games researcher have been outlined above, and better ways of dealing with them may be devised.

Second, although there is little evidence that students learn more when taught by games than by conventional methods, there is no evidence that they learn *less*. In fact, studies of cognitive learning point to "no difference" or "differences in favor of games that are not statistically significant." Hence, games seem to be at least as effective as other modes of teaching, and further studies may show yet more significant results.

Third, one must keep in mind the *general* scarcity of demonstrable evidence in social science and in education. Where, for example, is there good evidence that students learn when teachers lecture and class members take notes?

Fourth, a "data-based decision" to use games for teaching obviously should not be dependent upon evidence in support of *all* the claims outlined. If, for example, it were found that games stimulated interest and motivation and did nothing else, their use to get students involved and questioning might well be warranted.

Despite the general paucity of evidence, then, it is to be hoped that games will continue to be employed, and that they will receive closer scrutiny by researchers and educators. I am hopeful that, in addition to trying to test the propositions offered above, those who design, utilize, and evaluate games will turn their attention to the following types of questions.

(1) How and when and how much should games be used, and what is the best way of integrating them with other modes of teaching? Rather than introducing simulations in an ad hoc manner into a class, we should seek ways to make them parts of units of study, tied with correlate readings and other experiences.

(2) As noted earlier, questions of use cannot be answered for games in general. Rather, we must ask which games (or which types of games) best meet specific instructional

goals, and thus how and when they should be employed. For some games, Bloomfield and Padelford's (1959: 1115) comment on political games seems highly appropriate:

> The political exercise is by no means a substitute for systematic rigorous analysis by teacher and student in either international or domestic politics. Indeed, the gaming technique should be used only after the student has had extensive reading and thorough instruction on the political process, the nature and operation of institutions and laws, group dynamics, and the substance of political problems themselves. Otherwise the gaming is likely to become artificial, pursued largely on the basis of hunches or intuitions rather than of knowledge and understanding.

Other games, however, require little advance information or knowledge, and their primary value is their motivating power. Their contribution seems to lie in heightening students' interest in the topic and in sensitizing them to the complexities and variables involved in the social structure or social process simulated. We must, then, attempt to develop a taxonomy of games to guide the user or potential user.

(3) If, as has been suggested, participation in simulation games leads to attitudinal change, serious thought must be given to the dangers as well as to the promises of the technique. What are the underlying assumptions and values of the models developed and used? What are we *deciding* to teach, and what are we teaching *inadvertently?* When does "attitude change" become "brainwashing"?

(4) What are some of the less obvious personal consequences of participation in simulations? For example, where involvement is high and is manifested by severe frustration, anger, depression, and so on, are there "costs" to students that must be weighed against the gains? And how do these costs compare to those stemming from the frustration, anger, and depression created when students are "talked at," kept still and obedient, and the like in the regular classrooms?

(5) Finally, if further testing lends greater support to the propositions listed earlier, we then must continue our investigations to attain a greater understanding of *how and why* games have these effects. Through this type of query, researchers can make a contribution not only to those interested in this educational technique, but to those in all fields concerned with the nature and process of learning.

NOTE

1. Happily, the number of studies is steadily increasing; hence any review such as this is quickly dated. The reader is urged to consult recent issues of *Simulation and Games* (quarterly journal published by Sage Publications, Inc.) for reports published since September 1971, when this paper was completed.

REFERENCES

ABT, C. (1970) Serious Games. New York: Viking.
ALGER, C. F. (1963) "Use of the Inter-Nation Simulation in undergraduate teaching." pp. 150-189 in H. Guetzkow et al., Simulation in International Relations: Developments for Research and Teaching. Englewood Cliffs, N.J.: Prentice-Hall.
ANDERSON, C. R. (1970) "An experiment on behavioral learning in a consumer credit game." Simulation and Games 1 (March): 43-54.
ATTIG, J. C. (1967) "Use of games as a teaching technique." Social Studies 58 (January): 25-29.

BLOOMFIELD, L. P. and N. J. PADELFORD (1959) "Three experiments in political gaming." Amer. Pol. Sci. Rev. 53 (December): 1105-1115.

BOOCOCK, S. S. (1967) "Life Career Game." Personnel and Guidance J. 46 (December): 328-334.

——— (1966) "An experimental study of the learning effects of two games with simulated environments." Amer. Behavioral Scientist 10 (September/October): 8-17.

——— and J. S. COLEMAN (1966) "Games with simulated environments in learning." Sociology of Education 39 (Summer): 215-236.

BRUNER, J. (1961) The Process of Education. Cambridge, Mass.: Harvard Univ. Press.

CARLSON, E. (1969) Learning Through Games. Washington, D.C.: Public Affairs.

CHERRYHOLMES, C. H. (1966) "Some current research on effectiveness of educational simulations: implications on alternative strategies." Amer. Behavioral Scientist 10 (October): 4-7.

COHEN, B. S. (1962) "Political gaming in the classroom." J. of Politics 24 (March): 367-381.

COLEMAN, J. S. (1967a) Academic Games and Learning. Princeton: Educational Testing Service.

——— (1967b) "Learning through games." National Education Association J. 56 (January): 69-70.

FELDT, A. G. (1966) "Operational gaming in planning education." J. of Amer. Institute of Planners 22 (January): 17-23.

GOFFMAN, E. (1961) Encounters: Two Studies in the Sociology of Interaction. Indianapolis: Bobbs-Merrill.

GOLDHAMER, H. and H. SPEIER (1959) "Some observations on political gaming." World Politics 12 (October): 71-83.

GREENBLAT, C. S. (1971) "Simulations, games and the sociologist." Amer. Sociologist 6 (May): 161-164.

INBAR, M. (1970) "Participation in a simulation game." J. of Applied Behavioral Sci. 6 (Spring): 239-244.

——— (1966) "The differential impact of a game simulating a community disaster." Amer. Behavioral Scientist 10 (September/October): 18-27.

MEIER, R. L. (1967) "Simulations for transmitting concepts of social organiation," pp. 156-175 in W. Z. Hirsch et al., Inventing Education for the Future. San Francisco: Chandler.

RASER, J. (1969) Simulation and Society. New York: Allyn & Bacon.

ROBINSON, J. A. (1966) "Simulation and games," pp. 85-123 in P. Rossi and B. Biddle (eds.) The New Media and Education. Chicago: Aldine.

——— L. F. ANDERSON, M. G. HERMANN, and R. C. SNYDER (1966) "Teaching with Inter-Nation Simulation and case studies." Amer. Pol. Sci. Rev. 60 (March): 53-64.

SHIRTS, R. G. (1970) "Games students play." Saturday Rev. 53 (May 16): 81-82.

SPRAGUE, H. T. (n.d.) "Using simulations to teach international relations." Simile II, La Jolla, California. (mimeo)

STOLL, C. S. (1969) "Player characteristics and interaction in a parent-child interaction game." Sociometry 32 (September): 259-272.

THE EDUCATIONAL EFFECTIVENESS
OF SIMULATION GAMES:
A Synthesis of Recent Findings

FRANK H. ROSENFELD

Douglass College of Rutgers University

In the beginning, claims of the effectiveness of simulation games as teaching tools were accepted largely on faith (Boocock and Schild, 1968). As the field of simulations evolved, skeptics arose, and the first empirical studies were done. The research was primitive, but results suggested that some games were not any more effective as teaching/learning devices than more traditional educational techniques. In the face of this information, game developers and users cast doubts on the validity of the studies and made new claims, especially in the area of increasing motivation. After all, was not the interest and enthusiasm of the participants prima facie evidence of the effectiveness of simulation games?

During the late sixties, claims of the efficacy of simulation games continued to expand in the absence of hard evidence to include not only factual learning and motivation, but also affective learning, attitude change, and improvement in the classroom (Greenblat, 1973). The number of research studies grew, too, but still they were plagued by such weaknesses as poor design and sampling, poor operationalization of concepts, generalizations from one game to another or to all games, failure to examine the context in which the game was used, and failure to consider game and participant variables (Greenblat, 1973).

The early seventies has been an age of relative enlightenment. Due to the efforts of the staff of the Academic Games Program of the Center for Social Organization of Schools at Johns Hopkins University and others, a body of accurate knowledge about the educational effectiveness of simulation games has been developing. Research techniques have become refined; not only are the subjective valuations of participants considered, but a host of more objective methods, including controlled experiments and pre- and post-simulation tests and attitude surveys are employed.

AUTHOR'S NOTE: *I would like to thank Dr. Cathy S. Greenblat for her help in preparing this article.*

This paper is a synthesis of research findings published in the years 1971-1973. A more complete picture of the effectiveness of simulation games is provided by taking these findings together with those from earlier research. However, the recent research does cover many of the claims made about games and can serve as a source of some descriptive propositions.

A word of caution must be introduced here. Practically all the studies cited here are based on one-game samples. Since simulations differ a great deal in structure, content, and procedure, and since there are great variations in the administration of any one game, general statements about simulation games should be taken as descriptive rather than definitive.

Borrowing the outline developed by Greenblat (1973), evidence of the educational effectiveness of simulation games will be divided into six general categories: (1) motivation and interest, (2) cognitive learning, (3) changes in the character of later course work, (4) affective learning with respect to the subject matter, (5) general affective learning, and (6) changes in classroom structure and relations.

MOTIVATION AND INTEREST

Although it is difficult to measure, there are many reports that simulation games significantly increase the motivation and interest level of student players (Taylor and Walford, 1972). The source of such data is usually participant reports, and it is possible that there are discrepancies between what the participants perceive and what actually happens. In one controlled experiment, it was shown that a simulation game (INTER-NATION SIMULATION) could produce significant gains in the attitude of students toward their class as a learning experience (Lee and O'Leary, 1971), and it has also been demonstrated that the use of a game (CONSUMER) could lead to decreases in the absences of disadvantaged students (Livingston et al., 1973).

There is also evidence to show that not all games cause such increases in motivation and interest. For example, a run of a political science game (DEMOCRACY) did not heighten students' interest in the legislative process (Livingston et al., 1973), and gains in interest in specific courses were not always followed by increases in interest in similar courses (Lee and O'Leary, 1971).

COGNITIVE LEARNING

One claim of simulation game advocates is that the games experientially teach students facts, concepts, and procedures. Implicit in this is that games are as effective as, if not more effective than, traditional teaching methods in doing this. Students do report that simulation games are valuable for learning relationships and that they feel that participants succeed or fail due to their own actions

(Edwards, 1971; game used: INTRODUCTION TO BUSINESS GAME). Games also have been shown to be more effective than traditional techniques in training people in certain specialized skills (Livingston et al., 1973).

Other research has shown that at least some simulation games may be no more effective in cognitive learning than are traditional methods (Livingston et al., 1973; games used: TRADE AND DEVELOP, CONSUMER). Role-playing in certain situations may be as effective in teaching facts as a full simulation (Livingston, 1972).

It has been suggested that simulation games may be particularly effective in teaching disadvantaged students. Students seem to be able to understand strategies they cannot verbalize, and learning might take place in the absence of well-developed language skills (Livingston et al., 1973). One study showed, however, that games have no special value for low-achieving students (Edwards, 1971).

There may be drawbacks in using simulation games to teach facts, concepts, and procedures. Attempts to test the efficacy of simulations suggest that they are only as effective a teaching/learning device as the simulation is accurate (i.e., reflects the critical aspects of the real world) (Boocock, 1972). Role-playing students may only be learning how they and other students play the simulated roles, not how the real role occupants act (Boocock and Schild, 1968). In general, though, simulations seem no less effective than other techniques for cognitive learning.

CHANGES IN THE CHARACTER OF LATER COURSE WORK

There is no recent research evidence which either supports or refutes the claim that simulation games create changes in the character of later course work.

AFFECTIVE LEARNING WITH RESPECT TO SUBJECT MATTER

Simulation games probably have the most potential in the area of affective learning; it seems logical to believe that experiencing the world of others would be more effective in increasing empathy and understanding for others than traditional teaching methods and might lead to changed perspectives and orientations. Recent evidence in this area, however, is ambiguous.

Political simulations (DEMOCRACY) have been shown to lead to both changes in political attitudes and to a greater acceptance of certain political practices (Livingston et al., 1973), and a social minorities game (GHETTO) has also produced in players new attitudes more favorable to the poor independent of the participants' factual knowledge about or previous experience with poverty (Livingston et al., 1973). Similar studies have shown that these attitude changes are not long-lasting.

Other studies have demonstrated that, although simulation games may change individual students' attitudes, there may be no consistent effect across groups. For example, a political science game (INTER-NATION SIMULATION) produced no consistent overall change in the participants' ideology in terms of feelings toward radical philosophies and tactics, traditional moralism, belief in people, cynicism about institutions, idealism versus pragmatism, and people's control over events (Lee and O'Leary, 1971). In the same simulation, students did not gain in appreciation for the "complex and difficult nature" of the leadership roles they played. One possible explanation is the finding that identification with the role is important for changing the attitudes of the players (Livingston, 1972; game used: DEMOCRACY); players who concentrate on analyzing strategies and who, therefore, try to objectively observe the simulation, may not be immersing themselves in the game and subjectively experiencing the roles they are occupying (Boocock and Schild, 1968). Even though some simulations may not produce changes in the attitudes of the participants, students still view them as valuable for "getting a feel for the real situation" (Edwards, 1971).

There is also some evidence to show that simulation games can have a negative effect on the participants. One run of a social minorities game (GHETTO) "seemed to induce pessimism with regard to human conduct which may have led to less positive attitudes towards people in general" (Kidder and Aubertine, 1972: 9). A student who finds that a certain improper strategy pays off within the context of the game (prostitution in GHETTO, for example) may impute that strategy to the real-world people represented ("All ghetto women are hustling."). Experiencing the "self-defeating nature of the ghetto" can produce a general cynicism about the urban poor (Kidder and Aubertine, 1972).

GENERAL AFFECTIVE LEARNING

Simulation games seem to be effective in the area of general affective learning; participants report both increased self-awareness and a greater sense of personal power. Several games (DEMOCRACY, INTER-NATION SIMULATION, INTRODUCTION TO BUSINESS GAME) produced with the players a greater sense of efficacy in areas related to the simulations (Livingston et al., 1973; Lee and O'Leary, 1971; Edwards, 1971). Students participating in simulations (INTER-NATION SIMULATION) felt more confident in their ability to make decisions and developed a greater tolerance for ambiguity and uncertainty (Lee and O'Leary, 1971). It has been suggested, though, that some simulations may negatively affect those participants who lack the ability to do well in the game, who are oversensitive to others' disapproval, or who generally cannot cope with the simulated situation (Taylor and Walford, 1972). This may be due to the tendency of students to view games as self-judging.

CHANGES IN CLASSROOM STRUCTURE AND RELATIONS

There have been no recent reports of studies analyzing the effect of simulation games on classroom structure and relations, although anecdotal reports suggest that games lead to better student-teacher relations and an improved learning environment.

As previously mentioned, early studies of simulation games suffered from several methodological weaknesses, including the failure to systematically examine game and participant variables. Recently, researchers have looked into the effects of post-game discussions (called "debriefings") and student variables on the game experience.

DEBRIEFINGS

An important part of experiential learning is generalizing from specific experiences, and post-game discussions are often seen as crucial for the development of generalizations (Livingston et al., 1973). Recent evidence suggests that the role of the debriefing may not always be critical for learning, though. One study found that post-game discussions seem to have a slight effect on attitudes and no effect on understanding the game (Livingston, 1973; game used: GHETTO). In an elaborate experiment to test the relative efficacy of simulations with discussions, simulations without discussions, discussions without simulations, and independent learning involving neither simulations nor discussions, it was discovered that, although participants in the simulations with discussions expressed more satisfaction with learning than those in the other groups, there was no significant difference in the amount of cognitive learning in the four test conditions (Chartier, 1972; game used: GENERATION GAP). These studies do not account for possible variations in the quality of the debriefing, though.

STUDENT CHARACTERISTICS

Characteristics of the student-participants can determine the efficacy of teaching with simulation games. The sex of the player can affect both the quality of the game and the quality of learning (Fletcher, 1971; game used: CARIBOU GAMES). This may be partly due to the subject matter or orientation of the game; games geared toward masculine or feminine activities may be more effective with one sex than another in terms of cognitive and affective learning. This does not say that male-oriented games, for example, are ineffective or unimportant for females; the knowledge gained and the situation experienced by them may be more salient than larger gains and greater experiences of males. The development of a rational strategy can be affected by both the sex of the player and prior acquaintance with other players (Livingston et al., 1973). The socio-economic status and race of participants can also affect the learning of factual material through simulation games (Livingston et al., 1973). Playing singly or in pairs or threes does not necessarily affect performance or learning (Livingston et al., 1973).

Recent research also shows that simulations generally are more effective with students of higher academic ability; although both low- and high-ability students are successful in learning winning strategies, those with higher ability are more successful in drawing analogies to the real world (Livingston et al., 1973). Factual and conceptual learning seem to be correlated with verbal ability (Livingston et al., 1973). Attitudes of players prior to the simulation can affect what they get out of the game, as can personality variables (Lee, 1971; game used: INTER-NATION SIMULATION). Finally, the participants' relationship to the roles they take on can determine the effectiveness of the simulation (Livingston et al., 1973; Dukes and Seidner, 1973).

CONCLUSIONS

Findings of recent research into the educational effectiveness of simulation games can be summed up in five propositions:

(1) Simulation games generally seem no less effective as teaching/learning devices than more traditional methods; they may be more effective.

(2) There can be vast differences between the assessment of the effectiveness of simulation games by the participants (teachers and students) and their actual effectiveness.

(3) Simulation games can result in unanticipated outcomes, including negative attitudes and lower self-esteem of participants.

(4) Variations in the characteristics of the game players and the simulation environment can produce variations in the effectiveness of simulation games as a teaching/learning device.

(5) There is a great amount of variation in teaching/learning effectiveness from game to game.

These propositions are based only on research reported in the past three years (1971-1973) and must be examined together with findings of prior studies when assessing the effectiveness of simulations in education. The field of simulation games is relatively new, and so too is the area of research into games. Hopefully, improvements in both will soon be forthcoming.

Uses for simulation games are still developing. Careful monitoring of the effects of games through adequate and accurate research will allow users to know their strengths and weaknesses and designers to produce better games. Then, simulation games may fulfill their promise and become powerful instruments in education.

REFERENCES

Boocock, S. S. (1972) "Validity testing of an intergenerational relations game." *Simulations and Games* 3 (March): 29-40.

Boocock, S. S. and E. O. Schild (1968) *Simulation games in learning.* Beverly Hills: Sage Publications.

Chartier, M. R. (1972) "Learning effect: an experimental study of a simulation game and instrumented discussion." *Simulations and Games* 3 (June): 203-218.

Dukes, R. L. and C. J. Seidner (1973) "Self-role incongruence and role enactment in simulation games." *Simulations and Games* 4 (June): 159-173.

Edwards, K. J. (1971) *Students' evaluation of a business simulation game as a learning experience.* Baltimore: The Johns Hopkins University (Center for Social Organization of Schools, Report No. 121, December).

Fletcher, J. L. (1971) "Evaluation of learning in two social studies simulation games." *Simulation and Games* 2 (December): 259-286.

Greenblat, C. S. (1973) "Teaching with simulation games: a review of claims and evidence." *Teaching Sociology* 1 (October): 62-83.

Kidder, S. J. and H. E. Aubertine (1972) *Attitude change and number of plays of a social simulation game.* Baltimore: The Johns Hopkins University (Center for Social Organization of Schools, Report No. 145, December).

Lee, R. S. with A. O'Leary (1971) "Attitude and personality effects of a three-day simulation." *Simulation and Games* 2 (December): 309-347.

Livingston, S. A. (1972) *The academic games program: a summary of research results (1967-1972).* Baltimore: The Johns Hopkins University (Center of Social Organization of Schools, Report No. 146, December).

Livingston, S. A. (1973) *Simulation games in the classroom: how important is the post-game discussion?* Baltimore: The Johns Hopkins University (Center for Social Organization of Schools, Report No. 150, February).

Livingston, S. A., G. M. Fennessey, J. S. Coleman, K. J. Edwards, and S. J. Kidder (1973) *The Hopkins games program: final report on seven years of research.* Baltimore: The Johns Hopkins University (Center for Social Organization of Schools, Report No. 155, June)

Taylor, J. L., and R. Walford (1972) *Simulation in the classroom.* Baltimore: Penguin.

PROBLEMS IN THE STRUCTURE AND USE OF EDUCATIONAL SIMULATION*

CHARLES D. ELDER

University of Pennsylvania

During the past decade, we have witnessed an amazing proliferation of simulations designed exclusively or in part for the purposes of education (for example, see Teaching Research, 1967; Klietsch et al., 1969; Zuckerman and Horn, 1970).[1] The scope of these activities now encompasses nearly every academic discipline and extends to all levels of instruction. These trends seem likely to continue, although perhaps at a somewhat abated rate as the novelty wears off and gaming becomes less faddish. In any case, we probably can expect both a growing repertory of educational simulations and new and more extensive applications of gaming techniques.

While the use of instructional simulation has become more pervasive, developments in the area have not been guided by much in the way of explicit pedagogical criteria. Despite more than a decade of experience with these techniques, relatively little attention has been given to the question of what makes for an effective simulation exercise. Interestingly, there has been considerable discussion of the relative merits of simulation vis-a-vis other modes of instruction (for example, see Abt, 1970; Alger, 1963: 150-189; Brown, 1969: 41-43; Coleman, 1966: 3-4, 1967: 69-70; Kraft, 1967: 71-72; Raser, 1969; and Snyder, 1963: 1-23); but little consideration has been given to the relative merits of one simulation versus another. In a sense, the latter question is logically prior to the former. In the absence of any criteria (save one's gut reaction) for comparing educational games, the question of the value of simulation relative to other instructional techniques seems a bit premature and meaningful only in the abstract. More immediately, the lack of comparative criteria means that there is little in the way of useful guidelines to direct future developments in the area and to promote the effective use of existing games.

Reprinted from **Sociology of Education**, Volume 46, No. 3, Summer 1973, pp. 335-354, with the permission of the American Sociological Association.

*Practitioners of simulation and gaming differ in their use of terminology. Some argue that all games are simulations, but not all simulations are games. For these persons, the distinction between simulations and games rests on the type of structure involved and the

Surely, there are lessons to be drawn from the experience that has accumulated over the past decade. These lessons could help in providing the comparative criteria that are needed. Certainly, the variety of existing games and the range of their application is sufficiently broad to afford some perspective on the factors that contribute to their relative effectiveness as educational vehicles. This paper represents an attempt to identify some of these factors and thus to specify some of the major burdens that the designers and users of educational simulations must bear.

The instructional value of a simulation is largely a function of three partially interdependent factors: (1) content, (2) structure, and (3) implementation. Because it is difficult to generalize about content, primary attention will be given here to the problems of the structure and use of educational games. In general, it will be presumed that the designer or user knows what he wants to convey via the exercise. We will be concerned primarily with how that content can be conveyed most effectively. In this context, considerable attention will be given to the role of learning objectives and how they relate to the design and use of gaming exercises.

While it is difficult to talk about learning objectives in the abstract, it is a subject too vital to be overlooked. Simulation is purposive activity; but without explication, such a statement is largely meaningless. In the absence of explicit learning objectives, it is impossible to assess or evaluate either the appropriateness or the effectiveness of the technique.[2] Moreover, without such learning objectives, the whole enterprise is likely to be found an aimless venture by all involved.

STRUCTURAL DESIDERATA IN EDUCATIONAL SIMULATIONS

Broadly speaking, instruction may be divided into two general types: (1) training and teaching specific skills and facts and (2) teaching general principles, concepts, and orientations.[3] Although any given period of instruction might not fit neatly into either one of these categories, as one moves up the educational ladder emphasis tends to shift from the former to the latter type of instruction. Simulations can be and have been used in both areas. For present purposes, the important point is that the difficulties in both the design and use of simulation exercises increase enormously as one moves from training to more generalized education. As a general rule, the more narrow the problem, the more manageable it is; the more manageable the problem, the easier it is to game. As the scope and

purpose for which that structure is used. Others discriminate on the basis of content, arguing that all simulations are games, but not all games are simulations. Thus, it is possible to speak of "simulation games" and "non-simulation games," the later being contrived activities that have no external referents. Here, the terms simulation and game will be used interchangeably to refer to activities designed as analogues of referent processes. These activities will involve student participation and are explicitly intended to provide instruction regarding the referent processes.

complexity of the problem increase, focus becomes more difficult, crucial variables harder to identify, relationships blurred, and our knowledge more sparse.

In training situations, one normally has a rather clear idea of what he wants to convey and fairly firm control over what is conveyed and how well. In short, for most training problems, there are "textbook answers" making criteria for instruction and evaluation less difficult to define and implement. As a result of training, the student is to be able to perform a specific task in a specific manner. The educator's task is simply to find the most effective training vehicle (in terms of performance level, costs, and time—both his and the student's). In some cases, such as the training of prospective pilots, simulations, e.g., flight simulators, have abundant and fairly obvious advantages over alternative modes of training. In other cases, e.g., training police officers in the prevention and control of civil disorder, the potential advantages of simulation are clear but perhaps less compelling from a costs-benefits point of view. In any case, practical training exercises in the form of simulations are fairly easy to develop and use.

As one moves to the teaching of general principles, simulations tend to become less practical exercises and more dynamic case material to supplement other materials and modes of instruction. Problems of design and effective utilization increase as do the critical capabilities and inclinations of the students. There are no quick and easy solutions to these problems. It is possible, however, to identify certain desiderata in the way an educational simulation is structured. To some extent, the desired characteristics or structural properties are mutually competing. Nonetheless, they point to critical areas that must be considered in the basic design and construction of the exercise, if it is to be an effective educational tool. While the burden posed by these structural questions rests heavily on the designer, they are also crucial considerations to be weighed carefully by the potential user in selecting from existing simulation exercises.

Credibility

Just as *validity* is the sine qua non of theory, it is the ultimate and overriding requisite for simulations used for research and policy guidance. In educational simulations, however, it is not so much validity but credibility that is the primary requisite. In fact, to a large extent, *credibility* provides the raison d'etre for educational gaming.

Assuming by validity that we mean a demonstrable pattern of correspondence between simulate or game processes and observable referent processes,[4] it might seem that validity and credibility are perfectly compatible and that validity would be the best assurance of credibility. This, however, is not necessarily the case. Validity and credibility constitute distinct evaluative criteria and may at times actually be contradictory. Validity is predicated upon external standards (viz., the referent or object system being simulated) and is independent of the participants in the synthetic system created by the simulation. Assessments of validity are appropriately made by a detached observer. Credibility, on the other hand, is dependent upon the participants and their perceptions and reactions to the game. Thus, we would heartily agree with Hermann (1967: 219) when he suggests that in instructional simulations the primary evaluate criteria shifts "from the observable universe to the effects on the cognitive and affective systems of those individuals which the operating model is intended to instruct." Credibility is not, however, simply a matter of ex post facto assessment; rather it must be a consideration that is built into the very structure of the exercise.

To insure credibility, the designer and user of an educational simulation must take cognizance of the experience and capabilities of the student-participants. Ideally, a simulation will allow students to draw upon and to extrapolate from their experience or knowledge. The realism and relevance of the exercise thus will be readily apparent. Unless the game allows the student to relate referent experience or knowledge, it will be exceedingly difficult for him to adjust to the synthetic environment created by the simulation. If the purpose of the exercise is to convey poignantly the meaning of "culture shock," this is fine; but otherwise, it will be destructive to the intended learning objectives.

The requisite credibility, of course, is intertwined inextricably with the learning objectives of the exercise. To forcefully and effectively convey a lesson, it may be necessary artifically to isolate certain processes and to exaggerate, simplify, or otherwise distort them. The presumption is that in general one knows the points that he wants to make and is using simulation to convey these points in an experiential way. As long as the contrived situation remains plausible and reasonable within the student's frame of reference, such distortion will not adversely affect the credibility of the game.

Unnecessary abstraction of structures and processes that exist (or are assumed to exist) in the referent or object system, as well as the introduction of game-specific terminology and jargon, will tend to reduce credibility. Both are to be avoided. While simulation may be used to encourage students to view complex systems from an abstract or theoretical point of view (see, Alger, 1963: 162 and Verba, 1964: 494), this is best accomplished by incorporating concepts and variables of general applicability rather than game-specific ones.

Symmetry

What the participants learn from a simulation exercise is necessarily tied to their experience in that simulation. The more diverse and diffuse the experience of the participants, the more heterogeneous and uncertain the learning will be. It is not surprising then that simulations which place different students in structurally different situations and expose them to markedly different stimuli tend to produce mixed results in terms of student learning (for example, see Boocock and Coleman, 1966: 229 and Cohen et al., 1964: 263-264). If a simulation is used as a vehicle to achieve certain learning objectives, it seems only reasonable to provide the students with a common experience consonant with those objectives—common in the sense of coping with the same variables, relationships, parameters, and starting conditions. Simulations which place student-participants in different types of roles and/or in different types of decisional units create an asymmetry of experience that parallels the asymmetry of the game. Many of the more widely known simulations and games suffer to varying extents from such asymmetry (for example, see Guetzkow and Cherryholmes, 1966; Industrial College of the Armed Forces, 1969b; Cohen et al., 1964; Scott, 1966; Thorelli and Graves, 1964; and Weinbaum and Gold, 1969).

Symmetry in the structure of a game facilitates the accomplishment of the learning objectives desired of it. Unless this symmetry is built-in, iteration of the game with a reassignment of roles each cycle is perhaps the only way to achieve some symmetry of experience. To some extent, structural symmetry may fly in the face of the need for realism. However, a number of games have managed to treat this problem quite successfully (examples include Coplin et al., 1971; Summit, 1961; and Industrial College of the Armed Forces, 1969a, 1970). Asymmetry becomes most troublesome in complex, more comprehensive simulations, suggesting that from an educational point of view, simpler and more limited games may be desirable (for example, see Coplin, 1970: 412-426 and Stitelman and Coplin, 1969).

Synchronization

Another problem that often becomes acute, particularly in more complex and comprehensive games, is the problem of *asynchronization*. One of the virtues of simulation is that it allows for the condensation of time. However, in complex games, multiple time frames may be introduced that operate simultaneously but call for different reckoning scales, i.e., there are different scales for converting to real time. A certain amount of asynchronization is probably inevitable, since human participants will always be thinking in real time while the events of the simulation may be preceding along a generally condensed but largely arbitrary game time. However, when interactive elements are added wherein members of participant teams are allowed to interact, negotiate, and bargain with one another on a face-to-face basis, yet another time frame is introduced.

This asynchronization is further compounded by the fact that the distortions in time will not be experienced uniformly by all participants. This gives rise to a confusing type of temporal asymmetry wherein some participants may be operating in one time frame, while others are acting in quite a different time-reckoning system, neither of which may be in tune with the relentless beat of the clock used in the administration of the game. Thus, the problems of asynchronization are particularly acute in any simulation involving sequential dependence and dynamic change geared to a specific time schedule.[5]

The problem of time is indeed a perplexing one; and as Guetzkow (1963: 148) observed some years ago, it is a problem that deserves much closer attention. One thing seems clear, as asynchronization increases, the credibility of the game is diminished and frustration quickly can replace learning. Ideally, a simulation should have a single referent time span and uniform time-reckoning, i.e., equivalent conversion ratios for all simulated activities. This means that in some way the designer and user must attempt to reconcile game emergent time with the structured time frame of the game periods and make sure that both of these are compatible with the real time required for the mental operations of the participants.

This does not mean that condensation of time is impossible or that the simulated time frame must be consonant with real time. Rather, it simply means that the demands placed on the student-participants must be consonant with the temporal constraints imposed by the game. There are a variety of ways of reducing the synchronization problem. Perhaps the most obvious is to restrict the scope of activities being simulated to those that are amenable to uniform time-reckoning. This logic may be extrapolated readily to more complex games that are broader in scope by simply sequentially segmenting activities such that there is uniform time-reckoning with game periods (or segments thereof) but not necessarily across periods. While the credibility of exercises that encompass activities that are not sequentially dependent upon one another may suffer from this strategy, it has been employed quite successfully in simulations that involve activities that have an inherent chronology (examples include the AEROSPACE BUSINESS ENVIRONMENT SIMULATOR and the DEFENSE MANAGEMENT SIMULATION).

Because asynchronizations tend to be most acute in games where there are multiple decision-making units (be they teams or individuals) that are allowed to interact on a face-to-face basis, the problem may be mitigated by reducing the number of teams to two or by proscribing face-to-face interaction. Prohibiting face-to-face communication and bargaining does not necessarily preclude the inclusion of important interactive elements in the exercise. Interaction can be programmed into the structure of the game by making the results of decisions made by different participants interdependent (see Barton, 1970: 56; the MANAGEMENT DECISIONMAKING EXERCISE is an example of this type of simulation). This procedure allows (and may even promote) tacit bargaining and communication, while constraining the degree of asynchronization that can develop.

Manageability

That simulations can place students in relatively complex environments is surely an advantage. Such complexity can afford challenging experience that fosters an appreciation of the interdependence of various processes and the interface of multiple problems. Because of this, simulation can provide a more integrated or holistic view of phenomena than is generally possible through other educational techniques. Such complexity will be counter-productive, however, if it is more an artifact of the game than a reflection of the phenomenal complexity being simulated. A good example of such complexity (or perhaps it should be called complication) is the proliferation of forms that characterizes so many simulations. The student-participant literally can be overwhelmed by paper and paper work that bears little relevance to the main objectives of the game. Understandably, this can lead to great frustration.

Elaborate and complex games also can inhibit learning by obscuring cause and effect relationships. If it is difficult or impossible for the student to see the impact of the decisions he makes or if this impact is remote or miniscule in comparison to that of other variables over which he has no control, the feedback that he gets is unlikely to have much learning impact. This is one of the problems that has limited the value of TEMPER as an instructional exercise (see Industrial College of the Armed Forces, 1968). Similarly, learning may be impeded, if the student is forced to make so many decisions and to cope with so many facets of a game simultaneously that he has no time to analyze, deliberate upon, and think through his decisions. His behavior in such circumstances will be predictably nonsystematic, and his experience largely unrewarding.

The problem of *manageability* essentially means that the task that a student is asked to perform at any point in time must be limited but consequential to the outcome of the exercise (see Abt, 1970: 112-118). The factors that the student is to consider and the decisions he is to make should bear a direct relationship to the learning objectives established for the exercise. Busy work and tangential activity dilute the learning impact and obscure the purpose of participation.

If the learning objectives established for a game are multiple and sufficiently ambitious to require an elaborate and fairly complex environment, manageability perhaps can be achieved most readily through a progressively developing simulation. Here new elements are introduced and the environment changed or enriched as the game proceeds. This allows the student to see the effect of increasing complexity while preserving specific relational lessons. It also enables him to cope more effectively with complexity through progressive mastery of component aspects of the simulated environment.

Progressively developing simulations may take one of two forms: (1) one built up from a simplified situation which is gradually compounded by additional elements that add new dimensions to the original problem (see Sachs, 1970: 165) or (2) one which is segmented and sequenced so that a problem is dealt with in a piecemeal fashion in a manner consonant with the life cycle of the problem.[6] The decomposition and re-integration of situations through simulations of the first variety are somewhat artifical, but nonetheless this can be a potent way of conveying information. Simulations of the second variety may be segmented in a more realistic way but presume that the problem being simulated has itself a natural life cycle, e.g., the development and production of a specific product.

Another way in which a simulation exercise can be made more manageable (and thus more comprehensible to the student) is through provision for student interrogation of the model, i.e., exploration of the synthetic environment created by the simulation.[7] Such interrogation will not only enable the student to make reasonable projections of the consequences of his decisions, but more importantly it will allow him to find a degree of predictability in the synthetic environment that is essential for meaningful action.

On-line exploration (i.e., exploration occurring during the game) of the simulated environment also speaks to another perennial problem in instructional simulation; namely, the problem of acclimating students to the artificial environment created by the simulation. The time spent in getting the student attuned to the environment in which he must operate is a kind of overhead and is largely a waste in terms of specific learning objectives. This time can be reduced by providing the student with additional vehicles for familiarizing himself with the simulated environment and allowing him to explore it further as the game proceeds.

The primary purpose of such interrogative vehicles should be to provide the participant with only that information he reasonably could expect to be available in a referent type of situation. While most simulations are designed as exercises in decision-making under uncertainty, that uncertainty should be bounded. Insofar as the uncertainty under which the student must operate is merely an artifact of the game, it is destructive to its educational purpose.

In most decision-making situations, search activities are an important part of the overall process. Providing students with the capability of interrogating the model and exploring the game environment thus has the additional advantage of allowing this vital aspect of the decision-making process to be incorporated into the structure of the simulation. Thus, it can serve to promote the analytic and decisional skills of the student-participants by encouraging planning, deliberation, and anticipation.

Ease of Administration

Just as simulations can become so complicated and involved that they are unmanageable for the participants, they can become impossible to administer. Unless the game administrator can keep up with and coordinate the flow of events, the exercise will disintegrate, the students will become frustrated, and the game will degenerate into an exercise in futility. The *demands on the administrator* must be minimized. Generally, the more manageable a game from the participant's point of view, the easier it is to administer. Many of the things that facilitate manageability also will facilitate administration of the game, e.g., the fewer the game forms the participants have to work with, the fewer the administrator will have to worry about. In addition, the use of computers to perform involved, tedious, or lengthy calculations and collation will aid the administrator greatly. The use of remote, desk-type time-sharing terminals is particularly advantageous. They not only facilitate the effective and accurate administration of the game through quick and almost continuous access, but they also make feedback processes to the students more rapid and thus help to insure the continuity of the exercise. In general, the use of computers allows the simulation to be richer and more "realistic" (i.e., more credible) without placing additional or unnecessary burdens on either the students or the administrating faculty.

More Continuous Decisional and Feedback Processes

Reflecting in part the state of the art, expediency, and the facilities that were available when educational simulation first began to be used, most simulation exercises are broken into arbitrary playing periods with actual decisions coming at a single point in time. Such *point decisions and point feedback* often belie the dynamics of referent processes, detract from the credibility of the exercise, and contribute to the asynchronization and unmanageability that is so frustrating to the student-participants. As the end of a period approaches, there typically is a mad rush to meet arbitrary deadlines. Considerations of priorities, logical sequence, and relationships, along with any deliberative processes, give way to the relentless demands of the game clock. While time-pressure may be a variable that one wants his students to come to appreciate, it is doubtful that arbitrary deadlines will foster an appreciation of anything but the "non-reality" of the game or simulation.

Insofar as possible, deadlines should flow logically from the sequence of activities or events being simulated. Decisional and feedback processes should be both logical and continuous. The student should not be left to operate in an informational void and forced to make a series of decisions without the benefit of the kind of prior information that a person in an analogous referent situation reasonably could expect to have. Yet point decisions and point feedback tend to place the student in this very position.

To some extent, the problem is mitigated by provision for student interrogation and exploration of the model. However, this is only a step in the right direction. With the computer capabilities currently available, there is no reason why we should be satisfied with the discrete, punctuated time-slicing and haltering decision-making and feedback processes that always have characterized educational gaming. More continuous simulations, wherein the participants make decisions and get feedback more like they could expect in a referent situation, would allow us to do greater justice to the avowed claim of revealing to students the dynamics of the referent process or system.

EFFECTIVE USE OF EDUCATIONAL SIMULATIONS AND GAMES

In addition to the variety of structural problems that can detract from the effectiveness of educational simulations, there is a whole array of problems that the user can create for himself. By resorting to a variety of expedients, the most enthusiastic advocates of instructional simulation at times have imperiled the success of their own cause.

Innovation without Oversell

The inertia present in any institution makes the introduction of anything new and unfamiliar difficult. This problem is particularly acute in the case of instructional simulation, especially if computers are involved. Considerable skepticism continues to surround the use of simulation to say nothing of the problems of fitting new costs into already tight budgets. Confronted with these challenges, there is perhaps a natural tendency for the proponents of innovation in this area to resort to *exaggerated claims and promises.* Inordinate claims and promises, however, can breed disillusionment (among both students and administrative authorities) and can come back to haunt, if not destroy, a genuinely productive but less than spectacular effort.

Simulation is not (and should not be promoted or embraced as) a panacea for the ills of education. It is simply another method in the repertory of methodologies available to the educator. There is little hard evidence to suggest that, in general, educational gaming is any more or less effective than other teaching methodologies, all of which have their own strengths and weaknesses (see Robinson et al., 1966: 53-65; Boocock and Coleman, 1966: 215-236; Cherryholmes, 1967: 4-7; Boocock, 1968: 107-133; Baker, 1968: 135-142; Wing, 1968: 155-165; Fletcher, 1971: 259-286; and Lee and O'Leary, 1971: 309-347). Impressionistic evidence strongly does suggest that for certain purposes, simulation may be more useful than other educational methods (see Abt, 1970: 61-78); but by the same token, there are undoubtedly many instances where it is less appropriate.

The decision on which method will be employed is one the educator cannot escape. It must rest on considerations of time, facilities, staff assistance, and most importantly, the objectives of the instruction. In most circumstances, a mixed strategy probably is the most appropriate (Deutsch and Senghass, 1970: 37-40). Not least of the potential advantages of educational simulation is the interest and enthusiasm it can generate among students. This can do much to relieve the tedium often associated with more conventional modes of instruction.

Accurate Estimation of the Costs Involved

Just as advocates of simulation at times have been guilty of exaggerated claims and promises, they also have tended to *underestimate the time, effort, and material costs* required for the effective use of these techniques (see Shubik, 1968: 629-660; Cohen, 1962: 367-380). This again is to invite disillusionment. Interestingly, even those persons and institutions that conduct instructional games regularly have tended to play down these requirements. A smoothly operated and effectively administered simulation may belie the enormous amounts of time, energy, and effort that must be devoted to its preparation and conduct. While perhaps a tribute to the modesty and effectiveness of the user, it can be most deceiving to the unwary observer.

The costs of simulation in both time and money are not easy to calculate. Both types of costs have several dimensions. With respect to direct dollar expenditures, things like the costs of publishing game materials and the costs of computer time readily come to mind. This sort of accounting can be misleading, however. A game that can be conducted at one institution for a relatively modest sum might cost another many times that, owing to the disparity in assistance and facilities available to one user and not the other or to the costs of converting a computer program to make it compatible with a different machine.

With respect to time, one's cost calculations must include not only the time required to conduct the exercise but also the enormous amounts of time required to prepare the materials and set up the exercise, the time spent in learning to administer the game (or to train others to do it), the time spent to orient the student-participants, and the time the students themselves have to spend simply in learning the mechanics of the game. When all of these time expenditures are added up, we are talking about a considerable number of man-hours that must be put in before even the simplest game can start.

In mentioning these cost considerations, my intent is not to discourage the use of educational simulation and gaming. These are simply problems that must be reckoned with. Simulation and gaming can be an expensive enterprise, but there are ways of economizing in terms of both time and money. Careful design and discriminating use of simulation exercises can reduce the preparation time required of both the students and administering faculty. Mutual cooperation among institutions and the sharing of facilities also can reduce costs and staff requirements.

Integration into Broader Educational Experience

While novelty is perhaps one of the most appealing characteristics of simulation, it should not be viewed as a complete and independent form of instruction. *It cannot stand alone.* Simulation may be used to summarize, integrate, and illuminate previous instruction or to provide a common experience for reference in subsequent instruction. Without other instruction, however, participation in a simulation is an incomplete learning experience. For effective utilization, simulation must be meshed with other modes of instruction and integrated into the student's larger educational experience. Simulation is most effective when it is an integral part of a multifaceted and mutually supporting educational strategy.[8]

A number of implications flow from this. First, adjustments must be made not only in a proposed simulation but also in the curriculum or course that it is intended to support. Second, the enthusiastic support of other faculty members is likely to be needed. These persons should be thoroughly familiar with the exercise and have a firm understanding of the formal model or core assumptions upon which the exercise is predicated. Third, both the user and cooperating faculty must be prepared to point out the lessons to be drawn from the synthetic experience provided through the simulation and to show how these lessons may be linked to other instruction. Facts, be they real or synthetic, do not speak for themselves. Thus, without assistance, a student may fail to perceive or appreciate the larger implications of his experience. Clearly articulated learning objectives, a good orientation, and a thorough post-game review will aid immeasurably in getting the most out of a simulation, particularly when it is linked to other instruction.

Appropriate Substantive Qualifications and Precautions

Just as students are sometimes inclined to mistake written words for truth, there is sometimes a tendency for the neophyte to assume that simulations, particularly computer-assisted ones, embody proven principles and sound knowledge. While this may be a tribute to the credibility of an exercise, it is a sadly mistaken assumption. Most simulations, particularly those designed to simulate broad social or economic processes, often are predicated on little more than informed hunches. Moreover, it is often difficult, especially in large-scale games, to control the sort of impressions that the students extract from their experience. Clearly, one can mislearn as well as learn from an instructional game (Smoker, 1972: 322). Insofar as a game reinforces questionable or erroneous assumptions or serves as a "mechanization of popular mythology adding to it a dollop of scientism" (Shubik, 1968: 635), it is clearly counter-productive to the educational enterprise.

To avoid these dangers, it is important that faculty and students alike appreciate that *the simulation of phenomena does not transform ignorance into knowledge.* To foster this appreciation, the students should be invited to study and discuss the formal model or assumptions upon which a game is based (Deutsch and Senghaas, 1970: 39-40). If at all possible, they should be encouraged to grapple with the problems of model design and improvement. Perhaps this can best be accomplished by having them design and exercise small models of their own (Sachs, 1970: 164). However it is accomplished, it is vital that students become aware that simulation models are neither substantively neutral nor necessarily based on more than highly tentative assertions.

Definition and Implementation of Specific
Learning Objectives

Generally speaking, educational simulations are not conducted simply for the sake of simulating. A residual objective may be to familarize students with the technique, but the primary objectives are likely to be more substantive. In articulating these objectives, a statement of the general learning potentials attributable to the technique is a frequent but unsatisfactory substitute for a statement of the specific purposes of participation in the exercise.

The *learning objectives* associated with a simulation exercise should be sufficiently clear and specific to provide meaningful guidance to the student-participants. These objectives should be made clear at the outset and dwelt upon during the initial orientation. The post-game review should be structured around these objectives, which may be discussed in light of the simulation experience. Without the discipline imposed by such specific learning objectives, the results of the exercise are likely to be nebulous and disquieting no matter how well it is structured and administered.

STUDENT-ORIENTED LEARNING OBJECTIVES[9]

Before asking how a student learns from an educational simulation, we must be able to answer the question of what is to be learned. In fact, this question is logically prior to the questions of structural design (or selection) of a gaming exercise. Properly defined learning objectives can provide valuable guidance for the designer and prospective users of an instructional game. They also aid students as they prepare for and participate in the exercise. They facilitate the administration of the game and provide guidelines for the post-game review. Moreover, good learning objectives also serve as criteria for evaluating both student performance and the overall adequacy of the exercise, pointing to fairly specific ways in which both can be improved.

Insofar as possible, student or participant-centered objectives seem desirable. These objectives will describe what the student is to learn, not the scope of the phenomena being simulated nor what the simulation looks like. An objective will tend to be more meaningful if it provides an answer to the question: What am I to know, understand, appreciate, or be able to do as a result of having participated in this exercise? To describe the content of what the student is to learn as specifically as possible, multiple objectives probably will be required for most instructional games.

In addition to indicating what is to be learned, i.e., the content, an objective should indicate the type of learning (cognitive or noncognitive) and the level of competence that is expected. With respect to cognitive learning, any of three levels of competence may be sought: (1) general familiarity with a subject, (2) understanding or knowing specific substantive facts or principles, or (3) the ability to apply certain knowledge or to perform a specific task, e.g., be able to use trade-off curves or set up a linear programming problem. The second type of learning, noncognitive, relates to conceptions of or attitudes toward something. A learning objective of this variety indicates that the simulation is intended to help the student acquire an appreciation of a problem or problem area, e.g., the role pressures of a political decision maker, the complexity of an issue, or an adversary's perspective on a conflict.

This is only one of many ways of defining learning objectives. The most important thing is that they be defined and explicitly articulated. The discipline they impose upon both the designer and user of instructional games is likely to have its payoffs in student learning.

SUMMARY AND CONCLUSIONS

A number of problems in the structure and use of educational simulation have been identified and reviewed. I have suggested that credibility rather than validity is the prime requisite for an effective instructional simulation. Credibility, however, must not be sought at the expense of symmetry in the structure of a game. Asymmetry can diffuse the learning experience to the point that much of the instructional value of the game is lost. The inadvertent introduction of multiple time frames also can dilute the value of a game. It can lead to a severe asynchronization of activities, which detracts from the credibility of the exercise and tends to breed frustration.

Perhaps above all else, a game must be manageable from the student's point of view. Unnecessary busy work and cumbersome activities should be designed out. Ease of administration is also an imperative. The more difficult a simulation is to administer, the greater the probability of distracting, if not devastating, administrative errors and lapses of control.

For the sake of both credibility and manageability, more continuous decisional and feedback processes seem desirable. The artificiality of point decisions and point feedback can give rise to the simultaneous problems of an informational void and decisional overload. As a result, many decisions may be more artifacts of flaws in the structure of the game than the products of processes analogous to referent processes.

Moving from problems that are essentially structural in nature to those that arise more from the way simulations are introduced and used, a number of potential pitfalls were identified and suggestions made for avoiding them. In introducing a simulation exercise, it was suggested that care be taken to foster realistic expectations regarding the results it is likely to produce. These results are not likely to be dramatic and oversell invites dissatisfaction. Similarly, in entertaining the use of a simulation exercise, care must be taken not to underestimate the time and material costs it is likely to involve. The value of a game may be lost, if one is ill-prepared to conduct it.

It was further argued that a simulation exercise should not be viewed in isolation from the student's larger educational experience. Ideally, the objectives defined for an instructional simulation should mesh with and reinforce those defined for other modes of instruction. It also was suggested that the student be cautioned not to mistake hypothesis for fact and that he be made aware of the tentativeness of many of the lessons that a game may convey.

Finally, it was suggested that the user of instructional simulation is obliged to provide guidance to student-participants in the form of clear learning objectives.

Insofar as possible, these objectives should be student-centered and identify both the kind and level of learning that is sought.

The problems I have discussed pose a heavy burden for the designer and user of educational simulations. Although these demands may appear large in comparison to other modes of instruction, I suspect it is a misleading comparison. The burdens are perhaps clearer but probably no more onerous.

In concluding, at least four general implications seem to emerge from my analysis of problems in the structure and use of instructional simulation. First, the early hope of general purpose simulations that could adequately serve the purposes of education, substantive research on referent processes, and policy analysis was perhaps overly optimistic. While a number of games exist that have been used for all of these purposes, it is becoming increasingly clear that the best instructional simulations are those specifically designed for that purpose. Second and similarly, the prospects for general participant educational simulations designed to provide a valuable learning experience for participants regardless of their level of competence in a field appears limited. Once again, there are exercises that allegedly do this, allowing more sophisticated participants to find ever more subtle lessons in the game. However, considerations of credibility and manageability, to say nothing of content and associated learning objectives, suggest the need for simulations targeted at different levels of student competence. Third, despite the glamour and snob appeal of more complex and comprehensive games, small-scale simulations of limited scope seem likely to play a more prominent role in the future. Fourth and finally, insofar as large-scale simulations are required to meet certain learning objectives, the use of the computer to reduce the burdens of participation and administration is likely to be found essential. The increasing availability and convenience of relatively economical time-share facilities bodes well in this regard.

NOTES

1. The editors of Social Education, which provides fairly regular information on simulation and gaming, reported in 1969 that they received more articles on this subject than on any other (33, February, 1969: 176). New developments also are reported regularly in Simulation and Games, a journal created explicitly for this purpose.

2. This problem is not peculiar to simulation and gaming. It is a problem that has received insufficient attention with respect to almost all educational techniques (e.g., lectures, discussion, case studies). Perhaps this is one reason why research has been so inconclusive regarding what constitutes a good teacher, method, etc. and what the relative effectiveness of different methods is in different situations and with different subject matter. See Boocock, 1966: 1-45.

3. This distinction roughly reflects the distinction Boocock makes between "cognitive learning"—the acquistion of some new body of information—and "noncognitive learning"—a shift in values, attitudes, interests, or motivation (1966: 2).

4. The meaning of validity in simulation and gaming has received extensive attention. For example, see Guetzkow, 1968: 202-269; Hermann, 1967: 216-231; Powell, 1971; Raser, 1969: 137-160; and Smoker, 1969: 7-13.

Most definitions center on the degree of correspondence between referent and simulate processes and how these are to be assessed. Isomorphism, or a one-to-one correspondence, is not necessarily required for the validity criterion to be met. Hermann (1967: 219) suggests that what I am calling credibility may be considered one of many validity critieria, the appropriateness of any one being determined by the purpose for which the simulation is constructed. Hermann further identifies "face validity" as a possible criterion. Here surface or initial impressions of a game's "realism" are used to assess validity. It seems preferable, however, to distinguish validity and credibility, since they are predicated on different standards, the former supposedly on objective assessments of a detached observer, the latter on the subjective reactions of participants. This seems to be the tack taken by Barton (1970: 58) when he speaks of "verisimilitude" in man-machine simulations and by Abt (1970: 114-115) when he talks of "realism" in serious games. Also see Kardatzke (1969: 179-180).

5. These problems are fairly common in simulations like the WORLD POLITICS SIMULATION (Industrial College of the Armed Forces, 1969b).

6. The DEFENSE MANAGEMENT SIMULATION is structured in this way (Industrial College of the Armed Forces, 1970).

7. The MANAGEMENT DECISIONMAKING EXERCISE (Industrial College of the Armed Forces, 1969a) provides this capability through remote, time-share computer facilities. While computers may be used, interrogation need not be via a computer. It may be accomplished through written or verbal messages between participants and game administrators.

8. The use of the AEROSPACE BUSINESS ENVIRONMENT SIMULATOR (Summit, 1961) in courses in the School of Business at the University of Utah provides a compelling example of the use of this strategy.

9. I am indebted to Dr. Robert C. Andringa for many of the arguments contained within this section.

REFERENCES

Abt, Clark C.
    1970   Serious Games. New York: Viking.

Alger, Chadwick F.
    1963   "Use of the Inter-Nation Simulation in Undergraduate Teaching." Pp. 150-189 in Harold Guetzkow et al., Simulation in International Relations. Englewood Cliffs: Prentice-Hall.

Baker, Eugene H.
    1968   "A Pre-Civil War Simulation for Teaching American History." Pp. 135-142 in Sarane Boocock and E. O. Schild (eds.) Simulation Games in Learning. Beverly Hills: Sage.

Barton, Richard F.
    1970   A Primer on Simulation and Gaming. Englewood Cliffs: Prentice-Hall.

Boocock, Sarane S.
    1966   "Toward a Sociology of Learning." Sociology of Education 39 (Winter): 1-45.
    1968   "An Experimental Study of the Learning Effects of Two Games with Simulated Environments." Pp. 107-133 in Sarane Boocock and E. O. Schild (eds.) Simulation Games in Learning. Beverly Hills: Sage.

Boocock, Sarane S., and James S. Coleman.
    1966   "Games with Simulated Environments in Learning." Sociology of Education 39 (Summer): 215-236.

Brown, Fred R.
    1969   "Some Questions to Ponder." Perspectives in Defense Management (June): 41-43.

Cherryholmes, Cleo H.
    1966   "Some Current Research on Effectiveness of Educational Simulations." American Behavioral Scientist 10 (October): 4-7.

Cohen, Bernard C.
    1962   "Political Gaming in the Classroom." Journal of Politics 24 (May): 367-380.

Cohen, Kalman J., William R. Dill, Alfred A. Kuehn, and Peter R. Winters.
    1964   The Carnegie Tech Management Game. Homewood: Richard D. Irwin.

Coleman, James S.
    1966   "In Defense of Games." American Behavioral Scientist 10 (October): 3-4.
    1967   "Learning through Games." National Educational Association Journal 56 (January): 69-70.

Coplin, William D.
    1970   "The State System Exercise." International Studies Quarterly 14 (December): 412-426.

Coplin, William D., Michael K. O'Leary, and Stephen L. Mills.
    1971   PRINCE (Programmed International Computer Environment). Syracuse: International Relations Undergraduate Education Project, Syracuse University.

Deutsch, Karl W., and Dieter Senghaas.
    1970   "Simulation in International Politics." Perspective in Defense Management (March): 37-40.

Fletcher, Jerry L.
    1971   "Evaluation of Learning in Two Social Studies Simulation Games." Simulation and Games 2 (September): 259-286.

Guetzkow, Harold.
    1963   "Structured Programs and Their Relation to Free Activity Within the Inter-Nation Simulation." Pp. 103-149 in Harold Guetzkow et al., Simulation in International Relations. Englewood Cliffs: Prentice-Hall.
    1968   "Some Correspondences between Simulations and 'Realities' in International Relations." Pp. 202-269 in Morton Kaplan (ed.) New Approaches to International Relations. New York: St. Martin's Press.

Guetzkow, Harold, and Cleo Cherryholmes.
    1966   Inter-Nation Simulation Kit. Chicago: Science Research Associates.

Hermann, Charles F.
    1967   "Validation Problems in Games and Simulations with Special Reference to Models of International Politics." Behavioral Science 12 (May): 216-231.

Industrial College of the Armed Forces.
    1968   TEMPER (Technological, Economic, Military, and Political Evaluation Routine). Washington: ICAF.

1969a Management Decisionmaking Exercise. Washington: ICAF.
1969b World Politics Simulation. Washington: ICAF.
1970 Defense Management Simulation. Washington: ICAF.

Kardatzke, Howard.
1969 "Simulation Games in the Social Studies." Social Education 33 (February): 179-180.

Klietsch, Ronald et al.
1969 The Directory of Educational Simulations. Learning Games and Didactic Units. St. Paul: Instructional Simulations, Inc.

Kraft, Ivor.
1967 "Pedagogical Futility in Fun and Games?" National Education Association Journal 56 (January): 71-72.

Lee, Robert S., and Arlene O'Leary.
1971 "Attitude and Personality Effects of a Three-Day Simulation." Simulation and Games 2 (September): 309-347.

Powell, Charles A.
1971 "Validity Issues in Complex Experimentation." Los Angeles: University of Southern California.

Raser, John R.
1969 Simulation and Society. Boston: Allyn and Bacon.

Robinson, James A., Lee F. Anderson, Margaret G. Hermann, and Richard C. Snyder.
1966 "Teaching with Inter-Nation Simulation and Case Studies." American Political Science Review 60 (March): 53-65.

Sachs, Stephen M
1970 "The Use and Limits of Simulation Models in Teaching Social Science and History." Social Studies 61 (April): 163-167.

Scott, Andrew M., William A. Lucas, and Trudi Lucas.
1966 Simulation and National Development. New York: Wiley.

Shubik, Martin.
1968 "Gaming: Costs and Facilities." Management Science 14 (July): 629-660.

Smoker, Paul.
1969 "Social Research for Social Anticipation." American Behavioral Scientist 12 (July-August): 7-13.
1972 "International Relations Simulations." Pp. 296-339 in Harold Guetzkow, Philip Kotler, and Randall L. Schultz (eds.) Simulation in Social and Administrative Science. Englewood Cliffs: Prentice-Hall.

Snyder, Richard C.
1963 "Some Perspectives on the Use of Experimental Techniques in the Study of International Relations." Pp. 1-23 in Harold Guetzkow et al., Simulation in International Relations. Englewood Cliffs: Prentice-Hall.

Stitelman, Leonard, and William D. Coplin.
    1969 American Government Simulations Series. Chicago: Science Research Associates.

Summit, Roger K.
    1961 Aerospace Business Environment Simulator. Sunnyvale: Lockheed Missiles and Space Company.

Teaching Research.
    1967 Instructional Uses of Simulation. Portland: Oregon State System of Higher Education.

Thorelli, Hans, and Robert Graves.
    1964 International Operations Simulation. New York: Free Press.

Verba, Sidney.
    1964 "Simulation, Reality, and Theory in International Relations." World Politics 16 (April): 490-520.

Weinbaum, Marvin G., and Louis H. Gold.
    1969 Presidential Election. New York: Holt, Rinehart and Winston.

Wing, Richard L.
    1968 "Two Computer-Based Economics Games for Sixth Graders." Pp. 155-165 in Sarane Boocock and E. O. Schild (eds.) Simulation Games in Learning. Beverly Hills: Sage.

Zuckerman, David, and Robert Horn.
    1970 The Guide to Simulation Games for Education and Training. Cambridge: Information Resources.

Part IV

OTHER ARENAS AND APPLICATIONS

PUBLIC POLICY APPLICATIONS:
Using Gaming-Simulations
for Problem Exploration and Decision-Making

RICHARD D. DUKE

THREE USES FOR GAMING-SIMULATION

T he primary use of gaming-simulation for the past quarter-century has been in an academic context. Immediately following World War II, various schools of business administration adopted the technique (borrowing it largely from the war gamers) and incorporated it into their curricula. Significantly, many special uses were employed for the "adult-education" of businessmen in the field, in a variety of short-course formats. Both on and off campus, "real-world" business managers were given the opportunity to run some mythical corporation. Much in vogue during this era were arguments, pro and con, about possible effects that might be transferred from the game to the operation of a real firm. Since some of the business games were run for well established businessmen under very exotic conditions, the question of carryover was not entirely academic. In the paper by Cohen and Rhenman, presented in the preceding section of this volume, some of these questions have been explored in depth.

Schools of education and then of the various social sciences were close on the heels of the business schools in their adoption of use of gaming-simulation, resulting in a profusion of games, covering a wide variety of subject matter and technique. Their range of sophistication is enormous, and their correspondence to reality ranges from the purely abstract to the iconic. There are at least three published catalogs which attempt to document these materials.

Inevitably, these gaming-simulations, like the earlier business games, found their way helter-skelter into nonacademic use (although almost without exception, the gaming instruments were originally prepared in some academic context). And again, questions were raised about their potential carryover effects.

There appear to be three central circumstances in which gaming-simulation can be successfully employed:

(1)   for pedagogic purposes;

(2)   as a communications device by an interdisciplinary team engaged in sophisticated research;

(3)   for some direct public policy purpose—specifically to influence the decision-making of the voter, the civil servant, or the elected and/or appointed official.

Of these three purposes, only the first two have been pursued very assiduously. Attempts in the third area have been largely hit-or-miss forays spawned by the opportunity of the moment, using the resources at hand, with little attempt at measuring the impact of the effort. The question remains: Can gaming-simulation be effectively employed in actual public policy applications?

A positive response would seem to be justified; the historic use of gaming-simulation by the military is a case in point, as is the use of gaming-simulation in certain foreign policy applications. But can gaming-simulation be employed in less exotic settings? Both military and diplomatic considerations are sufficient to justify expensive technique. As the scale becomes less global, the problem more specific, the audience less sophisticated and more fleeting . . . is there a pragmatic, day-to-day potential for the technique?

Probably so, at least in the management of our urban centers, and these, of course, have a profound impact on the daily life of most Americans today. The Meier-Duke article examines urban complexity and sets the stage for the effective use of gaming/simulation in planning for urban regions.

THE AUDIENCES DEFINED

There are three applications in the urban arena for the productive use of games: first, directly with the voter; second, with the hired, appointed, or elected individuals who act for the citizen in the management of urban affairs; and third, with those individuals whose private actions have a profound impact on urban development—bankers, real estate men, large landholders, and other businessmen. It is useful to distinguish among the three as potential targets for the gaming technique because the circumstances of application vary considerably.

The citizens (for brief definitional purposes defined here as those eligible to vote) viewed as a gaming target may exist as individuals, members of loose coalitions, or members of a permanent or semi-permanent organization dedicated toward specified goals. However they may be found, they are generally motivated toward questions of public policy on an *issue* or *problem* basis. As the problem or issue emerges, coalitions form in response; as this issue is resolved, the coalition dissolves, to be replaced by another formed to meet the next crisis. The membership overlaps from group to group, and some stalwarts are to be found as more or less permanent fixtures in the process of societal dialogue. But in large part the citizen is a fleeting target for the gaming-simulation technique.

This implies that gaming-simulation will be successfully employed with the citizen only if certain conditions can be met:

—simple game devices are employed; emphasis will be on frame games, where the basic format can be assimilated, but content can be readily changed;

—the instrument is used in narrow, problem-specific applications; this implies new material being developed issue by issue;

—the gaming instruments are simple, requiring neither exotic paraphernalia nor trained personnel to either develop or to run the exercises;

—the central theme of the gaming exercise will be to convey the "big picture" at least to the extent that a correlation can be made for the citizen regarding the issue at hand and the central factors impinging on a knowledgeable decision.

If gaming-simulation is effectively employed, it will increase the power of the citizens, making them more knowledgeably effective in public affairs and less dependent on the actions of either their appointed/elected/hired representatives or the decision makers of the business community.

The second target group—the hired/elected/appointed—requires a different mode of operation for the successful application of gaming-simulation. This group is much smaller in number, more specifically defined in terms of responsibility and perspective (role) in community affairs, and considerably more knowledgeable than the average citizen, at least about their specific area of interest. Further, although their mode of decision-making is basically issue to issue, they are charged with a broader responsibility in at least two dimensions: they are obliged to consider not only a given issue or problem, but an array of these as appropriate at a given moment; further, they are obliged to consider the impact of any given decision through time (e.g., the impact on the municipal budget ten years from now. Given these considerations, the gaming-simulation vehicle(s) employed must meet several constraints:

—they must be reasonably complex; they must be sophisticated enough to meet the linkage of many dimensions of urban life;

—they must be valid; they must include or provide reasonable access to, specific data sources as required;

—they must provide for the perspective of the major urban roles, either by accommodating real-world decision makers directly in the game and/or through the inclusion of stereotypes in the gaming model;

—they must permit the thoughtful review of a given issue or problem in terms of both its own merits and its impact on other issues.

If gaming-simulation is effectively employed to increase communication among elected/appointed/hired urban decision makers, at least three advantages should occur: the quality of decisions on individual issues should be improved; the impact of individual decisions on future time should be less destructive; and, finally, the inevitable conflict between departments should be minimized or at least rationalized.

Finally, the major decision makers in the private sector are a different audience. They are, first of all, profit-oriented. They are reasonably expert in their area of interest. They are narrowly oriented toward a particular project or interest. They cannot be expected to respond to "community interest" or considerations beyond the scope of their private interest, except perhaps, to pay lip service. Gaming-simulation, to be effective with this group, must be:

—problem-specific; each gaming situation must explore a particular circumstance where tangible private interests are at stake;

—sophisticated in technique; the presentation must be at a level appropriate to an expert audience;

—focused on tangible considerations; municipal, law, the power of the electorate, administrative power must be stressed rather than ethereal notions of the public good;

—designed to permit an exchange, a dialogue, between experts representing both private and public interests.

Gaming-simulation of this style may be employed to obtain a better integration of projects by trading off private self-interest against the public interest and the clout of an informed electorate.

SOME RECENT APPLICATIONS

Gaming-simulation efforts to date for municipal policy purpose can be conveniently divided into the use of simple and complex exercises. For this discussion, "complex" implies the use of a computer.

To date, "simple" games devised for academic purposes have been employed for use with citizen groups. These include, but certainly are not limited to, a variety of field applications for STARPOWER, CLUG, METROPOLIS, POLICY NEGOTIATIONS, and many others. The use of these games has generally been with citizen groups with some degree of cohesion (e.g., League of Women Voters) and in a mode that comes closer to "adult education" than to a precise use of gaming for bona fide public policy purposes (i.e., actual decision-making). Some examples of such use are described in the article by Greenblat in the readings section.

A second use of simple games has been the development of special problem-specific games for explicit public policy use. A variety of public agencies have developed such games. One of the more recent examples is WALRUS, developed by Allan Feldt for use in Traverse City, Michigan, to improve public understanding of the activities conducted under the Michigan Sea Grant Program. Although quite successful, this game does not go the full route on the continuum of public education-public policy game use. That is, WALRUS is probably most effective in conveying a general situation; it is of limited value to decision makers in dealing with a variety of specific issues which might emerge in the region.

A third context for "simple" games is programmatic use. This implies the deliberate effort to employ gaming as one tool for public dialogue on an emergent issue basis. The only example known to me is in the Monterey Bay region of California (although some of the materials being prepared are being used elsewhere). The Association of Monterey Bay Area Governments and the Council of Monterey Bay, under funding from the National Science Foundation, have a project to develop gaming for programmatic citizen use. A series of frame-games has been developed (IMPASSE?; AT ISSUE!; and CONCEPTUAL MAP-PING . . .) and is currently being tested.

Complex (computer) games are less in evidence, partly because they are more cumbersome to use, but largely because of the expense involved. There are several such exercises in existence and fairly widespread use. In almost all applications, they are prototypical; their primary purpose (in nonacademic usage) is to illustrate the ultimate potential of the technique.

METRO-APEX

METRO-APEX is one well-known example commissioned by the Tri-County Regional Planning Commission of Lansing, Michigan, under Housing and Home Finance Administration funding; it was originally intended as a device for familiarizing the various planning commissioners with the dynamics of the region. Originally completed in 1968, it has subsequently undergone more or less continuous revision and is now in use at many American universities. METRO-APEX has served as a prototype to researchers in several other countries (the game has been translated from English and now "speaks" German, French, and Spanish). While METRO-APEX is prototypical of sophisticated gaming for public policy purpose, its use to date has not generally been in a "real-world" context.

METRO-APEX serves as a vehicle for both urban management training and urban policy research. It is a micro-environment for decision makers in which an abstracted, simulated metropolitan area is represented via gamed roles and in-computer models and data systems. There are a variety of uses for METRO-APEX: teaching of professional urbanists, social scientists, planners, and administrators; research on a wide range of computer-oriented urban models calibrated to a common data base; and research on political and/or planning decision-making in a realistic small group laboratory setting.

METRO-APEX is still evolving; the current version stresses training of adult groups in problems inherent to "comprehensive" and fiscal planning to urban areas. Policy problems are represented in a condensed time frame, to give rapid feedback of consequences of decisions.

New policy problems, such as air and water pollution, public health, transportation, etc., have been incorporated for particular occasions of use. METRO-APEX attempts to synthesize a coherent view of the city as a whole, to replace a narrow technical perspective with a broadened world view capable of grasping problems in their entirety. METRO-APEX starts with a *behavioral* as opposed to a technical model of urban systems, with bargaining emerging as a central phenomenon.

Perhaps the key distinguishing characteristic of METRO—APEX—in contrast with conventional urban simulations—is the stress on integrating a whole family of models (of only moderate complexity individually) to give a comprehensive representation of urban structure, rather than the progressive refinement and elaboration of specialized models.

The critical problems of urban regions are the ways in which decisions are made, with different views of the world leading to conflict, bargaining, and transactions. Too little is known or understood about these processes, so gaming instruments will always require continuous refinement. Playing out complex decision processes in gaming is the only device available for representing them, for either teaching or research, for mathematical models are a long way from handling the complexity (and the enormous permutational variety) they generate—even simple games have alternate solutions expressed in googols. The interaction of sets of decisions by players in multiple roles is the unique contribution of gaming-simulation to sensitizing researchers and future decision makers to the possibilities that can emerge in typical situations. Since all possibilities can never be evaluated, such devices are all that we have to develop, or to prepare for the developing of, contingency plans and strategies of intervention to promote plans.

The article by Allan Feldt in the readings section states the need for more effective tools for communicating the nature of "pure" simulations to decision makers who operate in the applied, nontechnical world of policy formation.

THE LABORATORY COMMUNITY

Even more futuristic is the concept of the "Laboratory Community." The idea is to have a continuing three-way interaction between modelling (which generates ideas for *what* to measure, as well as where solutions to problems may lie) and data collection and analysis (which may significantly alter our concepts and theories, as well as give parameter estimates for models) and public policy proposals (which will force the above two to be more realistic and less purely academic). One may conceive of the whole process as a spiral of increasing refinement of models, and of data collection and measures, and, hence, of policy ideas, over time. The Laboratory Community concept thus envisions work by a team of researchers from many disciplines in a single metropolitan area, over perhaps a decade. The attempt would be to establish a long-term study monitoring a broad spectrum of urban process for a single urban area, to generate time-series data that will be fed into and modify interdisciplinary policy-oriented urban models.

The research effort would be integrated through the development of a new METRO-style gaming-simulation of the Laboratory Community—with far more sophisticated data and models than now exist for any urban area. The next generation METRO, built around a Laboratory Community, would be sophisticated and detailed enough to really aid policy makers in a specific urban area in making better decisions. It would thus serve as an efficient device for disseminating the results of the Laboratory Community to both researchers and policy makers.

General advantages accruing from the development of the Laboratory Community would be that:

—it would be valuable to guide data collection for urban research by explicitly formulating models;

—it would serve as an integrative device for research; at the moment data and theories from different cities, different time periods, and different disciplines are very difficult to integrate into a common set of urban models;

—it would provide a better understanding of urban dynamics and of the real leverage points in the social structure.

The Laboratory Community remains a hope. It is further defined in the article by Miller and Duke.

GAMING-SIMULATION AS A TOOL FOR SOCIAL RESEARCH

CATHY S. GREENBLAT

In previous sections, the available research on learning, attitude change, etc., resulting from participation in gaming-simulations, has been reviewed. In the present section, the focus shifts from research on games to games for research—that is, to the question of what utility gaming-simulations may have for the social researcher.

Despite some early articles containing arguments that games might be useful to the social researcher (see, for example, Coleman's "Games as Vehicles for Social Theory" in Section IV), little has been written to explain just what the promise of these materials is. A perusal of the literature reveals a sprawl of ideas—some from work done, some speculative—but little in the way of organized thought outlining or describing modes of utilization. In this paper, an exploratory outline for the researcher is attempted. The framework derives from combinations of responses to the questions: (1) What is the researcher's purpose? (2) What kind of gaming-simulation does he employ? (3) What is the researcher's role? (4) What is the participant's role? These yielded four basic research modes, each with two variations based on the type of game used. These are illustrated in Figure 1 and are listed below. The body of the paper will include a more detailed explanation of each type, including real and hypothetical examples.

The eight modes of doing social research using gaming-simulations are listed below. The researcher:

I   A.   operates an existing game with "player-subjects." He counts or measures behavioral or verbal units to generate data to test hypotheses set up in advance.

I   B.   designs a new game and operates it with "player-subjects." He counts or measures behavioral units to generate data to test hypotheses set up in advance.

II   A.   designs a new game and operates it with "player-subjects." He observes and abstracts from the behavior and verbal messages and uses this to refine his theory.

320

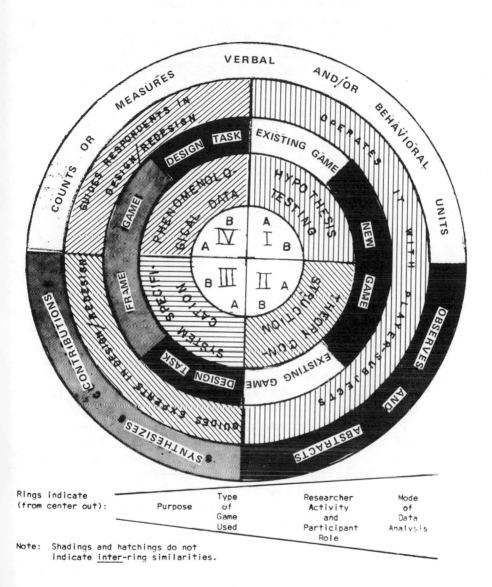

FIGURE 1. Four Basic Research Modes Using Gaming Simulations

II   B.   operates an existing game with "player-subjects." He observes and abstracts from the behavior and verbal messages and uses this to refine his theory.

III   A.   guides experts in designing a new game. He synthesizes their contributions and the product represents a specification of the system by experts.

III   B.   guides experts in redesigning a frame game. He synthesizes their contribution, and the product represents a specification of the system by experts.

IV   A.   guides respondents in the redesign process of a frame game to create a game about some system of which they are a part. He counts or measures the elements they include. The data thus generated are phenomenological—they represent the system as the respondents view it.

IV   B.   guides respondent-designers in the process of designing a new game about some system of which they are a part. He counts or measures the elements they include. The data thus generated are phenomenological—they represent the system as they view it.

Let's now see what these entail.

RESEARCH MODE I

In form A, the researcher has a set of hypotheses to be tested. He operates an existing game in a laboratory-like setting in which players serve as "subjects." He counts or measures behavioral or verbal units to generate data to test his hypotheses. The basic procedure followed is familiar to the social psychologist, except that a game is the basis of the activity that takes place in the laboratory environment.

The variables to be investigated or manipulated may be social-psychological or social-systemic. For example, the researcher may investigate sex-role differences in problem-solving style, the relationship between personality characteristics and behavioral styles, bargaining methods, coalition formation, types of leadership patterns, etc. He may, in such a laboratory setting, examine such things as the effects of alternative social structures on individual behavior in the situation by altering communication patterns, forms of social organization, etc. Through this systematic alteration of game parameters or the variation of player characteristics, causal hypotheses can be investigated.

The game-laboratory thus involves the creation of a social environment more natural than that in the usual laboratory experiment, yet less complex than the "real world." The complexity of the stimuli is greater than in the traditional laboratory, so one can observe more complex interactions; participants' motivations are likely to be higher; and players-subjects' actions may be more "natural" than the actions of traditional "subjects."

It is true that subjects may perceive the rules or the content of the simulation game as unrealistic or unrepresentative. However, due to their experience with other games in which certain aspects of reality have been modified, suspended, or ignored, it is felt the subjects in simulation game experiments will be less likely to respond to the experimental situation as if it were meaningless. Thus simulation games exploit the subjects' willingness to suspend disbelief in game-like situations, a willingness that appears not to be present in subjects in more traditional types of psychological experiments [McFarlane, 1971: 155].

The combination of the game and the simulation aspects allows the subjects to justify their actions after the fact in either serious or unserious terms; that is, the subjects can always say afterward 'it was only a game' *or* 'it was a very meaningful experience' [McFarlane, 1971: 154-155].

Gaming-simulations in the laboratory-type mode of use allow researchers to

(1) abstract only the processes in which they are interested; (2) replicate many times social systems which are alike or different along specified dimensions; (3) speed up or compress time; (4) safely investigate potentially dangerous or costly situations; and (5) provide a situation for players which offers its own rewards for participation [Dukes, 1973b: 4].

Comparing simulation games to both participant observation and to traditional social-psychological laboratory methods, McFarlane (1971: 150) cites as advantages of the gaming:

(1) an optimum combination of control and structure versus freedom and innovation with respect to experimental control of the subjects' actions; (2) a setting more likely to be perceived as "realistic" by the subjects participating in the experiment; and (3) a setting which allows the researcher more information with respect to complex, mutually contingent sequential interactions upon which he can perform his analysis.

Examples of studies done in this mode include the works of Vinacke (1959), a study by Clarice Stoll and Paul McFarlane (1969) and one by Stoll and Inbar (1970); some projects described by Raser (1969); and work by Terhune and Firestone (1970), Hermann (1969), Burgess and Robinson (1969), and Druckman (1968) in international relations. In several recent papers, Richard Dukes has described utilizations and potential utilizations of STARPOWER as a research tool (1973a and 1973b), and Coplin (1970) presents a general discussion of games as experimental devices to study human behavior in complex environment.

Form B of Research Mode I is very similar to form A; here, however, the researcher designs a NEW game, rather than employing an existing one. He then follows the same procedure: again he counts or measures behavorial or verbal units to generate data to test hypotheses specified in advance. The gaming-simulation thus again serves as the base for a social-psychological laboratory-type experiment.

Little is reported in the literature of attempts by researchers to create new games to test hypotheses. Among the only available examples are studies done by Murray Straus, who has designed a series of simple games for his research on family interation patterns. He has had family units, for example, play a game in which success rates for children and parents are manipulated, degrees of conflict and competition varied, and the ways in which these were dealt with by the family observed and analyzed. In this way, hypotheses regarding cross-class and cross-national differences have been tested. A clear example of the technique used in these studies can be found in his article "Methodology of a Laboratory Experimental Study of Families in Three Societies" (Straus, 1970).

RESEARCH MODE II

A look at Figure 1 will reveal the basis of the major difference between Modes I and II: the purpose in the latter is theory construction rather than hypothesis-testing. This mode is for more exploratory research studies in which the researcher is not seeking precise, prespecified data, but rather is interested in interactions which yield insights to help refine his theoretical formulations. Rather than counting or measuring the game behaviors, therefore, he observes them and abstracts from them; subsequently he may be able to use the resultant insights to generate hypotheses for testing in a real-world situation. The game thus serves as a heuristic device in theory-development.

In form A, the researcher designs a new game based on his theory, turning postulates into rules. After the initial design period, he finds players, observes their interactions, and abstracts from these. Often he redesigns the game or alters some of the parameters, and repeats the procedure of observation and abstraction until he has succeeded in generating a set of outcomes which replicate the outcomes seen in the real world.

This methodology contrasts to other techniques commonly employed by the theoretician-researcher. James Coleman, in "Games as Vehicles for Social Theory" (reprinted in this section), summarizes some of the differences as follows:

> In contrast to survey research and observations in natural settings, it depends on the creation of special environments, governed by rules that are designed precisely for the study of the particular form of organization. In contrast to experiments with their experimental probe or stimulus and the consequent response, the principal element in game methodology is the construction of rules which can elicit a given form of social organization [1969: 3].

A large number of topics can be examined using this procedure, including elements of social order and social organization, system linkages, the development of norms, social-psychological processes, and possible outcomes of alternative strategies. Some such uses are described in articles reprinted in this volume: William Gamson (1971), in the selection in Part II, describes his attempts to understand major elements of social order through design of SIM-SOC and subsequent modification upon modification of it due to initial lack of congruence between player behavior in the game and behavior in the real world. In this section, in "Collective Decisions," James Coleman (1966) illustrates how social theory can be translated into a gaming-simulation, and then observation of the play of the game may lead to refinements which are fed back into the theory. Philip Ennis' description of his experiments with a simulation of the artistic system demonstrates the creative use of the design methodology to refine theoretical understanding of social structures and processes.

Others have expounded suggestions that the development and observation of a simulation might lead to better insights into what *might* be. Smoker's "Social Research for Social Anticipation" (1969) and Boguslaw et al.'s "A Simulation Vehicle for Studying National Policy Formation in a Less Armed World" (1966) demonstrate such creation of hypothetical situation through game design and parameter manipulation. Questions of concern to social psychologists, such as processes in the development of consensus or the construction of shared perceptions, might be examined by creation of a "multiple realities" game and careful analysis of behavior in it. See Part II, Greenblat's "Sociological Theory and the 'Multiple Realities Game.' "

The advantage of this procedure over other modes of theory construction is that the researcher is forced to be explicit about his assumptions, to be concrete, yet he is able to avoid the semantic precision needed for a verbal model and the quantitative precision needed to construct a mathematical model. Such precision may be *emergent* from the model rather than a precursor to its development. Additional strengths are described by Druckman in "Understanding the Operation of Complex Social Systems: Some Uses of Simulation Design" (1971). Some of the costs and rewards of the method are described by Coplin in "Approaches to the Social Sciences Through Man-Computer Simulations." (1970)

Form B is the same as A, with the exception that the researcher, instead of *developing* a game, employs an existing one. Any of a large number of existing games could be used as is or with modifications to study any of a large number of social-psychological or social-systemic factors in an exploratory fashion. For example, METRO-APEX, SIMSOC, STARPOWER, and several other games present fruitful and rich sources of interaction which, if observed and analyzed, could yield abstractions about behavior in the social context simulated. Thus, they would be useful in the attempt to formulate more specific ideas about role conflict, emergence of norms, leadership patterns, relative deprivation, exchange mechanisms, etc. Bredemeier (1973: 77) suggests such a style of usage in urging:

Rawls deals with the matter of procedures for realizing the principles of justice in a provocative, if necessarily thin, discussion of some 200 pages comprising the middle of the book. There he discusses such matters as 'toleration of the intolerant,' civil disobedience, the branches of government and their functions, majority rule and 'participation,' among other institutions. . . . Are those the principles all rational persons with the qualities Rawls postulates would choose? The question might be nicely explored. Let students play the simulation games 'SIMSOC' or 'STAR-

POWER.' Let them experience the travails, restrictions on liberties and inequalities ordinarily experienced in those game settings. Then, stop the game and announce that it will be resumed later, with roles to be allocated all over again by lot. We now have a not-bad simulation of 'the original position.' Let the weeks (months) in the meantime be spent by the players in devising basic principles that will govern the game-playing when it is resumed. Will they arrive at Rawls' Principles? Without reading Rawls and the Utilitarians? With reading them? Under other experimental-pedagogical conditions?

Hypothetical conditions could similarly be investigated. The manual for THE MARRIAGE GAME, for example, includes instructions for the alteration of parameters concerning equality of the sexes, sexual freedom, etc. Play with such altered parameters should aid the researcher in generating fruitful hypotheses dealing with such conditions.

The "yield" from Research Mode II utilizations might take any of a number of forms, including lists of possible factors, consequences, obstacles, etc. For example, an undergraduate student of mine with little prior knowledge of educational systems played the basic version of POLICY NEGOTIATIONS and generated the following list of impediments to educational change:

   (1)   opposition to change by some system members;

   (2)   tactical difficulties in bringing about change;

   (3)   large amount of time required to implement proposed changes;

   (4)   lack of verbalization of intended outcomes, leading to misunderstanding;

   (5)   failure to understand sources of past failures;

   (6)   lack of understanding of sources of power and points of leverage;

   (7)   lack of understanding of change processes.

She planned to follow these leads by looking at real-world school systems to see if the same impediments blocked effective change.

As yet, little research of this sort has been done, although those interested in international relations have done a number of such investigations using Guetzkow's "Simulations in the Consolidation and Utilization of Knowledge about International Relations" (1972).

MODE III

A look at Figure 1 will reveal that there is a basic split in it between Research Modes I-II and Research Modes III-IV, for a number of the defining parameters of the research modes change as one goes from the righthand side of the figure to the lefthand side. In addition, the first two modes of utilization have been discussed in the literature, albeit in a limited fashion, and example of them can be found. The latter two modes of doing research utilizing gaming-simulations are more speculative; to the best of my knowledge they have not been employed and have not been described in the literature. The examples offered here, therefore, are hypothetical and simply indicative of the *types* of research problems that could be dealt with using gaming-simulation in these ways.

In Research Mode IIIA, the researcher's purpose is to gather from experts in a field their perceptions of the major dimensions of a system. Rather than interviewing them utilizing a standard questionnaire or interview guide, they are turned into game-designers and, under the guidance of the researcher, asked to express their ideas in the format of a game. If the system of concern is a prison, the experts would be penologists, and they would be asked to build a gaming-simulation of a prison; if the system of interest is an educational institution, scholars in the field of education would be called upon to serve as expert-designers creating a school or university game. The researcher's additional task is then to synthesize the definitions of the roles, goals, resources, constraints, and contingencies of the system as specified by the experts. The game-design task thus functions somewhat as an in-depth interview guide; it is a set of questions (see Duke, 1974, and the design section of this volume). If joint work sessions can be arranged, it may also serve as a vehicle to permit the equivalent of a "group interview" with experts. The task will generate dialogue and confront-ation, for the game design process inevitably entails arguments and debates among designers about what the system is *really* like. The advantage of this technique over more traditional modes of interviewing of experts derives from the fact that to design a game the experts will of necessity be confronted with the need to be specific about assumptions, to deal on various levels of abstrac-tion, etc.

In order to understand Research Mode IIIB, we must introduce a new concept—that of the "frame game." Probably the simplest example is the crossword puzzle. Its basic framework consists of a matrix of black and white squares, a numbering system, a set of clues, and a feedback mechanism in the form of answers for the player who wishes to check on his success or learn what he has missed. Once the basic framework is understood, different examples can be created by varying the content.

There are a number of gaming-simulations available which, like this, have standard frames into which content can be loaded. They consist of a framework, a sample set of content (often called a 'priming game'), and instructions for redesigning the game to make it content-specific to the user's needs. These have been widely used for teaching purposes and for training, but their potential as "questionnaires" to gather system-specification data from experts has not been exploited.

Such frame games as IMPASSE?; AT ISSUE; and the CONCEPTUAL MAP-PING GAME; NEXUS; POLICY NEGOTIATIONS; and COMMUNITY DIS-PUTES provide frameworks of varying degrees of complexity and embodying different types of models into which content can be loaded. If one of these (or another frame game) embodies an appropriate model, the researcher will find that the process of loading content into an existing frame proves far easier and less time-consuming than the process of starting from scratch to design a new game. It may, therefore, be more plausible to think of getting experts to take the time to engage in this process rather than in the full design procedure.

The prison system has been offered as simply one example of what could be done in the way of research in this mode. IMPASSE? versions have already been designed to deal with environmental and ecological problems, urban planning, health care delivery systems, etc. NEXUS versions deal with local authority finance problems, fire department operations, developing countries, etc. POL-ICY NEGOTIATIONS has been redesigned to deal with hospitals, social work agencies, etc. For Research Mode III purposes, however, the versions must be designed by those who are experts in the field rather than by game-designers with more limited understanding of the content area.

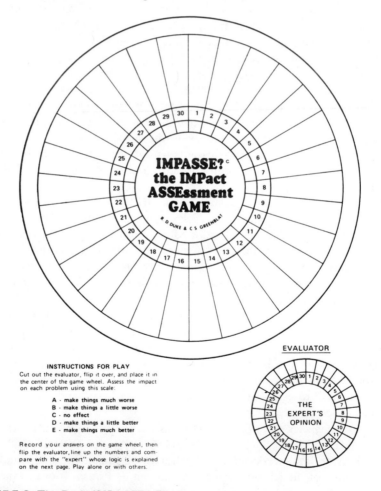

INSTRUCTIONS FOR PLAY

Cut out the evaluator, flip it over, and place it in the center of the game wheel. Assess the impact on each problem using this scale:

A - make things much worse
B - make things a little worse
C - no effect
D - make things a little better
E - make things much better

Record your answers on the game wheel, then flip the evaluator, line up the numbers and compare with the "expert" whose logic is explained on the next page. Play alone or with others.

FIGURE 2. The Basic IMPASSE? Frame

1-(E) Improved viability of the central business district would result in higher land values.

2-(D) More active business climate would result in higher tax derived from business.

3-(E) A successful, advanced rapid transit system will spawn other projects requiring federal aid.

4-(A) Basic changes in transportation capability will result inevitably in secondary costs for roads, sewers, etc.

5-(C) Some welfare recipients will be better off, but others will arrive to replace them.

6-(B) The existing tendancy of industry to decentralize will be encouraged.

7-(E) Populations will shift as land use patterns adjust to transit capability, affecting wards.

8-(E) The very magnitude of a rapid transit system requires discussions, perhaps agreement?

9-(B) The existing tendancy of the middle class to leave the city will be encouraged.

10-(B) Populations will inevitably shift; construction will intrude on existing neighborhoods.

11-(E) A viable rapid transit system inevitably makes a city a more viable "central place".

12-(A) Many actual improvements (low-cost transport, new jobs) will be offset by new indigents.

13-(B) Construction side effects as well as improved mobility will result in shifting populations.

14-(A) A more active, viable central area will discourage street crime.

15-(D) Better transit gives better access, more opportunity to reach a variety of facilities.

16-(B) Construction of this magnitude inevitably causes damage, some of which is permanent.

17-(E) Improved mobility brings a greater area of access to residents; more people moved in a given space.

18-(E) Existing pressures for change will have a better chance for success.

19-(E) No rapid transit system will inevitably lead to more sprawl and deterioration of the city.

20-(D) Entrepreneurial response to a new transport system is dramatic, perhaps too dramatic.

21-(E) The new transport mode will make large areas more accessible to the city.

22-(E) In the long-run, more-dense land uses will locate near the terminals; A more European pattern will result.

23-(E) Assuming proper integration (!) more people will commit to public transport.

24-(C) Expressways are here to stay; rapid transport is a complimentary system.

25-(D) Some improvement is to be expected, however the auto is always with us.

26-(D) A rapid transport system is a major component in regional growth permitting improved planning.

27-(E) Growth can be expected to concentrate at the terminals of the rapid transit system.

28-(A) New shopping centers can be expected at the nodes or transit terminals.

29-(E) The central business district will be more readily accessible and therefore more viable.

30-(E) Growth will be channelled by the transit system, planning decisions will be more orderly.

Our "expert" for this game is Dr. William Drake, Assoc. Dean for Research, School of Natural Resources, the University of Michigan. Dr. Drake is director of the Ann Arbor Transportation Authority, which has successfully pioneered in the use of "Dial-a-Ride" mini-buses.

Should your perceptions differ (either with regard to the problems in the impasse wheel, the "expert's" values as assessed, or the brief explanation of his choice) drop a note to the editor marked "Rapid Transit Impasse".

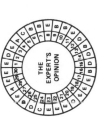

THE EXPERT'S OPINION

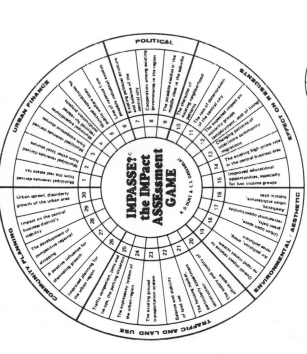

"RAPID TRANSIT" IMPASSE?

IMPASSE?
the IMPact
ASSEssment
GAME

R D DRAKE & C S GREENBLAT

EVALUATOR

As mayor of a major city, what impact would you expect an advanced rapid transit system to have on your town in the decade following its opening in terms of:

INSTRUCTIONS FOR PLAY

Cut out the evaluator and place it in the center of the game wheel. Line up the colors. Assess the impact on each problem using this scale:

A - make things much worse
B - make things a little worse
C - no effect
D - make things a little better
E - make things much better

Record your answers on the game wheel, then flip the evaluator, line up the colors and compare with the "expert" whose logic is explained on the next page. Play alone or with others.

FIGURE 3. An Example of a Loaded IMPASSE? Game

Figures 2 and 3 show (in reduced scale) the framework for IMPASSE?, the simplest of the frame games mentioned above. Figure 2 presents an example of the game materials loaded to deal with assessment of the impact of introduction of a rapid transit system, and Figure 3 presents the empty frame elements. Using IMPASSE?, for example, those concerned with a prison system could ask experts to put into the large wheel the variables of a prison system that should be taken into account in assessing any potential or proposed policy or action. Next they would be asked to outline a series of changes, actions, or proposals that might be brought to bear in such a system; and finally the experts would be asked to indicate on the small wheels ("serial evaluators") their assessments of the impact of the given policies on the system parameters. Whether or not the resultant game(s) were ever used (i.e., played), they would present statements of the system specifications as seen by these experts and hence be of value for research-theory building.

NEXUS employs a somewhat similar recording device for indicating the variables that characterize a system. The NEXUS center card is filled in with system variables and indications of the effects of various actions on each of the variables. These cards are later laid in the center of a board composed of scales for recording the cumulative changes in the state of the system resulting from the sequential actions. (The similarity between IMPASSE? and NEXUS derives from their common ancestry in Olaf Helmar's FUTURES GAME.) Experts could be asked to program the NEXUS cards in order to elicit their views of the system.

If a more interactive model were needed, the researcher could locate several experts and have them engage in the steps of redesign of POLICY NEGOTI-ATIONS. As will be noted, the steps require a quite elaborate set of system specifications.

The type of data collected by Olaf Helmar and Dennis Little via Delphi techniques to develop (respectively) the FUTURE STATE OF THE UNION GAME and STAPOL (State Policy Simulation) would be extremely useful to a researcher. These games could be used as frame games by wiping out the figures given by Helmar's and Little's respondents and having experts present their assessments of impacts. Again the game would serve as a sort of "questionnaire." In summary, then, the frame game in Research Mode IIIB is used as a device to elicit ideas of experts on the characteristics of complex systems: constituencies, roles, issues, impacts, problems, constraints, contingencies, etc.

MODE IV

The fourth Research Mode using gaming-simulations again involves utilization of a frame game. Here, however, our designers are not scholar-experts, but people who *live* in the system in question and thus serve as respondents to the games' questions, providing phenomenological data. In form A, for example, we would ask prisoners to design a prison version of POLICY NEGOTIATIONS rather than asking penologists to do this. The resultant model would reflect the biases inherent in the perceptions of those in one role in the system, but as such it would be a sound reflection of the prison system as seen by prisoners.

In this mode, then, the researcher guides designer-respondents in the redesign process of a frame game to create a game about some system of which they are a part. Having thus elicited the data, he counts or measures the elements they include. The data thus generated are phenomenological as they represent their views of the system.

Again IMPASSE? could be used in a variety of ways: A predesigned version could be played by a group (such as prison inmates) and their assessments tabulated and analyzed like survey data. Inmates or staff of a prison could be asked to define the elements that should be included in the large wheel to make a PRISON IMPASSE? game. The researchers' task would be to determine the frequency with which various aspects were mentioned. Students might be asked to express their views of their university by creating a version of POLICY NEGOTIATIONS reflecting their university. They would be guided by the researcher through the steps of redesign, and he would derive a picture of their views from the new games they created. In a more elaborate research design, students, junior faculty, senior faculty, and administration at the institution might be asked to engage in the redesign process. Comparisons could be made on a series of game parameters (e.g., amount of influence allocated to each of the major constituencies, impact of passage of particular policies, etc.) to compare the perceptions of those in differing roles in the system.

Form B is almost the same as form A. It involves, however, the creation of a new game rather than the redesign of a frame game. Here the researcher guides designer-respondents in the design process of a new game. For example, prisoners would be guided in the design of a prison game, students in the design of a university game, etc. In this manner, they would again present their perceptions of the system, generating phenomenological data for the researcher to count, measure, or otherwise analyze.

CONCLUSION

The utilization of gaming-simulation for social research has largely been ignored in methods texts and guides. Those that refer to the technique at all discuss only computer simulation or refer only to what is here described as Research Mode IA: utilization of an existing gaming-simulation for hypothesis testing in a social-psychological laboratory (cf. Phillips, 1971: 171-190.) Those in the infant field of gaming have also neglected in their writings the research potentials of the technique. It is hoped that in the future the ideas and suggestions offered here will be more fully explored and essayed to learn more of the potential contribution of gaming to the research endeavors of social scientists.

REFERENCES

A. Literature

Boguslaw, R., R. H. David, and E. B. Glick
    1966 "A Simulation Vehicle for Studying National Policy Formation in a Less Armed World." Behavioral Sciences II (January): 43-61.
Bredemeier, Harry C.
    1973 "Justice, Virtue and Social Science." Society (September/October); 76-83.
Burgess, Philip and James A. Robinson
    1969 "Alliances and the Theory of Collective Action: A Simulation of Coalition Processes." Pp. 640-653 in James Rosenau (ed.) Foreign Policy and International Politics. New York: Free Press.
Coleman, James S.
    1966 "Collective Decisions." Sociological Inquiry 34 (Spring).
    1969 "Games as Vehicles for Social Theory." American Behavioral Scientist 12 (July/August): 2-6.
Coplin, William
    1970 "Approaches to Social Sciences through Man-Computer Simulations." Simulation and Games I (December): 391-410.
Druckman, Daniel
    1968 "Ethnocentrism in the Inter-Nation Simulation." Journal of Conflict Resolution 12 (March): 45-68.
    1971 "Understanding the Operation of Complex Social Systems: Some Uses of Simulation Design." Simulation and Games 2 (June): 173-195.
Dukes, Richard L.
    1973a "A Research Package for STARPOWER." Boulder: University of Colorado, mimeographed.
    1973b "Symbolic Models and Simulation Games for Theory Construction." Delivered at the annual meetings of the American Sociological Association.
Gamson, William A.
    1973 SIMSOC. New York: Free Press.
    1971 "SIMSOC: Establishing Social Order in a Simulated Society." Simulation and Games 2 (September): 287-308.
Greenblat, Cathy S.
    1974 "Sociological Theory and the Multiple Realities Game." Simulation and Games 5 (March).
Guetzkow, Harold
    1972 "Simulations in the Consolidation and Utilization of Knowledge about International Relations." Pp. 674-690 in Randall L. Schultz (ed.) Simulation in Social and Administrative Science: Overviews and Case-Examples. Englewood Cliffs, N.J.: Prentice-Hall.
Hermann, C. F.
    1969 Crises in Foreign-Policy: A Simulation Analysis. Indianapolis: Bobbs-Merrill.
McFarlane, Paul A.
    1971 "Simulation Games as Social Psychological Research Sites: Methodological Advantages." Simulation and Games 2 (June): 149-161.

Phillips, Bernard S.
   1971 Social Research: Strategy and Tactics. New York: The Macmillan Company.
Raser, John
   1969 Simulation and Society. New York: Allyn & Bacon.
Smoker, Paul
   1969 "Social Research for Social Anticipation." American Behavioral Scientist 12 (July/August): 7-13.
Stoll, Clarice and Paul McFarlane
   1969 "Player Characteristics and Interaction in a Parent Child Simulation Games." Sociometry 32 (September): 259-271.
Stoll, Clarice S. and Michael Inbar
   1970 "Games and Socialization: An Exploratory Study of Race Differences." Sociological Quarterly 2 (Summer): 374-380.
Straus, Murray A.
   1970 "Methodology of a Laboratory Experimental Study of Families in Three Societies." Pp. 552-577 in Reuben Hill and Rene Konig (eds.) Families in East and West. Paris: Mouton.
Terhune, K. W. and J. H. Firestone
   1970 "Global War, Limited War, and Peace: Hypotheses from Three Experimental Worlds." International Studies Quarterly 14 (June): 195-218.
Vinacke, W. E.
   1959 "Sex Roles in the Three-Person Game." Sociometry 22: 343-360.

B. Gaming-Simulations

AT-ISSUE!; CONCEPTUAL MAPPING GAME; IMPASSE? Designed by Richard D. Duke and Cathy S. Greenblat. Published in Game-Generating Games: A Trilogy of Issue-Oriented Games for Community and Classroom from Environmental Simulation Laboratory, University of Michigan, Ann Arbor, Michigan.

COMMUNITY DISPUTES. Designed by Armand Lauffer. Published by Gamed Simulations, Inc.

FUTURES. Designed by Olaf Helmar. Published by Kaiser Aluminum and Chemical Corporation.

FUTURE STATE OF THE UNION. Designed by Olaf Helmar. Published by Institute for the Future, Santa Monica, California.

THE MARRIAGE GAME: UNDERSTANDING MARITAL DECISION-MAKING. Designed by Cathy S. Greenblat, Peter J. Stein, and Norman F. Washburne. Published by Random House, Inc., 201 E. 50th St., New York, New York.

METRO-APEX. Designed by Richard D. Duke. Published by Environmental Simulation Laboratory, University of Michigan.

NEXUS. Designed by R.H.R. Armstrong and Margaret Hobson. Published by Institute for Local Government Studies. University of Birmingham, Birmingham, England.

GAMING-SIMULATION FOR
URBAN PLANNING

RICHARD L. MEIER and RICHARD D. DUKE

University of Michigan

T here is a time in the evolution of every discipline when its milieu, and the phenomena with which its practitioners must deal, grow in complexity, and responsibilities multiply beyond the capacity to cope with them. That time is at hand for planners.

It is reassuring to recall that the medical profession went through such a transition in the 1920s, when advanced chemistry, physiology, and microbiology were added to the basic training. Physics encountered one such transition with quantum mechanics in the late 1920s and another of quite a different kind when physicists became immersed in politics and secrecy after the advent of nuclear weapons. The chemists were the first group to face up to the knowledge explosion, making important beginnings toward the organization of basic knowledge in the 1920s, while engineers are still in the throes of adjustment to the expanding body of technical knowledge affecting design and production.

The immediate pressures upon planning result from a recognition that traditional measures—principally civic design, the control of land use, and the allocation of subsidies—do not cure the ills for which they were prescribed. The treatments for depressed areas and slum-ridden cities will have to affect something more fundamental than appearances.[1] New powers are being offered to the responsible agencies, providing them additional means of treating the conditions that create slums and other metropolitan disorders.[2] Programs are expanding at a bewildering rate, and new decision processes are being recommended in planning agencies. In the near future data banks will broaden still further the spectrum of possibilities.[3]

Increasing complexity presses a profession toward simultaneous development in two directions: one, specialization of technique and subject matter into slightly overlapping cells of sufficient simplicity for individual decision-making; and two, the elaboration of new models and devices for achieving an overview. This division of labor within the field of planning permits one to address

Reprinted by permission of the Journal of the American Institute of Planners, Volume 32, No. 1, January 1966.

specifically the difficult task of achieving a comprehensive overview of metropolitan regions. Simulation is of greatest value for this purpose in that it provides a means of coping with complexity; its primary value is that it enables one to abstract the critical structure or process involved and to specify the components and their interactions.[4]

A model is essentially a set of criteria for selecting significant data; it provides rules for applying logic to the distilled information.[5] The larger the body of data from which the selection is made, the more information one must incorporate. Hitherto, site planners have prepared three-dimensional scale models for proposed physical improvements on major sites, and have charted the sequence in which these changes could be achieved. Such iconic models left the optimization processes to the intuition of the senior planners. Mistakes frequently were made because the costs and benefits to society arising from implementation of the model were rarely assessed. Scale models are less useful when traffic or social planning is crucial. For public utilities, traffic, and communications-based activities, early approximations are obtained with arithmetic calculations of flows, and more careful estimates are achieved with simple mathematical models operated alongside the three-dimensional models.

UNDERTAKING SIMULATION

Much planning already revolves around the budgeting process and the capital improvement program. Over the next decade the various regional development organizations, whether metropolitan, state, or watershed-oriented, will be faced with the preparation of very sizable, integrated capital programs. The outcome may fit into a plan for transportation and land use, as with the Penn-Jersey enterprise. The plan may also produce thoroughgoing urban redevelopment, as in present-day Boston, or redistribute water, as in California's statewide program. When the planning effort is large and long-range, the projects will be sponsored by a variety of interests. The governmental units serving these interests will retain a concern for planning procedures, usually in proportion to their financial contributions. From one of these sources, or their consultants, a suggestion will come, "Why don't you simulate?" The stimulus will almost always come from agencies that are more concerned with capital budgets than with space or time budgets.

At the present time, simulation is a technique often discussed, and sometimes introduced, by the avant garde in planning. Although the basic idea is very old and a matter of common sense, the technique has now been encumbered with social science terminology and computer technology.[6] A simulation worthy of the name is no longer to be taken lightly.

During the 1960s, at least, none of the chief planners or their office heads is likely to have had an opportunity to study simulation techniques at a university, and very few have had practical experience with simulation. Buying the know-how from consultants carries a high price tag; in addition, certain risks are run in trusting technical specialists on matters so closely affecting strategy and policy. The barriers to adequate communication are evident, and misunderstandings are prevalent. Therefore, it is not unreasonable for the responsible practitioner to reject simulation in favor of the planning tools he knows how to use. However, requests for construction of simulations are increasing in frequency and intensity, and they come from the most unexpected quarters. Within planning staffs, there is a junior contingent who view simulations, along with monorails and malls, as ultramodern and, therefore, very fashionable, but there is also a group who have learned to do simple computer programming and who have been exposed to courses in systems analysis. For the latter, to undertake a simulation represents the path to prestige and promotion.

When new projects are planned, particularly when property must be condemned, conflicting views must sometimes be contested in court. The outcomes from a simulation with explicit decision criteria may be highly persuasive in court as compared with plans derived by more conventional methods. New members of commissions and advisory boards, if they have undertaken simulations in their own organizations, are likely to offer support for the attempt to simulate. Once raised, therefore, the simulation issue is as likely to be settled on grounds of political and administrative convenience as upon the heuristic utility of the technique. Then the burning questions will concern *what* should be simulated and the *style* or form the simulation should take.

Before simulation is adopted as a procedure, the reported good features will be weighed against the foreseeable limitations. This will require a procedure similar to that of firms pioneering the idea of computerization,[7] where a committee is given the task of familiarizing itself with the new technique and its application. Learning about simulation by hearsay and consultation teaches more about costs and risks than about methodology; however, it does introduce the vocabulary.

This background for the decision to simulate is important because features of it are often responsible for the difficulties encountered later. In the preliminary internal plans, for example, and even during the first full-scale trials, a natural tendency toward perfectionism is rarely kept under control. Mathematically trained persons have a bias in favor of elegance and rigor which can delay the attainment of results far beyond the deadlines that are initially set. There is a strong chance that no results at all will be achieved if the working model becomes too cumbersome to operate. The expense sometimes borders upon the scandalous because computer charges and the use of skilled manpower can mount up very rapidly. Thus the learning process is necessarily an expensive one and the costs are borne most heavily by the pioneers.

A decade or so hence, when simulation is counted as one of an extensive repertory of skills mobilized by planners to explore alternative courses of action, some of the agencies that adopted simulation early will use their experience as a foundation for the synthesis of still more complex systems of control. No names have yet been assigned to the data sources and the feedback loops in subsequent stages of development, but the ideas are already being generated at the academic level.[8]

STYLES OF SIMULATION

Although two quite different simulation procedures relevant to planning are readily identified, at least three styles can be detected. One procedure treats regions as an interacting system blindly responding to a series of externally applied forces. The relevant simulation style in that case is mathematical or a "pure" computerized technique. The second procedure recognizes the competition for scarce resources in the region and explores the strategies open to the various contestants. It emphasizes role-playing and the possibilities for integrating the respective roles into some system or institution at a higher level.[9] The third style is eclectic, since it borrows from both of these approaches, adds any extra tools and components that seem useful, and synthesizes an ad hoc, hybridized procedure for problem-solving.

The term *style* has been used here to imply that the methods used for problem-solving are chosen to suit the personality of the director and the experience of the organization. Situations and facts must be reported in a terminology and format that communicate both up and down the hierarchy of the agencies that the planning group serves. Therefore the most important single determinant of the procedure to be adopted for simulation is the style of operation that has been established previously by the director and his staff. Aside from these considerations, however, the characteristics of the core problem may influence the style of the simulation undertaken.

When organizational needs do not determine style and a choice must be made, it is important to assess the sources of information from which the bulk of the plan will be constructed. Thus, the proposition:

If: the basic data are produced by natural phenomena or a large number of small, independent transactions and the final design of the plan must produce a balanced network of central services (water supply, drainage, housing market, regional retailing, electrical power distribution, traffic, banking services, schooling),

Then: sampling procedure can be developed which fits into an outline description of the total system (that is, a model of the system) that can be manipulated with the aid of a computer. The inputs to the model containing an initial state of the system are new facilities and regulations, while the output is a sequence of resulting states of the system.

The computerized model, once it has been set up, should suggest the proper combination of new construction and regulation so that the greatest returns to the public are achieved. Such models have apparently served best thus far in electrical utilities, where the main task has been that of choosing the best network of principal routes for distributing power. Traffic and banking may well have been aided by a quite different set of models based upon trips and transactions. The faults in computerized models are rarely reported in the literature (few investigators have the fortitude to undertake the post mortem analysis of a failure with which they have been associated). Our own impression is that conditions change so rapidly any assumptions introduced will appear unrealistic by the time the implications are successfully ground out by the debugged program. Thus the range of possibilities usually has been inadequately explored at the time a model is selected, or the model chosen was too elaborate for the task it was assigned. It is therefore necessary to adjust the findings to the conditions known to exist at the time the decisions are made. These experiences suggest a second proposition:

If: the problem seems to be that of acquiring insights into organizational behavior under conditions in which resources are truly scarce, so that competition becomes intense, and the essential data are qualitative or subjective (housing types, budgeting choices, demand in marketing, or support in politics),

Then: a gaming procedure can be developed which allows surrogates of the chief competitors to test the principal strategies open to them and so discover what new and unexpected situations may arise. These anticipations should suggest the use of interventions of various kinds which prevent the worst from happening and increase the likelihood that a more desirable outcome will eventuate.

If the planning of regions, metropolitan and otherwise, were to be begun, it would first be necessary to identify the policy instruments available to the planners for the guidance of long-term developments. For example, the agency sponsoring the planning could have a great deal of influence over (a) zoning and land use controls, (b) capital budget, (c) building permits and inspection, (d) the choice of routes or areas for improving municipal services, (e) the standards for quality of services, and (f) information acquisition and release. The application of such policies is poorly defined and not readily suited to detailed programming. The well-defined features of the environment are derived from census-type data to which is added extra countable uses of public facilities. Within the American mode of government and American patterns of settlement, a few computational routines can be found which need very little adaptation as they are transferred from one locale to another. One of the best examples is population forecasting, combined with school enrollment projection, using the cohort survival technique. Within a few years a number of standard programs and procedures should be in existence which can be utilized virtually anywhere as "subassemblies" for simulations designed to fit specific regions.

DESIGNING COMPUTER SIMULATIONS

To date, simulations have not had a very salutary history. The most common error seems to be that of overcommitment. As people learn more about what computers can do and get more deeply involved in systems analysis of the operations of their organization, they become intoxicated with the potentials. The experience of administrators and planners with data collection, coding, storage, model-making, and debugging has been too limited as yet to enable proper appraisal of the amount of work involved in completing a computerized working model at the level of sophistication desired. The error is uniformly an underestimate of the amount of time and labor actually required, with the result that the scheduling of completion dates has been over-optimistic. Consultants tend to make mistakes in the same direction because the lacunae in planning data and the judgmental features in the interpretation of these data seldom come to light until the final stages of the effort. Planners are rarely conscious that they quite dogmatically apply rules of thumb which are difficult to justify to the outsider, so the careful consultant will need to make an independent appraisal. Specific traps in the use of simulation techniques are not yet signposted in the way that statisticians have mapped out their methods over the past two decades; statistical workers have raised helpful warnings for the novice wherever the risks of incorrect inference were significant.

A more fundamental cause of failure than underestimation of time and expense, however, arises from an inability to answer the philosophical question, "When is a simulation a good enough replica?" Every staff member feels the impulse to introduce detail up to the degree that its deviation from empirical data can no longer be discriminated by the decision maker when his attention is focused upon a given subsystem or process. Yet to operate at a grosser level exposes the simulation to criticism for its crudity. Where shall the compromise be made?

Physical planners have faced a similar problem in their approach to the use of three-dimensional models for site planning and land use. In models prepared for the educated public they now place painted wooden blocks, the volume and height of which are scaled according to plan, on the respective map sites, but the architectural details that distinguish these buildings from each other are not added. Existing buildings are given a grey or tan tone while proposed buildings are highlighted in an appropriate hue. Viewers accustomed to museum-like replicas, accurate to the resolving power of the eye, are often disappointed because such a model seems drab. Nevertheless the features essential to planning-type judgments are retained.

If public decision makers are to participate freely in design decisions, they must either learn the referents for the adopted abbreviations, or proposals must be translated into more understandable models.

A simulation for regional development must compress time as well as space. When computers are used, it is best to plan upon reducing an annual cycle to a few minutes of computer operation, the exact amount depending upon the capacities of the computer employed and the size of the budget available for simulation. When people become used to "gaming" the cycles, a year is normally reduced to about one hour. Much of this time is required for communication of relevant information, for comprehending crucial shifts in relationships, and for bargaining between players. Human capacities therefore determine the degree of reduction. The typical planning horizon—five to seven years before project initiation—can then be reached by the end of a single day of gaming. It allows a small amount of time for a review of what had transpired and the raising of new questions to be considered. The ratio of time compression—one year to one hour—is about ten thousand to one.

By coincidence, the reduction in spatial coordinates for the maps which encompass a whole metropolitan region or small watershed for use in the game is also about ten thousand to one. However, since this scale applies to both the North-South and East-West dimensions, the reduction in area is about a hundred million to one. It is meaningless to discuss the size of a map inside a computer, so the comparison between computer simulations and gaming simulations cannot be extended into this dimension.

A further compression occurs in the number of actors. A sampling rate of about one in one hundred is feasible for a computer simulation; it is normally necessary to maintain a population of 2000-3000 households, or firms, or voters in order to have confidence in the outcomes. Each of these simulated "actors" is highly simplified, rarely possessing more than ten attributes, and seldom is each permitted to make more than three different kinds of decisions. In that instance a great deal more is shorn away, but the amount of simplication or compression is difficult to measure.

For gaming simulations the same approach for measuring reduction will apply. We use people as actors and accept all their cultural, social, economic, and political attributes, but when we do so only a few (three to twenty) persons can be efficiently engaged as players. The average number may be in the neighborhood of ten persons. Moreover the rules of the game and the issues imbedded in the model restrict the attention of the players to a few sectors of the total range affecting the course of public affairs in the real city. An alternative approach would observe the dynamics of interaction in the group and count the number of completed transactions (including all the trips, calls, purchases, lessons, letters, votes, decisions, and so forth, recordable in the real world). In that case, it is estimated from observation that the players normally can complete several hundred transactions during an intensive hour of play that may represent a year, while the city itself produces many billions of such transactions per year. Thus the reduction, viewed in this manner, is on the order of 10^7 fold (billions divided by hundreds). This represents a notable degree of simplification. Only a few natural scientists (principally those working with modern telescopes, microscopes, spectroscopes), investigate phenomena with this degree of reduction, magnification, or amplification in methodological improvements than have thus far been plowed into planning-related modeling. Even today, years after their introduction, the big scientific instruments are known to produce many meaningless outputs for each one that is good enough to incorporate in a published report. A truly top-rank scientist is required to distinguish which outputs are important enough to integrate into the exisiting body of knowledge.

Several conclusions follow immediately from these observations:

(1) The principles for gaming simulation must be capable of reducing the number of urban transactions by a factor of at least ten million to one and yet retain the pattern discernible for the aggregate; they do this by paring away the least relevant transactions and collapsing many others into a single bundle that can be treated as a unit.

(2) Those cyclical events are retained which make up essential links in several causal chains; the gaming events, including decisions, should have the same degree of influence over subsequent events as their counterparts in the real world.

(3) A number of crucial decisions or turning points in development (elections, transport changes, settlement of industries) should be illuminated by the provision of maps, statistical series, and indexes which summarize the trends in certain aggregates of transactions (for example, sales, traffic, educational levels, mass media use) so that rational calculations are possible.

(4) Allowance must be made for the utilization of typical strategies known to decision makers, so that they can be tested at the level of simulation.

This set of requirements places extraordinary demands upon the design of a simulation. The working model demonstrates that it meets these design requirements by producing a good facsimile of the past. Among technical specialists the crucial tests for a good simulation focus upon its ability to obtain reliable performances when redoing the recent history—recognizing that critical turning points cannot be reproduced in detail, but aggregates, ratios, permissible directions of growth, and the like can be confirmed. This kind of reliability is rarely adequate for regional planners because it is calculated to obtain the best hindsight at least cost. However, hindsight is highly correlated with foresight only at equilibrium, a condition where forecasting and planning are not really needed![11] Therefore the planner must suspect the models that are prestigious among the specialists whose expertise is derived from a close study of the specialty in the recent past. Are they tuned so as to prevent the most costly errors of the past? If so, they cannot help very much in dealing with the crises of the future.

The models for planning must be tied to certain foreseeable periodic decision processes, such as, one, reconsideration of the capital improvement program and the budget recommendations following that review, two, the elections and their repercussions, and, three, the foreseeable crises in employment, health, or welfare. Most of the crises fall into a category of unique threats, at least for the history of the region being modeled. Often they fit into a class where demand for a service rendered or the introduction of new rules for the allocation of the scarce services is out of focus. The planning models should be flexible enough to allow the introduction of a variety of these crises, as well as strategies for meeting them.

This mode of learning about simulation, that is, learning by doing, introduces a new insight. Investigation of about a dozen or so instances where a body of data is available which describes changes in a system over the passage of time, and where various approaches to simulation have been attempted, suggests that people should not expect to discover a get-rich-quick formula through the synthesis of simulation model. Its primary function is to improve the judgment of ordinary people with limited experience who must make decisions even though they may not be aware of all the relevant factors. In other words, it is a concept-enriching, loss-minimizing technique, similar in effect to other forms of "broadening" education. An "art of simulation" must be evolved which introduces ways to expose the player to new experience at a rate which presses him but stops short of causing confusion. The sophisticated person should be able to perform more effectively than the normal outcome from any policy suggested by experience from play because he knows much more than can be put into a simulation model.

A city is an extraordinarily complex institution as compared to those that have hitherto been reduced to working models. The diversity of the city and its range of activities force us to seek new principles in model-making. Existing simulations for social institutions, the best known of which depict systems of firms, nations, voters, or travelers, fail to provide adequate analogs for representing whole cities progressing through time. Obviously a great deal of borrowing of previously developed concepts and insights is possible, but the central mechanism that organizes a city, inducing it to live and adapt, cannot be found in this prior work. *It is asserted here that this focusing element is the production of income, opportunities, and other rewards that are held to be most important by the mobile, migratory, marginal elements in human society who have choices to make about where they will work and live.*

Thus the ultimate criterion in decision-making for a city in the era that is now upon us must be the *quality* of the urban environment produced in that city as a consequence of the decision. Therefore enhanced quality must also be the aim of the game. Quality is to be achieved within the constraints of the geography, history, laws, conflicts, wealth, educational endowments, physical equipment, and governmental capabilities. Then, with a lag of only a few years, a better-than-average balanced performance in operating a city should result in population growth greater than its principal competitors. Thus the quantitative index of success in the middle run is ecological more than anything else. It is the same for a metropolis as for a biological community—enhancement of the numbers being supported—but growth is not achieved through accommodating distress migration from elsewhere as has been the explanation in the past. A better urban environment provides a quality of life that discriminating people appreciate and will seek out. Such immigrants bring most of their assets with them, and are willing to put their special skills to work in new wealth-creating enterprises operating in the vicinity of the new home.

THE URBAN DEVELOPERS

What urban roles become most significant in this new competition between cities? How are decisions about the quality of the urban environment reached?

For a gaming simulation we are forced to identify the central roles, simplify them, and finally cast them into a game structure that can handle the most typical decision processes.

The defining of such roles is influenced by the conclusions reached in the foregoing pages. At different stages in the analysis it was argued that (a) the middle- to long-range future should be in sharpest focus, (b) the capital improvement program is the most suitable administrative instrument for identifying public commitments to growth and development over this period, and (c) the criterion of success should be an increase in overall attractiveness of the city as viewed by mobile urban residents. The amount of social theory applicable to urban development is quite small and provides little guidance. Exploratory experiments in game construction are more helpful.

Five categories are proposed here as crucial: *Politicans* settle conflicts in the demands for space and other scarce resources in urban development. *Businessmen* create enterprises which guide settlement patterns and secure new commercial and industrial growth: their managers are expected to make profitable use of the improvements in the urban environment. *Administrators* and *planners,* taken as a single category because their efforts very much overlap, enunciate specific goals for growth and generate the projects for civic development. *Educators* superintend the major investments being made in human resources. *Judges* rule on the otherwise insoluble conflicts, determining the rate that the interpretation of constitutional provisions ("rules of the game") may evolve, and accommodate the pressures for structural change.

Politicians are given top billing in our cast because each significant internal crisis in a city is quickly brought to their attention. Politicians can do something about this maelstrom of issues only if they play a private game of their own that of seeking and maintaining power In the American city of the recent past the urban issues were those of accommodating the rural influx and developing the economic base that would provide jobs for these people. The civil rights and housing problems of great cities that stem from the same era of immigration will be demanding attention for decades to come, but they will be incorporated increasingly into the arguments for improving the quality of the urban environment. A new kind of urban politician is coming into being that is willing to cope constructively with these new problems. Perhaps the type to rise to the top most frequently in the future is best epitomized by Mayors Lee of New Haven and Cavanaugh of Detroit. The game they must play is essentially that of mobilizing support for continuation in office by guiding the flow of benefits from improvements of various parts and functions of the city.

Politicians must not only seek power, they must spend it wisely. Although the central city, which is ordinarily kept most closely under control, may contain a predominant share of the low-income and poorly educated population, the metropolis as a whole must be made a desirable place for the middle class, which is coming to control the votes. Therefore the current crop of strong mayors is intent upon gaining the initiative in bargaining with the largely middle class suburbs in the course of determining transport, recreation, water, and pollution prevention policy. The fragmentation caused by the insulated metropolitan communities on the periphery, each dedicated to the preservation of a special style of life, is declining in importance for these issues. Urban politicians of the future, we believe', will dispose of metropolitan issues and depend upon metropolitan support, regardless of the immediate constituency. The ascendancy of the concept of a metropolitan data bank offers an important confirmation of the trend, and there are others. A model of the politician's role which largely neglects the frustratingly complex effects of polyarchy may therefore be increasingly realistic.

A chief politician, such as a mayor or a boss, holds court in a manner reminiscent of the reigning monarch. Various other responsible roles have access to him, funneling in information and opinions while transmitting outward both queries and decisions. Businessmen hold a crucial place in this inner circle because they make a variety of contributions to urban development.

The range of productive activities is most often extended by businessmen, and the increasing variety of jobs they stimulate may provide greater opportunities for the labor force. The businessmen participating in government do so by supporting the election campaigns and special publicity programs of the politicians. They hope to obtain informal vetoes over proposals that seriously diminish the expectations of profit and viability of the firms. Politically effective entrepreneurs are usually too well established to be innovators themselves. Nevertheless the businessmen exerting influence hold strong sympathies for the builders of new businesses and serve as spokesmen for the whole occupational group.

Businessmen on the whole cannot pay too much attention to politics because they are forced to play a business game in order to survive. A minority component is active in the urban land market, buying land, promoting structures, and shifting the uses to which specific properties are put. Another group is operating commercial establishments whose prosperity is sensitive to the physical organization of the city, especially the accessibility of customers to sites. A growing group is dependent upon educational, recreational, and cultural investments made in the name of the public. However, they all measure personal success in terms of profits, and security in terms of a dependable share of the market. A condensation of their behavior is to be found in the CORNELL LAND USE GAME.[12]

Planners propose components of the capital improvement program, working out the balances that are needed for the orderly development of the city. Administrators, almost interchangeable in this sense with planners in various parts of the continent, have a strong influence in the formulation of the respective projects and in the construction of the annual budget. Since the private game they play seems to be primarily that of professionalism, where competence and achievement on the job are rewarded with respect by fellow professionals (later by fellow players), eventually this respect must be confirmed by salary increases and promotions. Briefly stated, these bureaucrats are pressed by their circumstances to maximize professional prestige. The accumulation of prestige is a means of acquiring security, since if political conditions worsen and a man is being forced to operate contrary to his principles, respected professionals can always move elsewhere to a job as good as or better than the one presently held. Professionalism, even if it produces little security and no extra salary, explains much of the budget-to-budget behavior of planners and administrators that is independent of the urban crises with which they must deal. They compete for prestige points while working out and implementing programs for improvement. Sociologically speaking, the professions are by nature quite conservative and tradition-bound, so it is not uncommon that a good program for a given city requires compromise with professional aims, thus creating a typical bind for the responsible bureaucrat.

Educators are more difficult to analyze. For a long time their private game was separated from urban politics, because education was not compatible with the normal level of corruption. Education remained under the control of middle-class elements in the population—the professionals, proprietors, and managers. Separate school districts with their own taxing power emphasized the independence of education. Health services, the other major facet of social welfare that conserves human resources, were distributed between private, cooperative, charity, religious, and public institutions, and therefore often excluded from the capital improvement program.[13]

Times are changing rapidly, however. The quality of public schools has become a prime feature in the attractiveness of a city. Most tracts of land on the periphery can be developed only if new schools of sufficient capacity and of good teaching standards can be assured. Some of the desirable new industries refuse to move into a community that has been reluctant to improve its schools. The citizens engaged in voluntary action for community improvement and social reform generally place education at the top of the list of priorities. Increasingly the school elections and bond issues are carefully meshed with those affecting sewers, streets, and parks, and school locations are linked closely to land development plans. The top educators need continuous access to the locus of power, and must participate in the formulation of development plans.

Successful educators reorganize old school systems, and start new institutions. They set standards of instruction that are respected by other educators delegated to inspect and review their performance. School board chairmen (often independent professionals or businessmen) act as members of a team which usually has school quality improvement as a goal.[14] The team does not attempt to maximize knowledge, or understanding, because these outputs of an educational system still defy measurement, but it aims for respectable schools as gauged by educators.

Judges are rarely considered as actors in urban development. Nevertheless, as has already been illustrated, changes in values, goals, and role responsibilities in urban development have been accelerating. Cities are the arenas of conflict, as every case study testifies. Eventually it is the judges who determine whose interests shall be sustained. Their responsibility is to the law and the maintenance of public order. New precedents are established primarily to make the governmental system work without the outbreaks of riots or the open use of coercion if it is at all possible. The judge tries to find the rule that will stand the test of appeal.

In a gaming simulation the role of the judge is played by the designer of the game as it proceeds through a series of improvements. The person supervising the play, sometimes formally given the role of referee, serves as the equivalent of the judge of an appellate court. It is their combined responsibility to keep the simulation going as long as the participants wish to play; their rulings are made with minimum likelihood of overruling a prior arbitration.

The simplest urban gaming simulation is a scenario with a director and four types of roles. The stage is a land-use map for a city and the backdrop a wide variety of socioeconomic and historical information about that city. Each scene is a decision-making cycle closing with the setting of the tax rate and the determination of what goes into the capital improvement program. Each act in the play terminates with an election. The most important audience is the players themselves, because they see nuances in the systems of which they were previously not aware. But there must be a much larger hypothetical audience built into the script; it is made up of residents and sojourners. Will they like what happens to their city? Their approval must be built into the election simulation. Beyond this audience is a population receiving information about the results of the play. Do the reports describe the development as a more interesting and rewarding environment than their own? If so, they are impelled to change environments. When the box office prospers the players are rewarded with bonuses. This, in outline, is the system that emerges: it is a play with lines composed by the actors themselves according to rules carefully designed to model the real world.

This exposition began with a very scientific-sounding definition of the reduction process in model-making. The technical challenge was that of retaining a modicum of verisimilitude while selecting less than one part per million of the public transactions. Nevertheless the description of the elements of a workable model seemed to fall back upon the dramatic arts. The connection is closer than that of an analogy. In large part this is because playwriting is dependent upon a variety of techniques of compression, and the dramatic medium has educated a large public to the point where many of its conventions for summarizing action are readily comprehended. The heuristics for gaming simulation are to be found in the fashioning of successful drama. Recognition of these precedents greatly accelerates the trial-and-error search for acceptable modes of condensation.

However, if a game is to be used to improve decision-making, there must be a plot that faithfully represents the metropolis being modeled. The plot is fixed only as long as it reviews the recent past. When the game is launched into the future the players have freedom to decide how they will act, but the rewards and penalties are calculated more finely than can ever be accomplished by measuring applause. After each scene players must be told what action in the immediate past drew the respective payoffs. Each player is therefore rapidly conditioned to stay within reasonable limits of role behavior. Together the players are then able to explore one of the more likely futures of the metropolis, progressing year-by-year.

A game we have constructed, METROPOLIS, fits these specifications. It has been played about thirty times by students, professionals of various kinds, and citizens filling responsible community roles. Although the kind of experience acquired in a gaming simulation is not readily measured by an objective test, the students' test scores were clearly superior to those of a control group learning about community process and structure in a normal fashion. This game has demonstrated its flexibility in permitting from three to thirty to play, and it can be operated so the intercycle computations are handled either manually or with a computer. Also, unusual crises that may possibly affect a metropolis of the size and structure modeled (an inland city without satellites in the 100,000 to 500,000 category) have, on occasion, been introduced and played out.[15]

THE HYBRIDIZATION PROCESS

An interesting model will mutate, evolve, amalgamate, and proliferate. The process has already begun to occur in our case.

In the community model we have created only five kinds of computations which are required. This degree of simplicity is required because a schedule of payoffs connected with explicit decisions can very quickly exceed the capacities of manual calculation. The outputs from cycle include:

(1) On the basis of role-to-environment interactions, the roles are rewarded for the expertise the players bring to them or have acquired as they proceeded;

(2) role-to-role interactions are a function of success in gathering current information and in bargaining;

(3) the collective effects of role actions on urban prospects must be summed (a capital budget that maintains balance between the wards of the city and between the various categories of service, while at the same time not raising taxes, is optimal), enabling calculation of the overall change in attractiveness of the metropolis;

(4) a feedback from change in attractiveness results in additional rewards and penalties for the players;

(5) a partial record of the sequence of play is retained in order to expedite subsequent discussion and analysis.

A carefully constructed block diagram showing information flows, decision points, value flows, and feedbacks is readily converted into a computer program. Standardization of operations in the game up to that level may be achieved after ten to twenty exploratory manual runs. The programming effort changes the model in some details and, it is hoped, the use of a computer will eventually speed up the play somewhat. Figure 1 represents the fourth version that was formulated, and the second programmed for computer-assisted operation.

Once a debugged program exists a few of the crudities of the original model become salient. It is evident that some subroutines synthesized for describing features of the city will retain their basic structure regardless of any reformulation of strategies in the game. Some of the information that would go into a metropolitan data bank can be synopsized and kept available. A sample of the household sector provides one body of such information, and the distribution of land use another. With these subroutines at hand the payoffs to businessmen investing in property can be rendered more realistic. School enrollment projections are also sufficiently routinized to be introduced as a computer-simulated subsystem.

The most practical use of a gaming simulation is a for operational gaming. This involves data from a real city introduced in order to throw light upon the implications of policy alternatives. The technique becomes feasible after several

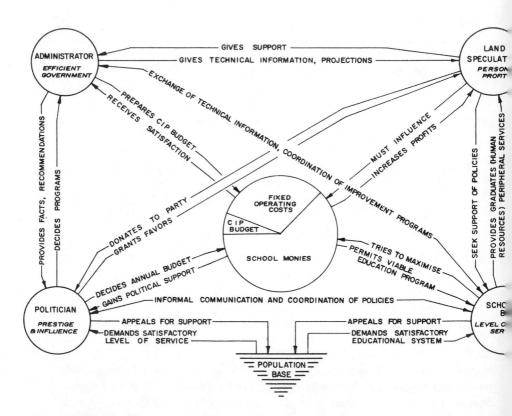

FIGURE 1.

subsystems have been successfully joined to the basic model. An attempt to construct such an operational gaming simulation was undertaken early in 1965 for the tri-county area surrounding Lansing, Michigan.

When a gaming simulation is backed up by these routines, and the feedbacks are refined to make the extra capabilities worthwhile, a hybrid simulation has come into being which can be used for testing unusual policies that would not be risked under ordinary circumstances. Under ideal circumstances a run should be mountable in about a week. A day or two per run would be needed for each operation of the game.

Over the years that a hybridized gaming-simulation is being constructed, however, much self-education will have occurred. What is systematic in the decision-making for the city should have been teased out of its operations. The artistic, intuitive, qualitative, often highly political features would have been noted. During this period many of the most crucial turning points in development would have been explored with the crude model of the game to the point that policy makers became familiar with the predictable consequences. Middle-rank professionals and new additions to the staff would have been trained to expand the range of considerations taken into account in planning and budgeting. The fully assembled hybrid would constitute a tool that is already quite well understood, and is not likely to be misused. The cost is moderate, and the benefits are continuous.

The "pure" computer simulation, in contrast, is a heroic task requiring years to assemble and test before it is ready to serve. Thus far they are much less flexible in structure, and are only gradually improving in this respect.[16] Although much can be learned about systematics in the course of preparing the original design of the computer simulation and collecting data to fill in the gaps, the processes by which the conclusions are reached are difficult to keep in mind. The results are likely to be fed directly into the decision centers with a minimum of interpretation and qualification. If no other institutional changes are made, it represents an authoritarian approach to planning, because the boss, in conjunction with his computer technicians and immediate staff, would be able to reach the only informed decision. The rest of the staff would collect and prepare the relatively microscopic data required to feed the computer simulation. Some cities presently function in an elitist manner that is compatible with this process for developing a working model and the decision makers would be more comfortable taking such an approach. It is much closer to the style of modeling that is now popular in business and the defense industries, and therefore has a reservoir of talent to draw upon that is not nearly as suited to the hybrid form.

This evaluation of the present state of knowledge is based upon the communicated output of no more than fifty persons who have expended about a hundred professional man-years on the development of simulation techniques over the past six years or so. This is small compared to the investment of time and funds which could be expedited by games over the next half-dozen years. Metropolitan data bank organization alone will shortly require the expenditure of that amount of effort per year. It seems highly likely than in the not-too-distant future the electronic data processing subsystems will handle all local problems with definite solutions, the kind that decision makers find are quite easy to handle.

Will the planning and administrative task become easier when the routine work becomes automatic? Systems engineers promise that it will be, but their arguments flow from an analysis of the job of decision-making in the past. A behavioral analysis arrives at the opposite conclusion. Administrators and planners derive much of their personal satisfaction from exhibiting that their training and experience allows them to find a good solution quickly. However, the relatively insoluble issues unaffected by data remain to occupy their attention and soon almost full time will be devoted to them by decision makers.

Extra effort applied to "insoluble problems" is expected to be unrewarding to the office holder, unless he is masochistic. Thus the role of the urban decision maker will be more stressful than it is now. The anticipated result, then, is that top positions in cities will become virtually intolerable and will be unoccupied for long periods at a time.

Bargaining between representatives of conflicting forces may, as a further consequence of this result, become even more important than at present. Role-playing is an accepted technique for conflict resolution. The metropolitan gaming simulations may therefore evolve into quite a differnt kind of instrument in the 1970s and 1980s than is visualized here. But that form would be the offspring of the hybrid.

SUMMARY

Systemization of accumulating urban data, resulting eventually in the establishment of comprehensive data banks for metropolitan areas, together with the introduction of new treatments for the ills of regions, have expanded the range of planning considerations available to be reviewed at a given time. Project synthesis, program formulation, budgeting, policy-making, and long-range planning are activities rendered more complex by the extra information and the introduction of more explicit performance standards. The possibility of making a serious miscalculation or oversight when using traditional methods of planning is increasing rapidly. Thus the need for a simplified working model that tests the feasibility of proposed policy combinations is frequently expressed.

Simulations are primarily intended to cope with the modern complexities of regional and metropolitan planning. Simulations are inappropriate, however, for most planning situations today. Either the planning staff is unprepared, the data base is inadequate, or the usefulness of the results is dubious. Nevertheless, for reasons that often will be political in origin or merely because new fads are gaining acceptance, simulation will be undertaken with increasing frequency over the next decade. The challenge lies in exploiting the technique to fullest advantage.

Three different styles of simulation models can be identified. If data are numerous and optimization of the balance between standard services is desired, a computerized model is indicated. If the range of relevant considerations is imperfectly known, and essential data deal with quality of services (about which there is partial consensus), a gaming model will be more informative. If planning includes budgeting and other routinized procedures along with a variety of poorly defined situations, a hybrid form of simulation, combining gaming with computer programs, should be most effective.

In the development of computer simulations the greatest risks arise from overoptimism about time-of-completion, costs, and explicitness of outcomes. Metropolitan activities are not as well defined and bounded as federal departments and business routines, so that attempts to forecast trends in metropolitan activities trip over unanticipated lacunae in the information base. Filling in these data gaps is expensive and time-consuming. Delays can often be traced to conflict between elegance of study design and the flexibility of outlook required by decision makers. In addition, the "learning by doing" phenomenon, especially the increasing sophistication of the planners involved in the task of replication-in-miniature, often results in major program shifts that delay the production of results.

Recent efforts to perfect a model reveal that an overall performance index for large cities is needed in order to determine the utility of various developmental strategies. In the near future it will be the relative livability of a city that must guide urban development policies. Decision units will be rewarded in the long run if they exploit opportunities for making the city more attractive to new value-creating organizations and to the educated populations that initiate and operate such organizations. The decision units must create attractiveness in their city more rapidly than their competitors in other cities. Thus the ultimate goal of planning at the regional level is to raise the quality of the total environment—social, economic, cultural, and physical—as judged by residents and the sojourners that sample it.

Gaming simulations must identify the key roles for producing urban values. We believe the simplest city model must include politicians, businessmen, administrator-planners, and educators, each of whom plays a subgame of his own with explicit measures of utility. Thus politicians accumulate and spend *power*, businessmen are similarly engaged for *profits*, the administrator-planners seek *influence* of a kind that results in prestige for themselves and the educators seek *respect* for the performance of their school system. These are representative of a set of "selfish interests" which compete with public interest in decision-making. Another role, that of conflict resolution mainly through *consistency* or fairness in rule-making and enforcing, is implicit in gaming, but it is assumed by the designer of the game and the referee of the run.

Hybrid models will best simulate metropolitan situations in the long run. They can achieve the most thorough compression of the real world while still retaining the flexibility for exploring widely divergent futures. They should evolve as "teaching machines for institutions" so that a staff can acquire a common image of their organization and its environment which expands upon their specialist training. Planning agencies for metropolitan centers should not only use such models to prepare for complicated transitions, but should make them available for training personnel in other agencies. Schools would find them even more useful than the "business games" that have already been hybridized and the "inter-nation political games" which are about ready for this step. The primary value of these models is to improve human judgment; thus their preparation is essentially an educational enterprise.

NOTES

1. No strongly developmental effects have been achieved through urban renewal as presently conceived (cf. A. H. Schaaf, "Public Policies in Urban Renewal: An Economic Analysis of Justifications and Effects," *Land Economics,* XL (February, 1964), 67-78; Roger Montgomery, "Improving the Design Process in Urban Renewal," *Journal of the American Institute of Planners,* XXXI (February, 1965), 7-20; David A. Wallace, "Renaissancemanship," Ibid,. XXVI (August 1960), 157-76.) Criticism of the welfare outcomes of these programs has been assembled by Herbert J. Gans, "The Failure of Urban Renewal," *Commentary* (April, 1965), 29-37. Planned regional development has also yielded little net impetus (cf. John Friedmann, "Regional Development in Post-Industrial Society," *Journal of the American Institute of Planners,* XXX (May, 1964), 84-90; John E. Moes, *Local Subsidies for Industry* (Chapel Hill: University of North Carolina Press, 1962) unless one employs a variety of extra techniques for stimulation and guidance such as were, for example, employed in Puerto Rico. R. L. Meier, *Developmental Planning* (New York: McGraw-Hill, 1965).

2. President Johnson's "Message on Cities," March 2, 1965.

3. A review of the present expectations was conducted at the Systems Development Corporation, New York University symposium on *Electronic Data Processing Systems for State and Local Government,* September 30 to October 2, 1964. For an abstract see Systems Development Corporation Magazine (November, 1964). The evidence strongly suggests that Los Angeles and Puerto Rico are setting the pace in the introduction of EDP.

4. Many workers in the simulation field may object that this ascription of value is too limited. In his introduction to the symposium, "Urban Development Models: New Tools for Planning," *Journal of the American Institute of Planners,* XXXI (May, 1965), 90-95, Britton Harris adds the calculation of optima and the evaluation of alternative development strategies, When the kinds of optimization and evaluation that are feasible are reviewed, we discover there is little difference of opinion except on the proper terms with which to express modesty.

5. H. Guetzkow, *Simulation in Social Science* (Englewood Cliffs, N.J.: Prentice Hall, 1962); Ira S. Lowry, "A Short Course in Model Design," *Journal of the American Institute of Planners,* XXXI (May, 1965), 158-66.

6. Ithiel D. Pool, "Simulating Social Systems," *International Science and Technology* (March, 1964), 62-70.

7. Recommendations for this kind of procedure based upon recent experience with computerized models are now being provided. Cf. Britton Harris, "Organizing the Use of Models in Metropolitan Planning," presented to the Seminar on Metropolitan Land Use Models, Berkeley, California, March 19-20, 1965, mimeo 41 pp.

8. R. L. Meier, *A Communications Theory of Urban Growth* (Cambridge: M.I.T. Press, 1962).

9. H. Guetzkow, et al., *Simulation in International Relations* (Englewood Cliffs, N.J.: Prentice-Hall, 1963).

10. The evolution of the images of political units from that of distant figures is provided by David Easton and Robert D. Hess, "The Child's Political World," *Midwest Journal of Political Science,* VI (August, 1962), 229-46. The properties of these images in the adult

world are provided by Gabriel Almond and Sidney Verba, *The Civic Culture* (Princeton, N.J.: Princeton, 1963).

11. The nature of scientific explanation enters very strongly at this stage in the argument. Cyert and Grunberg faced similar questions in their work on simulation of the firm, viz. R. M. Cyert and J. G. March (eds.), *A Behavioral Theory of the Firm* (Englewood Cliffs, N.J.: Prentice-Hall, 1963), Appendix A, pp. 298-311.

12. Allan Feldt, "The Cornell Land Use Game," Department of City and Regional Planning," Cornell Univiersity, Ithaca, New York, 1964, mimeo.

13. A major attempt at model-making for health has been stimulated by the needs of public health education. "Dixon-Tiller County" is depicted in a series of reports from the Hypothetical Community Training Unit, Community Services Training Section, Communicable Disease Center, Atlanta, Georgia, Bayard F. Bjornson, Chief. The vital statistics, history, and geography strongly resemble a small Georgian metropolis, but politics, race and controversy have been expunged!

14. Normal D. Kerr, "The School Board as an Agency of Legitimation," *Sociology of Education,* XXXVIII (Fall, 1964), 34-59.

15. Richard D. Duke, *Gaming Simulation in Urban Research,* Institute for Community Development, Michigan State University, East Lansing, Michigan, 1964.

16. In the long run it is hoped that all-purpose simulation languages will exist which greatly reduce the programming effort and increase flexibility. But it will be some time before they have sufficient capacity for the simulation of a city. H. S. Krasnow and R. Merikallio, "The Past, Present and Future of General Simulation Languages," *Management Science,* XI (November, 1964), 236-67.

POTENTIAL RELATIONSHIPS BETWEEN
ECONOMIC MODELS AND HEURISTIC
GAMING DEVICES

ALLAN G. FELDT

My comments are addressed to those problems which arise in attaining effective utilization of economic models by persons charged with making strategic policy decisions based upon the outcomes and predictions generated by such models. It seems quite apparent that in recent years enormous strides have been made in our ability to adequately model and represent a great variety of complex organizational and behavioral processes through simulation and related techniques. If we have not yet attained a fully satisfactory stage of development in our actual construction of such models, it does seem safe to assume that the time is not far off when a number of highly general, precise and fully validated models of this type will be available for use in policy formulation. The generation of these models is not enough, however. They must also be employed effectively in policy decisions, usually by persons quite different from those who have worked on the development of the models themselves.

In general, it seems that potential users of models may be broken down into three basic types:

(1) Those who possess a high degree of training in mathematical, economic, and computer techniques and who fully understand the workings of the model and its output;

(2) those who understand nothing at all about the model and either distrust it because it is inhuman or are afraid of it because it seems to indicate that their own knowledge and skill is now completely out of date, and;

(3) those who have enough understanding of modeling procedures to appreciate that the construction of a model is possible and worthwhile and who are likely to believe anything which the model tells them because of the reverence with which they view the entire process of model construction and operation.

Reprinted from Vincent P. Rock (ed.) Policymakers and Modelbuilders: Cases and Concepts, 1966, with the permission of Gordon and Breach, Science Publishers.

A fair example of the latter class is those persons who will blindly accept any output from a cohort-survival population projection without giving any careful consideration to the morality, fertility, or migration assumptions upon which it is based. Members of the first class are rare, particularly in important policy-making positions. Unfortunately, it is persons in classes two and three who are most likely to predominate in any potential public of model users in the foreseeable future. Their lack of understanding has seriously inhibited the adoption of models and their results in policy-making and has occasionally led to inadequate public support of modeling activities.

Although the development and validation of a model may often be of great theoretical significance in proving or disproving any particular body of theory, their proliferation is of little public consequence until these models may be brought into normal usage in policy formulation and decision-making. It is to a potential solution of this dilemma that I would like to direct the content of this paper.

At the same time that advances in computer technology and operations research techniques have flourished and brought about our present relative sophistication in modeling techniques, another body of techniques has also been developing which is called gaming-simulation techniques—now more commonly known as either operational or heuristic gaming. Developments in this area with respect to urban studies are still relatively recent although the techniques have a much longer—if more prosaic—history in war games, business games, and a few related areas. Our experience with heuristic gaming devices in studying and teaching about urban processes and growth makes it clear that these techniques offer a very powerful device for communicating complex bodies of ideas and system processes to almost any audience at all, regardless of their degree of mathematical or economic sophistication. The stimulation and involvement of participants in a heuristic gaming exercise is great and almost anyone who has played a well-designed game at some length comes away with considerable insight and understanding as to the way in which the game model works no matter how complex it may be. Essentially, the game players learn the process through sheer experience in collapsed time. At the same time, it has become apparent to many of the persons involved in the development and use of heuristic gaming devices that their predictive power is low and their ability to provide suitable specific information for real world policy formulation is virtually nil.

Such games, however, are models of essentially the same type as the more widely known and appreciated economic models. They contain a set of inter-related factors which can only be handled within certain constraints set by the rules of the game. The sequencing of the processes follows a relatively rigid set of steps of play and development which is, again, set forth within the rules of the game. The only necessary difference between a heuristic gaming model and a more common simulation model is the degree of detail and precision contained within particular subroutines and the fact that within the gaming framework the players themselves make a number of the decisions which a computer would more normally make by linear programming or Monte Carlo techniques. Irrational decisions are also more common in a game since players are seldom able to collate and evaluate all the information which should bear upon a particular decision and they are often biased by their own perception of other players and events in the game. Although irrational by any objective criterion such behavior is at the the same time more real than some of the decision processes provided in simulation models.

We thus find ourselves with two related but distinctive devices in the growth of models. The first device promises a high degree of precision and reliability but suffers from its inability to be understood by the persons most likely to have need of its results. The second device offers low precision and reliability but seems clearly to be able to communicate its own nature to a broad range of potential publics and users. It seems abundantly clear that very serious consideration must be given to the possibility of combining the use of these two devices in such a manner as to allow each of them to compensate for the other's basic weakness. What I am trying to suggest is that any well-developed and proven simulation model should have a heuristic gaming device constructed to teach users the basic principles of the operation of the more complex model.

To illustrate the possibility of such a combination, I would like to describe briefly the operation of a heuristic game and a simulation model which, quite by accident, are fairly close analogies of each other. The heuristic game I will discuss is a device developed at Cornell University and now in fairly common usage in a number of universities, both in this country and abroad.[1] It also is being employed to at least a limited extent by a few professional planning offices and research groups. The counterpart simulation model is one developed for the Pittsburgh Community Renewal Program by the CONSAD Research Corporation based upon an earlier formulation by Ira S. Lowry.[2] This model is quite widely known and appears to represent a rather important step in the evolution of simulations of urban systems. Unfortunately, it has never become fully operational and is presently suffering from lack of support. The two models were developed completely independently of each other and their correspondence is only approximate at a fairly general level. That they are related to each other at all is entirely fortuitous, but the similarity is enough that one may begin to perceive how easy it would have been to have made them virtually identical.

The Cornell model is now evolving into a second generation of games of considerable complexity which are incapable of being handled without substantial computer assistance. The exact nature of these second-generation models is not yet fully known, and I will restrict my comments at this point to the more simple first-generation game which presents a set of economic and technological conditions affecting the growth of an uncomplicated and abstract community. This game can be handled without computer assistance if at least two persons are available to assist the players in the accounting procedures which the game requires.

The Pittsburgh model consists of a number of interrelated subroutines imbedded in a larger decision framework which appears to provide a reasonable picture of the overall decision process of a large city. The basic elements of the model with which we are concerned are the Industrial Spatial Allocation Model and the Time-Oriented Commercial and Residential Allocation Model. My familiarity with the Pittsburgh models is limited and based primarily upon the reports published a few years ago. The intention of this discussion is not to adequately describe any or all of these models but simply to show their relationship to the Cornell game model.

In the Pittsburgh models, a projection of industrial growth and employment is provided by subroutines external to the location models themselves. These projections are then fed into the Industrial Employment Allocation Model in the form of aggregate employment projections by type of industry for the city and other surrounding areas. This first locational model then allocates new employment to specific areas of the city according to site selection criteria such as availability of land, existing land use restraints, and assessment practices, accessibility to transportation facilities, proximity to previously existing employment clusters, and so on. Following the distribution of new employment centers within the city, the Time-Oriented Commercial and Residential Allocation Model provides for the distribution of the new residential units and commercial services within areas of the city to provide appropriate industrial and commercial employee households and related services generated by the new industrial employment centers. Households are located largely in relation to their required journey to work under the restraints imposed by availability of land, price of land, and the type of land use control in effect. Commercial establishments are located with some reflection of the previously established residential sites. Employment generated by the commercial establishments is then located in turn and some further provision for appropriate commercial service establishments to service these households is also made. A few rapidly converging iterations of this cycle leads to a full allocation of all the new households generated by the industrial employment and all the subsidiary land uses generated by this new employment. In both the industrial and commercial land use allocations, some restraints on the maximum and minimum sizes of units are imposed to avoid developing totally unreal sizes of units. In all cases, provision is made for considering the prices and quality of land being used as well as existing constraints on land development in each area. Failure to find an adequate location may be used to either suggest a relaxing of restraints on new types of land use in an area or to indicate a withdrawal of that type of land use from the area under consideration.

The operation of these two models rests upon the solution of a set of equations in a fairly large number of unknowns and the collection of a considerable amount of data on the properties of existing land uses and their futures within the city. The major delays in putting the model into operation have reportedly been encountered in problems of data collection rather than in the construction and mechanics of the model itself. I understand that some work is still proceeding on the development of the model in the Pittsburgh area although not at the same scale as was true a few years ago. I have also been informed that a model substantially congruent to the Pittsburgh model has been successfully developed for use in Lansing, Michigan.

The Cornell model provides for playing through essentially the same steps of site selection as in the Pittsburgh model, although with fewer variations in type of land use and less precision due to the scale at which the playing pieces have been constructed. The smallest residential playing piece in the elementary game represents approximately 4,000 persons with 1,000 of them in the employed labor force. As in the earlier Lowry model, there is no differentiation of household type provided for in the game. Industrial units are at a scale to provide for employment of either two or four residential units, i.e., 2,000 and 4,000 employees. Only three kinds of commercial establishments are provided for, each scaled at a level to employ a single residential unit. Several modes of transportation are provided within the community as desired by the model operators and costs on the various modes are simple multiples of the costs per unit distance on the most economical mode.

Players begin by purchasing land for development within an existing community which has been established according to the requirements of the game operator. When playing the game for the first time, an empty field is usually provided. For more advanced players, however, any desired pattern of land use can be used as the starting point. Players are provided an opportunity to buy land on a relatively free market basis and are then required to vote as a community for the provision of utilities to various plots of land within the community. Depending upon available capital resources, players may then construct any form of building they choose on any piece of owned land which has been provided with utilities. The type of building erected is generally determined by the amount of capital available and the economic opportunities which present themselves in terms of new industrial employment for residences, potential commercial market areas generated by new residences or industries, and so on. The growth of new industry is basic to community growth in this model and is determined almost entirely by available capital and competition for alternative forms of construction. Some experimentation with stimulating or retarding new industrial growth through factors external to the game has been made which would provide a closer analogy to the Pittsburgh model but it has not been found to be of sufficient relevance for our purposes to warrant its inclusion in the basic game framework.

Following construction of new buildings, players are provided a monetary input to the local economy through the industrial sector, which is in turn partially dissipated to the household sector by payments of wages to the players owning residential employee units. The household sector in turn makes payments to the several commercial sectors for the various types of goods provided and they in turn meet payrolls to their own employees. This step is followed by a levy of transportation charges upon each player for his costs incurred in shipping manufactured products to the terminal connecting to the outside market and for payments for journeys to shop and to work from the owners of the residences. Trying to minimize the amount of this transportation payment provides the major locational determinant for all land use decisions, mediated by the availability of land, its price, and the taxes levied in various areas when a differential tax policy is provided in the game. The determinants are not different in any essential way from those in the Pittsburgh model except for the level of detail provided. Following the payment of transportation charges, players are then presented with their tax bill for the utilities and municipal services provided on the previous round. These charges are usually made in a fairly normal fashion of assessing the value of land and buildings of each player and applying a tax rate voted by the players in the previous round. Some community indebtedness is allowed within existing constitutional debt limits and several alternative forms of taxation and assessment are allowed according to the desires of the players or the game operator. Taxation is followed in turn by land purchases and sales and the game then proceeds through the same sequence of operations for another round of play.

Characteristically, fairly careful players are able to generate a relatively mature city of about 250,000 population from an open field within about twenty rounds of play. Blight and related economic and natural catastrophies are provided for every five rounds by keeping track of the ages of all buildings and forcing players to roll dice upon their buildings with odds of losing them which are proportional to their age and supposed state of deterioration or technological backwardness. Buildings which are lost in this fashion are simply out of play for five rounds but still occupy the player's property and are still counted on the tax rolls of the appropriate players. Before rolling the dice, players are provided an opportunity to renovate buildings, thereby decreasing their possibility of loss; but the costs involved are often prohibitive. Patterns of ecological succession in areas undergoing a transition in land use are commonly observable in the form of very old buildings which are never renovated and are awaiting demolition or movement to some other area.

The basic steps of operation of this model do not depart in any major way from the heart of the Pittsburgh models. The locational factors follow a sequence of industrial location, residential location with respect to the industries, commercial location with respect to the residences, and finally more residential location with respect to commercial sites for their employees. Factors of taxation, land value, assessment, rebuilding and demolition, transportation costs, topography, soil condition, utilities, financing, and zoning also enter into the locational decisions, although some of these factors are only introduced in runs of the game operated for advanced and well-qualified players. In computer-assisted versions, we also provide for overloads upon the various utility plants located in the community and traffic congestion on certain major arteries. The models are in many respects identical, and full identity would have been relatively easy to achieve had we been more fully aware of the elements of the Pittsburgh model and the possibilities inherent in achieving identity.

Use of the game in a wide variety of contexts has made it clear that almost anyone can play it reasonably well and understand its basic operation after ten or twenty rounds. It has been used extensively at the University of Pittsburgh and has included as players some persons who were totally unaware of its inherent relationship to the Pittsburgh models themselves. The acceptance of the game model by some of the same persons who have not fully appreciated the more sophisticated Pittsburgh models offers wry testimony to the relative power of the two formulations in communicating basically similar ideas. It might even be argued that the Pittsburgh models would have enjoyed greater financial and political support if they had provided for the concurrent development of a heuristic gaming representation of the models themselves.

Experience in the construction of heuristic gaming models is still limited but a few general observations may be drawn at this stage. The development of a well-formed game reflecting basic economic or sociological principles is not a simple task and requires several man-years of input for its elaboration and testing. This time is shortened, however, if the basic theoretical structure is provided beforehand by an accepted mathematical formulation. The time involved to develop such a model seems great in relation to the development required for normal teaching devices but it is quite small in relation to the time and financial load required for the development of a traditional simulation model. It seems reasonable to assume that a heuristic gaming counterpart to almost any operating simulation model could be constructed for less than five percent of the cost of the basic simulation model itself. Finally, it seems likely that persons charged with constructing a full-scale simulation model would benefit greatly from playing a counterpart heuristic game even in its preliminary stages of development as part of the larger model construction project. Such benefit would be even greater if some of the opposing players in the game were made up from the potential users of the model including higher-level policy makers. Many of the problems in communication between these disparate sets of actors would be cleared up readily in such a context, and further developments of the model would be more fruitful and readily understood.

The basic thesis of this paper is, then, that closer attention to the potentialities of heuristic gaming must be given by model builders. These devices possess important qualities despite the unfortunate connotations of the word *game*. Their basic quality, the ability to communicate complex ideas in simple terms, may provide a means of solving the increasingly important problem of making the meaning and content of simulation models intelligible to the general public.

NOTES

1. For a somewhat more complete description of this game and references to additional sources on it see Allan G. Feldt, *Operational Gaming in Planning Education, Journal of the American Institute of Planners.* Vol. 32, No. 1 (January 1966): 17-23.

2. A general discussion of this model and references to the technical reports in which it is described more completely is given in Wilbur A. Steger. *The Pittsburgh Urban Renewal Simulation Model, Journal of the American Institute of Planners.* Vol. 31, No. 2 (May 1965): 144-150. See also Ira S. Lowry, *A Model of Metropolis* (The RAND Corporation: RAND Research Memorandum RM-4035-RC. August 1964).

GAMING:
A Methodological Experiment

ROY I. MILLER and RICHARD D. DUKE

University of Michigan

INTRODUCTION

J ust five years ago, government officials in nonmilitary agencies would shudder in disbelief at the prospect of using "games" in an applied (decision-making) context and would grimace with skepticism at proposals to use "games" in an instructional setting. Two years ago, this skepticism was replaced by curiosity as two major governmental agencies (HUD and DOT) circulated RFPs for the construction of gaming-simulations for training purposes. During this past year, the number of gaming-simulations under construction increased dramatically and the number of people using games, especially in educational institutions, sky-rocketed. Yet, it is fast becoming apparent that government expectations were not met, as much of the early curiosity has reverted to a healthy skepticism (try to fund research in gaming with federal resources). As with so many other innovative approaches to social system problem-solving (e.g., PPBS), enthusiasm for gaming-simulation appears to be going full circle—first none, then too much, and then none again. This should be neither surprising nor depressing. Rather, it should be a signal to people in the field to reassess their efforts and place them in a proper perspective. This paper is an attempt to do just that. It is the major thesis of this paper that, like those other innovative methodologies, gaming-simulation will find its place in society if it is allowed to mature and develop in the proper way. Furthermore, the impact of gaming-simulation on the political process can and will be dramatic.

WHAT IS GAMING-SIMULATION?

The temptation to launch a full-blown discussion of the nature of gaming-simulation is hard to resist. The ambiguity of the term in the literature and in

AUTHORS' NOTE: *This is a slightly revised version of a paper prepared for delivery at the 1972 Annual Meeting of the American Political Science Association, Washington Hilton Hotel Washington, D.C., September 5-9.*

government circles is great, and a new clarity of definition is needed. However, although we have tried and are still trying, we have found it difficult to arrive at the type of definition needed to add precision to the analysis and evaluation of gaming-simulation exercises.[1] Because a more standard and less precise definition will suffice for this paper, we, too, will be fuzzy. Unfortunately, we will not offer the definitive statement by which one might measure the value or appropriateness of using a particular gaming-simulation. In fact, our definition may even fall short of providing help in determining whether some "thing" is or is not a bona fide gaming-simulation.

The gaming-simulation methodology is a merger or hybrid of two distinct techniques: gaming and simulation. Although many authors, game builders, and game users use the terms interchangeably, there are games that are not simulations and, similarly, there are simulations that are not games. Although we make this distinction to contribute to the clarity of our definitions, much of what is said later applies equally well to games and gaming-simulations.

Simulation is the attempt to abstract and reproduce the central features of a complex system for the purpose of understanding, experimenting with, and predicting the behavior of the system. Usually, the abstraction or simulation utilizes some model or representation of the structure of the system and some mechanism for re creating the process by which the system operates. In modern social science, simulation often takes the form of a computer program; however, restriction of the term to computer modelling is overly constraining. Whereas one might resort to simulating river flows on a computer to predict the future state or gain understanding of the river system, so might one resort to simulating the flow of the river by building a small-scale physical model of the river and pumping water through it.

Games are more difficult to define. Simply, they are sets of activities performed by groups of people (groups of one are allowed) where a set of rules or conventions constrains or defines the limits of activity. The rules themselves may be dynamic and, therefore, subject to change during the play of the game. Generally, people playing roles perform the sets of tasks suggested by the rules. Interaction in the form of competition and/or cooperation is almost always required to enable players to perform those tasks. If this was the entire definition, practically any group activity could be called a game. In fact, some scholars have argued that all group activities can best be understood when viewed as games. Norton Long (1958: 3) argues

> It is the contention of this paper that the structured group activities that coexist in a
> particular territorial system [community] can be looked at as games.

We feel it necessary to restrict our definition to something that differentiates between "real-world," serious gaming and gaming for fun. In the latter case, people participate for fun and, when the activity becomes tedious, they quit. The characteristic of serious games that distinguishes them from fun games is that they are pursued even under adversity. The characteristic that distinguishes serious games from the real-world (the characteristic that probably accounts for the regrettable selection of that particular word as a name for this type of serious activity) is that the outcome—the winners and losers—are not final, nor are the participants held accountable for their behavior outside the game environment. (One might argue that the $20,000.00 won by each man on a professional championship football team is serious and final and that football is fun; however, this apparent contradiction of the definition is easily explained by acknowledging that professional football is a business, but touch football is a game played for fun.)

Gaming-simulation combines the two techniques: simulation and serious gaming. The set of rules governing the play of the game and the scenario setting the game environment act to constrain behaviors in the game so that the game activity simulates some more complex system. Quite often, computer-simulations are meshed with human decision makers in gaming-simulations but, again, it is overly restrictive to limit the term gaming-simulation to such man-machine systems.

These definitions or concepts of simulations, games, and gaming-simulations are virtually content- or subject-free. As defined, simulations, games, and gaming-simulations can (and do) appear in any and all fields. In short, we have defined a methodology, a way of looking at many kinds of systems in a variety of ways. To illustrate these concepts, let us offer a few examples.

It is possible to distinguish between physical systems such as climatological or hydrological systems, social systems such as cities or political parties, and logical systems such as mathematics or grammar. A group of non-simulation games have been developed that do not abstract physical or social systems. Instead of re-creating a physical or social system in a gaming-model, these non-simulation games re-create a system of logic or body of abstract knowledge and transfer it to participants through the activity of play. The Resource Allocation Games designed by Layman Allen and others are examples of non-simulation games.

> When the resources to be allocated in such games are symbols representing the fundamental ideas of a field of knowledge, the resulting activity can be a powerful instructional interaction. A learning environment can be designed to emphasize interacting peers creating and solving highly individualized problems for each other. . . . Players have something to do with the ideas that they are engaged in mastering; they don't merely hear them or see them expressed in print [Allen, 1972: 2].

Examples of simulations that are not games are more plentiful. Probably the best known political science simulation is the Simulmatics Project conducted to enable the Democratic Party to estimate the probable impact of different issues on the public during the 1960 election campaign (Pool and Abelson, 1962). Here, actual data were used to simulate the response to key issues of carefully selected voter classes reacting to cross-pressures created by opposing religious and party affiliations.

Again from the political sciences we draw an example of a gaming-simulation—the INTER-NATION SIMULATION (INS) (Guetzkow, 1968) The INS represents intranational processes through a computerized simulation relating selected prototypic variables and introduces international processes through gaming. Thus, it is a man-machine form of gaming-simulation.

WHAT DOES A GAMING-SIMULATION DO?

The three examples in the last section illustrate not only the distinction between games, simulations, and gaming-simulations but also the diverse range of application of *each* of these techniques. The Resource Allocation Games are primarily (if not entirely) education or training exercises. The Simulmatics model is used to make conditional predictions of election outcomes through pure simulation. The INS is most often thought of as a device to build theory or compare theories—that is, do research.

Quite frankly, we know of only a handful of vigorous scientific attempts to demonstrate that gaming-simulations do any of these things well—teach, predict, or coordinate research.[2] The claims of our colleagues are justifiable by example (all games have a stock of anecdotes to illustrate the wonders of games) or by intuition (all gamers argue that if you'll just try it, you'll like it). This lack of more scientific justification or proof of the efficacy of gaming-simulation should not be construed as a "cop out" or failure on the part of gaming-simulators. The simple fact is that games are attempting to replace or supplement conventional forms of teaching, speculation, and/or research by introducing new and intangible features to the more standard approaches; therefore, conventional modes of evaluation are inadequate. Furthermore, the single most important one of these features is the ability to "handle" complexity, a feature that is hardly understood and certainly not measurable.

To complicate matters further, many people in the field argue that the methodology can be applied in a fourth way—to aid decision makers by allowing such people to explore the present and/or the future through the play of gaming-simulations. This, of course, creates a new set of problems for proponents of the methodology and opens the door to a completely different type of skepticism on the part of its critics.

We feel that even though the uses of the methodology are varied (and, therefore, the forms gaming-simulation takes are many), there is a single common characteristic of games that helps clarify the field and aid in understanding its significance. The claims and counter-claims of the gamers and the skeptics might, at least, deal with the same substance if this characteristic is carefully identified. If people using games acknowledge this attribute of the methodology, the focus of application in each context (education, predictive modeling, research, decision-making) will sharpen, and the methodology will find its proper place in society. In short, *games communicate.*

METRO—AN EXAMPLE

Like so many of our colleagues, the logic of our position about gaming revolves around example and anecdote. With our colleagues at the Environmental Simulation Laboratory, we have worked on the METRO gaming-simulation for eight years.[3] METRO is a complex, computerized man-machine gaming-simulation of an urban area. Players take on the roles of key decision makers: politicians, planners, land developers, industrialists, environmental control officers, and news media specialists. They run a city for a period of five to ten years by making annual decisions that parallel those of their real-world counterparts. The decisions are fed to a series of simulations of natural and artificial phenomena (for example, an air mass model and a land market model) to determine the immediate effect of each year's decisions. The decision-making process is repeated on the basis of the new computer printout (status of the city) for each year in the five- to ten-year period.

The project has been "finished" at least twice only to be restarted again with fresh funding and new personnel. As with so many other major projects beset by funding problems and the associated discontinuities of thought and action, the emphasis and direction of the development of METRO has shifted several times to accommodate the most recent funding agent or client group. Yet, METRO has been nothing more or less than a vehicle to communicate the nature of the urban system to students, policy makers, citizen groups, and anyone else who "wanted to know." We quote from the Phase I report—a document published in January 1966, even before the first computer program had been written.

> The METRO project will attempt to close this gap [between the urban "planmakers" and the decision makers they attempt to serve], utilizing operational gaming techniques to develop a plan effectuation instrument. The instrument will have the capacity of demonstrating to appropriate human decision-makers the consequences of alternative decision chains on metropolitan growth patterns. This will be accomplished through the use of a simulated abstracted environment, employing a reduction of time span and dynamic inter-play of current decisions with fixed policies. The instrument is intended to simulate growth patterns which should occur naturally and enable their comparison with planned growth patterns.
>
> This technique will introduce a dynamic quality into urban "plan-making" activities and allow for a structured interplay between those concerned with plan design and the "decision-makers" upon whom the plan's ultimate implementation depends [METRO Phase I report, 1966: 5].

Although METRO has become known as a training tool, it was initially thought of as a dynamic supplement to conventional planning to further communication among decision makers, planners, and citizens involved in planmaking. As with *all* other large-scale simulations, METRO suffers from some major gaps and inaccuracies in the data and some weaknesses in the theoretical foundations of some of its components. Unlike some other supporters of simulation, the designers of METRO recognize these weaknesses and, therefore, emphasize the training use of the exercise.

As a result, in 1972, when we spoke of METRO as a prototype for a more sophisticated gaming-simulation to be used as a methodology for "dynamic urban planning," the response was that gaming-simulations are teaching tools or, at best, research vehicles. Yet, from the outset, our pioneering work in gaming-simulation has been moving toward a new planning concept—one that will upset the traditional political as well as planning processes. Our hope is to launch this planning concept in some city or region and demonstrate that it can, in fact, work. We call this anticipated first attempt the Laboratory Community.

THE LABORATORY COMMUNITY

Following the almost perverse penchant of people in gaming to select poor titles (witness the horror of the uninitiated at the thought of using games for serious purposes), we selected the title Laboratory Community for our vision. The intent is to provide a community with a set of gaming-simulation devices to be used in a laboratory by people in the community, in which they can register, explore, and play out their problems prior to any actions in the "real world." Unfortunately, too many people hear our title and think of some poor community being used as a laboratory by hard-hearted scientists out to prove they know how to run the world. Despite this, we exhibit the stubbornness and near fanaticism of our gaming-simulation colleagues and retain this title.

In our more fanciful moments, we envision some city procuring a series of gaming-simulation devices and setting them up in a large room in the basement of city hall.[4] In much the same way that the military chiefs in the movie "Dr. Strangelove" used their war room to thrash out key tactical and strategic problems, decision makers would use this "game room" to explore tactical and strategic problems for their city. Thus, the game room becomes the new political forum; however, it is a forum in which decisions are not final and the participants are not held accountable for their behavior when they go upstairs to council chambers. But when they do go up to council chambers, they will have a better understanding of the problems and a better feeling for the nature of the arguments for and against assorted policy options.

In order to appreciate how this game room might work we must explore the types of gaming-simulation tools that would be made available in the laboratory of the Laboratory Community. Each tool is there for a purpose and that purpose must be to further the communication to someone, from someone, or between groups and individuals.

The basic tool would be the METRO-style computerized gaming-simulation. This would be a sophisticated and accurate set of models to estimate the impact of key decisions on the urban system. The data base would be a living (easily updated), viable resource for the community to be used as part of the simulation and also as an independent entity. The purpose for building such an expensive and cumbersome gaming-simulation (we estimate that it would take our staff ten years at $500,000.00/year plus the strong support of the jurisdiction being simulated) is not so much to predict the future as to foster nonsophomoric discussion of the key current issues in the community. After all, the predictions generated by such a model would be conditional on many ceteris paribus assumptions no matter how complete the simulation; but, with the aid of the model, the discussion on the issues could take on substance and quality.

For example, consider the generally accepted principle that the fate of the city of Detroit rises and falls with the fate of the auto industry. So many decisions are made in Detroit with the impact on the auto industry serving as a prime argument for or against particular action. Yet, the actual magnitude of the effect the auto industry has on the economy is probably exaggerated, as is suggested by the recent failure of a boom in the industry to cure Detroit's unemployment.[5] Only in-depth exploration of the facts surrounding assumptions such as these can improve the quality of decision-making, and (only?) gaming-simulation will make the results of that exploration understood.

A second tool would be a NEXUS-type gaming-simulation questionnaire to communicate the opinions and feelings of citizens' groups to the decision makers and to communicate the professional expertise of the technician to both citizen groups and decision makers (Armstrong and Hobson, 1970). NEXUS is an information-generating and information-disseminating gaming-simulation that combines aspects of the DELPHI technique and survey research in a single gaming-simulation exercise. Players are asked to comment on (or estimate the magnitude of) the impact of various decisions on carefully selected quality of life indicators from the vantage point of their real-life role or some other assigned role. When specialists or "technocrats" play, the resulting consensus (if there is one) on impact represents expert opinion. When citizens play, the resulting consensus (or lack of it) indicates the intensity and direction of public opinion. When combinations of people play, these "consensus agreements" can be juxtaposed and fought over.[6]

As an example of how this tool might work in practice, consider the current method of resolving a request for a rezoning for a given parcel of land. Such a request would be announced in highly technical and unintelligible articles in the local newspaper. A public hearing is held in which citizens get five or ten minutes each to verbalize their feelings to Common Council. The planners present their recommendations to Council at the hearing and/or in a special report. Pressure groups might contact particular politicians to assert their influence. At no time is a clear statement of the impact of the rezoning and the reasons behind the decisions made on the request made clear. Suppose that all persons concerned had systematically responded to a uniform questionnaire via a gaming-simulation device. The results could be tabulated and analyzed routinely if the process of administering the game-like questionnaire was ritualized. The outcome of the play would not necessarily alter the advocacy process in any way but would substitute a uniform method of argument for a haphazard battle of words.

Other types of problem-oriented gaming-simulations would be available through the Laboratory Community. For example, a gaming-simulation might be designed to teach people how to purchase housing, or how to interpret a budget. These are educational devices—but what is education but a form of communication? Similarly, gaming-simulations might be built to solve the problems of communication between the police and the youth of a community[7] or the blacks and the whites.[8] In fact, the possibilities are limitless, subject, of course, to the constraints set by our own imaginations and willingness to experiment.

In sum, the combination of gaming-simulation techniques made available in the laboratory would serve the larger community in the following ways:

(1) Help decision makers identify new and potentially troublesome issues at the earliest possible moment (NEXUS-type game).

(2) Let competing groups reformulate and restate the issues to suit their perspective (NEXUS-type game).

(3) Help decision makers gain reasonable input from technocrats and citizen groups with regard to the probable impacts of decisions on key issues (NEXUS-type game).

(4) Allow concerned people (decision makers, technicians, citizen groups, interest groups, students, etc.) to systematically explore the primary, secondary, and tertiary ramifications of select policy options in the urban or regional system (NEXUS- or METRO-type game).

(5) Bring the power of the computer, a large data base, and related technical aids to those responsible for governing our urban and regional centers as well as to concerned citizens in those centers (METRO-type game).

(6) Provide the mechanism for creating a dynamic, living plan (through the games) to replace the standard maps of the "Comprehensive Plan" used today.

THE IMPLICATIONS OF THE LABORATORY COMMUNITY

The implications for the world of politics (and, hence, political science) of establishing and institutionalizing the Laboratory Community are many. The last section of this paper made repeated reference to multiple sources of input (technicians, citizens, etc.) to the planning and decision-making processes. Of course, these processes are highly political and bound up in tradition but, nevertheless, political obstacles to broadening the base of decision-making and planning must be overcome by today's establishment. In 1963, James Q. Wilson made this point quite clear.

Citizen participation in urban renewal [or other urban programs], then, is not simply (or even most importantly) a way of winning popular consent for controversial programs. It is part and parcel of a more fundamental reorganization of American local politics. It is another illustration—if one is needed—of how deeply embedded in politics the planning process is [Wilson, 1968: 223].

Donald N. Michael suggests that the widespread use of computers with legally enforced citizen accessibility provisions may be the method for reorganizing local politics to accommodate such participation.

> For those who want to participate in the political process, the opportunity to challenge the system or to support it on basis of knowledge *as* the government develops its own position and then to monitor and criticize *continuously* the implementation of whatever policy prevails should be a heady incentive for extensive use of such computer facilities [Michael, 1968: 1190].

He then offers this warning.

> But the approach proposed here involves no casual laying-on of minor modifications in the conduct of urban government. Opening up the information base of political decision-making would be one of the most painful wrenches conceivable for conventional styles of government [Michael, 1968: 1190].

Although the Laboratory Community as described here is a somewhat different concept than Michael's open-access, multiterminal, computerized information system, it will achieve the same end and, of course, pose the same threat to the conventional political process. We submit that one of the most serious potential impediments to the Laboratory Community concept arises out of this threat. Politicians and decision makers may well be afraid of the Laboratory Community.

There are several sources of fear, some healthy and others disturbing. First, policy makers are afraid of being replaced by the "machine." If a simulation were devised that could make accurate and objective predictions of future states given policy alternatives, this fear might prove to be partially valid. However, somebody must still design alternatives and evaluate future states of society—the policy maker. Realistically, the design of such a powerful simulation is highly unlikely. At best, simulation and gaming-simulation will give a feeling for the possible configuration of future states and allow people to communicate their fears and opinions about those states, but the policy makers will still have to interpret the validity and consequences of the assumptions underlying the gaming-simulation when making their decisions.

A second source of fear is the widespread citizen input needed to make the game room viable and, if that input is secured, the consequences of grass-roots decision-making. The expense and inefficiency of maintaining the capability to mingle with citizens in the gaming-simulation environment is a legitimate concern—legitimate, but not wise. The political system in this country must find a way to systematically get at grass-roots solutions to the problems, whatever the cost. Failure to do so is an invitation to social upheaval even more severe than that already experienced by our larger cities. Furthermore, once the cost of initiating the Laboratory Community has been met, the cost of maintenance in time and money should not be too great if the process is routinized. The fear of the consequences of grass-roots decision-making has even less foundation. The elitist view that technocrats know best has been proven wrong. Technology has run away from the reins of society and must be balanced by citizen control.

A third source of fear is the most depressing from the perspective of the advocate of the use of gaming-simulation in a public policy context. A scientific gaming-simulation Laboratory in the basement of city hall, open to all, will erode the power base of those who currently have the information and clout and return the bureaucracy to a position of service rather than dominance. We argue that this erosion must take place despite the fear and opposition of the bureaucracy to such erosion, and, because gaming-simulation has the potential for causing that erosion to begin, it ought to be pursued.

NOTES

1. This problem is treated in detail in a paper presented at the Eleventh Symposium of the National Gaming Council at Johns Hopkins University, Baltimore, Maryland, in October, 1972.

2. Possibly the best attempt to perform controlled experimentation on the teaching capability of games is reported by Margaret Monroe (1968). Other research concerned with the validation of games (as opposed to evaluation) is going on at the University of Southern California (Russell, 1971; Boocock, 1971).

3. A full set of manuals and the computer program for METRO-APEX, the most recent version of the game, are now available from several sources including the Environmental Simulation Laboratory of the University of Michigan, the COMEX Project at the University of Southern California and the Office of Manpower Training of the Enviornmental Protection Agency.

4. As far-fetched as this may sound, we are currently working with the University of Wisconsin at Milwaukee in setting up such a game room on their campus. Although the room is going to serve students at the university in the early stages of the program, the intention is to include citizens' groups and the establishment in the operation of the Laboratory as the gaming-simulations housed therein begin to pertain more directly to the problems faced in Milwaukee.

5. See the News Analysis by Ralph Orr entitled "Auto Sales Soar, But Not the Jobs" in the Sunday edition of the Detroit *Free Press* of June 25, 1972 for some comments on the impact of the auto industry on Detroit.

6. We had hoped to report on our experiences with this technique more fully in this paper; however, our time schedules for experimentation have been set back by funding delays. To date, we have prepared a NEXUS-type game, POLICY PLAN, to study housing problems in Rochester, N.Y., but have not finished testing this particular exercise. POLICY PLAN was designed by Larry Coppard and Mary Kay Naulin of our staff.

7. One such game, POLICE COMMUNITY RELATIONS or, more popularly, THEY SHOOT MARBLES, DON'T THEY, has been designed by Fred Goodman of the School of Education at the University of Michigan.

8. Several games have been designed on this theme. One of the better ones is URBAN DYNAMICS, designed by the staff of Urbandyne in Chicago.

REFERENCES

Allen, Layman E., "RAG-PELT: Resource Allocation Games: Planned Environments for Learning and Thinking." Presented to the International Conference on Simulation and Gaming (Birmingham, England: July, 1971)

Armstrong, R.H.R. and Margaret Hobson, *The Use of Gaming/Simulation Techniques in the Decision Making Process.* Prepared for the United Nations Interregional Seminar on the Use of Modern Management Techniques in the Public Administration of Developing Countries (Washington D.C.: August 31, 1970)

Boocock, Sarane S., "Validity Testing of an Inter-Generational Relations Game," in the Proceedings of the Tenth Annual Symposium of the National Gaming Council (Ann Arbor, Michigan: The Environmental Simulation Laboratory, 1971)

Guetzkow, Harold, "Simulations in International Relations" in *Simulation in the Study of Politics*, ed., William Coplin (Chicago: Markham Press Co. 1968)

Long, Norton E., "The Local Community as an Ecology of Games." *The Journal of Sociology*, 64 (November, 1958)

METRO: A Gaming Simulation. Report on Phase I, prepared by the staff of the Urban Regional Research Institute for the Tri-County Regional Planning Commission (Lansing, Michigan: 1966)

Michael, Donald N., "On Coping with Complexity: Planning and Politics," *Daedalus* 97, (Fall, 1968)

Monroe, Margaret Warne, *Games as Teaching Tools: An Examination of the Community Land Use Game.* Thesis presented for the Master of Science degree at Cornell University (Ithaca, New York: Center for Housing and Environmental Studies, Papers of Gaming-Simulation No. 1, 1968)

Pool, Ithiel De Sola, and Robert Abelson, "The Simulmatics Project," in *Simulation in Social Science: Readings,* ed., Harold Guetzkow (Englewood Cliffs, New Jersey: Prentice Hall, Inc., 1962)

Russell, Constance, "Validity Testing of a Social Simulation," in the Proceedings of the Tenth Annual Symposium of the National Gaming Council (Ann Arbor, Michigan: The Environmental Simulation Laboratory, 1971)

Wilson, James Q., "Planning and Politics: Citizen Participation in Urban Renewal," in *Urban Planning and Social Policy,* eds., Bernard J. Frieden and Robert Morris (New York: Basic Books, Inc., 1968) Article reprinted from the Journal of the American Institute of Planners XXIX (November, 1963)

GAMING AS APPLIED SOCIOLOGY

CATHY S. GREENBLAT

$\mathbf{I}$n the midst of a heated debate at the recent National Gaming Council meetings, one member of the audience began his attempt at reconciliation of two seemingly diverse positions with the statement, "I want to try to narrow the gap, so we can leap over it." This paper has a somewhat similar purpose: to propose a way of narrowing the gap between the things sociologists know about and do in the classroom or research center and the things they might do in the community.

Even those sociologists who are "sure-footed" in their research and teaching capacities often find the jump to the out-of-class community too big to allow them to land on their feet. Many questions present problems to those who want to take their ideas beyond the ivory tower: How do you get across an understanding of social organization, community development, problems of diverse subgroups, to an audience that is not as "captive" as students? How do you transmit an understanding of the dynamics of interpersonal and intergroup relations in terms that assist people in learning to assess their position and their opposition, to develop effective organization, and to exercise greater control over their lives? How do you get groups in conflict to become more aware of how the world looks and feels to those with whom they are embattled so some reconciliation can be approximated? How do you help people expand their horizons concerning available alternatives and their costs and rewards so they can make choices about their lives? The difficulties of response send many scurrying back to familiar territory in haste.

CHANGES IN GAMES AND PLAYERS

Simulation games are operating models of central features of real or proposed systems or processes. Scenarios are developed, roles are defined in interacting systems, and players are given goals, resources, and rules. Then they make the system work, trying out alternative strategies within the system constraints presented.[1]

War games have been played for centuries, not simply by those seeking "fun,"

Reprinted from Arthur Shostak (ed.) **Putting Sociology To Work**, published by David McKay, 1974.

but often by those who later would put what they learned in the exercises into practice in real-life military operations. Both the Japanese and the Americans used the technique extensively in the pre-World War II and during the war days, and the U.S. Joint Chiefs of Staff still "game" a number of military and diplomatic situations.

In 1956 the American Manufacturing Association developed the first business game—a training device to develop management skills through practice and subsequent analysis of various strategies where the costs of real-life errors were absent. Within a few years, more than two hundred companies were utilizing business games for training in marketing, production, investment, organizational decision-making, and other business problems; for the identification of potential management skills; and as part of their larger programs of general education for employees.

Movement of games into the schools followed somewhat later. In the early 1960s social scientists at Johns Hopkins University started work under James Coleman to develop game models for teaching sociological and political science concepts and theories; Abt Associates in Boston began developing a number of games with simulated environments; and project SIMILE at Western Behavioral Sciences Institute moved from early work using Harold Guetzkow's INTER-NATION SIMULATION (INS)[2] for research purposes to the development of simulation games for teaching. If participants in the research endeavors became so involved and excited during the experience, they reasoned, could not some of these interest and motivational outcomes be brought into the classroom via games?

Since that time numerous developments have taken place in the field of education. Individuals and companies all over the country are developing and disseminating games for teaching and training in a variety of disciplines.[3] Elementary and secondary-school teachers and college instructors are turning to the use of gaming as a means of developing interest and motivation, promoting cognitive learning, bringing about affective change concerning oneself and others, and changing the structure and climate of the classroom.[4]

The next step seems to be one of moving the gaming technique into the community, and this has already begun. Despite the fact that the purposes for which these endeavors have been undertaken are those sociologists speak of, most of those who have started using gaming in the community seem *not* to be sociologists. Rather, some community organizers have "discovered" gaming as an effective tool for transmitting understandings of urban problems and developing skills; numbers of ministers and church people have found them good ways of opening channels of communication; and others have found different gains. The city of Plainfield, New Jersey, hired a psychologist in 1970 to run a Community Relations Training Program for city employees and interested citizens; two or three days of the five-day workshop were spent playing William Gamson's SIMSOC[5] and discussing the questions it raised about the relationships among individual, local and national needs, the relationship between alienation and deviance, and a host of other issues.

My own work with games has been limited largely to the classroom, and discussions with others in the profession reveal similar limitations of locale. Surely some sociologists have made important strides in the use of gaming in the community. In exploring game use in the out-of-class context, however, I have found sociological enterprises, but not sociologists. What I shall try to do, therefore, in the rest of this paper, is to describe some activities of people whose work seems to represent "sociology in action," and then review the lessons these vignettes offer for sociologists interested in using games to "make a difference" in the off-campus world.[6]

LARRY McCLELLAN

Many of Larry McClellan's endeavors in designing and using games in the community have had the dual function of giving people more holistic notions of cities, communities, colleges, and other social systems, and of training people in the political skills of negotiating, strategy planning, forming coalitions, and the like.

Larry was an organizer in Cicero, a white working-class neighborhood in Chicago, and in several affluent white neighborhoods, when he and three fellow students in divinity school played Allen Feldt's COMMUNITY LAND USE GAME (CLUG)[7] in a course on metropolitan planning. The others had had experiences in Chicago working with poor blacks, running coffeehouses, organizing in other communities and so on, and they found CLUG an exciting experience related to their concerns. Some of the problems they were involved with, however, were not dealt with in the game, and they thought that a simple modification might bring these concerns into the play experience. Nine months later and with another recruit to the design process they completed URBAN DYNAMICS.[8] They report:

I think a large part of our impulse in putting the thing together was our frustration, particularly in dealing with affluent suburban whites and working-class whites, in trying to give them "handles" on the city. Most whites don't understand the internal dynamics, let alone the hassle that poor blacks go through—they just don't have a perception. We believed that blacks in Chicago are powerless partly because of overt racism, but largely because of certain kinds of structured, institutional relationships. So we wanted a game that would help people realize the institutional and group dimensions of poverty and other urban problems—about the relationships between individuals and systems.

Our original concern was with that teaching function. Ideally we'd really want to get whites spending six months working on the streets, but that's not a very practical alternative; and the game we developed seemed to create a very intense, dramatic situation in which people started making discoveries. Those caught up in the individualistic ethos began to realize that no matter how good you are, if you're caught up in a group with limited resources you have real constraints, which they hadn't recognized before playing. *Urban Dynamics* was based on the city of Chicago and the kinds of inequities that existed there between the different economic and racial groups, but as we played it elsewhere it became clear we had built a more generalized model. People in New York and Rochester and Cleveland were all saying, "Yes, this is really true, this is the way it is."

After using URBAN DYNAMICS a number of times, Larry and his colleagues in Urbandyne, a company formed by the five designers, began finding out that the process of playing had other consequences than the learning of content.

Players really struggled with questions of decision making, organizing, negotiating, getting things together, and for many of the groups that was the real value of playing: struggling with the problems of community organizing in a simulated urban context as a way of developing these skills for application in the real world. In doing so, we've had some groups that have really faced up to their own disorganization. For example, I ran one game for a planning group and they just couldn't get themselves together. I found out after the game that this staff group was going through a great deal of administrative difficulty, but nobody had been able to talk about it. In the game situation the same difficulties emerged. These problems were analyzed in the debriefing, and later the players were able to draw parallels to their actual organizational dilemmas.

We've also discovered that people started using the game as a broad kind of strategy tool. For example, they say, "What would happen if you divided the resources up differently?" and then do it and play and see what the consequences are. Also, we have had some groups who asked, "How could you equalize suburban-urban development?" and then used the game for exploring the possible alternatives.

Much of this work in using games for pedagogy and as deliberate organizing tools has proceeded with URBAN DYNAMICS. The Urbandyne staff, however, have developed and used several other games. EDGE CITY COLLEGE[9] is a simulation of a college. It developed from a concern about how to get people to understand the dynamics of a college campus, and has been used with high school students and incoming freshmen to illustrate some of the nonclassroom aspects of a university and the way it functions. De Paul University used it two years ago for orientation for approximately 500 incoming students, and last year as part of the orientation program for 700-1,000 incoming students. Larry reports that the president of the student body indicated she believes that playing the game politicized the students to the nature of the university, and that they have been involved in the life of the university in a far more significant way as a result. In a postgame questionnaire, 87 percent of the participants indicated that playing had bettered their "understanding of the structures and processes of the university." Other evaluations were also positive. I asked Larry if he thought the game's contribution was in increasing knowledge of what the system is like, or in increasing students' sense of efficacy within it, and he could not be sure.

Whatever the source of its effectiveness, the game was suggested as an essential ingredient of future orientation programs by the student participants.

LEARNING A&M

The Learning Activities and Materials Cooperative is an informal group of people—most of them students and faculty at the University of Michigan—who share an interest in games. Michael VanderVelde explains:

> The games that excite us are those that get people to share *themselves* as well as a common experience. We just finished a few weeks' work designing and running a gerontology game for institutes of gerontology all over Michigan. The game we developed was a loosely structured one used to get participants reacting to gerontology problems as they ordinarily do. Then through postgame discussion we tried to get them gradually to talk about the issues that are affecting them and their ways of dealing with them. It's the same philosophy of gaming coming through in all our work: the people bring whatever it is, but use the gaming device as the *vehicle* to get talking about things.

Bob Parnes points out that Learning A&M is primarily concerned with designing and disseminating games that are easy to construct, demolish, and redesign.

> Social-simulation gamelike exercises that's how we've come to define what we're doing. I think probably one of the things that symbolizes our group more than anything is that we don't think people learn very much from playing games, but that they learn a lot about themselves and the systems they're working in by taking a game and redesigning it for their own purposes. So some of the games we have allow the group to play many versions of it—to just play around *with* it.

One such game, developed by Fred Goodman, a professor of education and member of Learning A&M, is POLICY NEGOTIATIONS. [10] In the basic version the content concerns a series of educational issues that players must deal with through influence allocation. Variations have been developed and used as training and educational games for university students, church workers, police, social welfare workers, and others. When I spoke with Michael VanderVelde he was beginning work with the nursing staff at a local hospital, developing a variation to show how hospitals run, what the power structure is and how it works, who the people are and what kind of influence each would have on various critical issues. Then they'll have student nurses play. The important thing about POLICY NEGOTIATIONS is that it's a "priming game"; once it has been introduced and played, it can be revised easily by participants to reflect their own system's issues, constituencies, and agencies. Thus, instead of a game director redesigning the game, the participants can do so. This process of articulation and analysis of system components may have important consequences for heightening awareness and understanding of the system in which the player-designers live and work.

THEY SHOOT MARBLES, DON'T THEY?[11] is an experimental, experiential game designed by Goodman, in which players build a community. A few rules are given and participants make up the rest in the course of play, experimenting with different forms of government and of social organization. People from Learning A&M have "gone off with the game—some to schools, some to churches—using it with kids and with adults." Bob talked about Wes, who is interested in playgrounds and in kids' play, and wanted to get a feel for how kids would respond to the game.

> This summer every afternoon he would dress up in as "non-grown-up" clothes as he could possibly find and go out to playgrounds. It's funny, because Wes is about six foot six—a big black guy, a football player in college—and here he was carrying around his little "Marbles" game and trying to get these five-, six- and seven-year-olds to play. And it wasn't anything formal at all; he didn't ask for an invitation. He'd just walk onto the playground and get kids to play. Why? Well, it's fun to play and we think it's a great way to start discussions, and that's an important aspect of learning—people sharing experiences with each other.

R. GARRY SHIRTS

As director of SIMILE II, Garry Shirts has worked extensively with games for education. He has designed several games dealing with international relations, domestic politics, city problems, stratification, race relations, etc.; he runs workshops on simulation use and design for teachers, researchers, community workers, and interested others; and he often runs game sessions in schools. Some of Garry's work in out-of-class settings has entailed use of games for education; for example he has used games for teaching purposes with police groups throughout the state of California, focusing on such problems as the pressures and problems of minority groups.

Some of his other endeavors, however, do not involve games for transmission of information. Rather, games are utilized to facilitate communication between antagonistic groups of people who have difficulty talking and relating to one another. In these cases the games are not the end and the content is not relevant; the games are employed to set a climate, a stage, that permits other things to happen later. The game-play is thus task-oriented: its function is that of changing group dynamics and processes. Garry says:

> Let me tell you about two different experiences that we had. An organization had been trying to develop a program to get blacks and whites and browns communicating with one another. They had run a series of discussion sessions, but these tended to break down, to develop into encounter-type sessions which completely polarized the groups. The blacks ended up browbeating the whites and the whites felt guilty and then more guilty and then finally became angry and fought back. The sessions exacerbated racial tensions rather than improving them.

I received a call asking me to design a game to help facilitate communication. I told them that since they knew the specific problems better than I did, I thought it would be better if I came and taught them how to build a game themselves. They agreed, and put together a group which included members of the university community, the police, the white community, the militant minorities, and the moderate minorities. We set up a three-day program to help them design a game, and the program consisted mostly of having them participate in and discuss a variety of existing simulations.

The first night the militant black faction announced they wouldn't play. They said that their problems were real and they weren't interested in playing games, but wanted to talk about pressing issues. We asked them to go along with our plan for the one evening and they agreed. In the game that night, one militant got in a position of power and took advantage of the other groups. The other militants challenged him on this and said, "Here you are always complaining about the white Establishment and as soon as you get in power you do the same things they do. You're really no different than they are." That turned out to be a very real experience for him. From then on there was nobody who didn't want to participate and who didn't see the games as important. We played several SIMILE II games, including STARPOWER,[12] SITTE,[13] and NAPOLI.[14]

After the three days there was an incredible amount of rapport between the groups. Initially, when coffee and meal breaks came, the blacks went out with the blacks, the whites with the whites, the police with the police, etc. At the end, everybody went out together. One policeman said that for the first time in his life he was able to respond to a black person as a person rather than as a stereotype, and that type of feeling was expressed by several of the people.

It was an extraordinarily emotional experience for me, seeing the change from a group in desperate positions and highly antagonistic, to a group that communicated in a way that you see only after the very best of encounter sessions. We tried to analyze what had happened and why, and it seemed that maybe the games did several things. First, by participating on teams, players formed new in-group relationships. If you have people from groups A, B, C, and D on a team and they get a sense of working on a team, then they develop a sense of themselves as a group. They can communicate with each other instead of A only being able to talk with other As, etc. Second, participation allowed them to begin communicating with each other in nondefensive ways. They were able to learn how a person responded to authority, how well he was able to persuade, how well he spoke in public—so many things they were able to learn about each other in that amount of time. While they never found out about personal hangups and the other things you learn about in encounter groups, they did learn a lot about one another—the same kinds of things they would find out about if they were to work side by side for an extended period of time. The third possible explanation is that I think participation in games provided a common experience which not only helped to solidify the group, but helped them to communicate with each other and to make sense to each other. Previously when they talked, for example about "power," they all meant different things. We played a game about power, and then they had a common referent.

We got a chance to try these ideas out when we got a call from a California community with a serious problem. An interracial incident had polarized the community, and the work of the Human Relations Commission which was set up immediately afterward had exacerbated the situation. Every time they interviewed someone, the person was suspected of being racist, and charges and countercharges flew.

After three months of this, we were called in and asked to run a conference. We decided to try out what we had learned in the earlier experience, for it was clear that the community was torn with dissension and that the community leaders gathered for the conference would be unable to work together. We thus spent the first morning and part of the afternoon having them participate in a simulation (SITTE). Our aims were what we discussed before: to break up in-group relationships as the only mode of relating, to help them to become acquainted with each other, and to give them a common experience that would unify the group and help them develop a common language. After the simulation, we had a working situation in which we had them look at some broad goals for the community and devise ways to attain them, instead of looking directly at the racial question. They outlined a series of steps and assigned responsibilities for executing them. It went remarkably well, and one year later essentially all these goals were met and the racial situation had substantially subsided.

We can't be sure what effect the simulation game actually had, but before playing these people were unable to work together on anything, and afterwards they functioned beautifully as a group and accomplished the things they set out to do. Original in-group ties no longer hampered cross-group cooperation.

SOME OUT-OF-CLASS USES OF GAMES

The foregoing accounts suggest several uses of games in working in out-of-class settings.

First, games may be important tools for teaching in the community. The same kinds of outcomes cited for games in the school context may be found in these situations: learning of principles, processes, structures, and inter-relationships; development of empathy for and understanding of the predic-aments, pressures, and problems of real-world role players; a stronger sense of efficacy. Particularly with an audience not used to being "lectured at," the approach of teaching through play of a high-verisimilitude simulation game and postgame discussion and analysis may be a far more potent way of getting across information and ideas than is the more traditional lecture. Thus, those working to increase the information and understanding of general citizens' groups or special-interest groups may find gaming a valuable tool for transmitting ideas.

Second, gaming may provide a training device for those who wish to help develop skills. Businessmen and diplomats have been using games to cultivate abilities of persuasion, bargaining, and strategy planning; police, minority-group members, commissions, and the like may derive similar benefits from the technique.

Third, games may be good vehicles for social planning or "future testing." Using some games, players can try out alternative forms of social organization, resource allocation, communication, etc., within a simulated context to "test" the efficacy of their ideas, the costs and rewards of various options, and the difficulties of going from the present structure to the desired future one.

Fourth, games may be good vehicles for helping individuals and groups to explore values, ideas, and behaviors—for making explicit what has been implicit.

This is a general communication function—leading people to a better understanding of themselves and others, as has been discussed in the account of the activities of Learning A&M. In a somewhat similar vein, Peter Stein, Norman Washburne, and I have been exploring the uses of our newly designed MARRIAGE GAME [15] for counseling purposes. Although we designed it for teaching about marriage as a social system, discussions with marital counselors suggest that participating in the simulation (which involves much of the decision-making and interaction typical of real-life marriages) may prove to have heuristic functions. Analysis of game behavior may be a less threatening way to begin exploration of marital difficulties than "jumping right in" to look at the real-life, emotion-laden situations and behaviors. Again, the aim is to use the game to elicit perspectives about "the way the world is" and one's place in it, as a means to begin to develop new modes of relating and of problem-solving.

Finally, the fifth use is clearly exemplified by Garry Shirts' reports of games for dealing with problems of conflict resolution. Here the aim is to use games to facilitate communication between disparate, antagonistic groups so that other tasks can be undertaken more effectively. The game is employed as a vehicle for changing group structure and the work climate, opening new channels of communication; thus the game-play is for naught if other activities do not follow. Game use is preparation for more serious endeavors.

GAMING AS APPLIED SOCIOLOGY

How, then, can sociologists use gaming to "make a difference in the out-of-class world?" First, they can work on the *design* of games, particularly those to be used for teaching purposes, which require high verisimilitude. The translation of sociological theory and data into the game format would create new and better tools for use in teaching community groups. Games seem to have real potential for increasing people's understanding of the nature of alternative social environments and the consequences of living in them. If sociologists could translate their research findings into games, people could then play to learn about the costs and rewards of living under fascism, democracy, or socialism; in monogamous marital arrangements or in communes; in the mainstream of contemporary American society, in minority groups, or in groups that have deliberately relinquished some of the technological accoutrements. Thus the sociologist-game designer could give people greater opportunities to make decisions about their lives and their worlds.

Second, sociologists can use some existing games in their attempts to teach community groups. The most important part of gaming is the post-play discussion, and the sociologists' expertise could help to make post-play discussion and analysis a rich learning experience.

Third, those sociologists interested in helping create new relationships, new patterns of communication, new skills for those they are working with, might well consider the other uses of games described in this paper. In such ways, perhaps the community role of the sociologist will become more vital.

Gaming is relatively new in academic quarters, and as is typical of many such enterprises, research lags behind the innovation, Hence, although numerous claims are made for the efficacy of games for cognitive, affective, and behavioral learning, the empirical status of these claims is fairly low. Anecdotal evidence abounds—all of us who have used games in and out of the classroom can offer examples of significant happenings, important changes in individuals and groups, major insights gained into oneself, others, or social systems. Reports of those who have run or participated in games are almost always highly enthusiastic. The experience is involving for both participants and directors; the games raise many questions and present operating relationships rather than static elements for later analysis. The interaction during play creates new relationships between those who otherwise may see fellow students or workers simply as faces, names or numbers.

Unfortunately, most anecdotal reports are not yet backed up by research findings proving them to be prevalent outcomes rather than singular occurrences. If your decision to try gaming is attendant upon a body of "hard data" from research, therefore, this is not the time to begin. On the other hand, where is the demonstrable evidence that more traditional approaches bring about these results? The *general* paucity of evidence in social science and in education must surely be kept in mind in looking for evidence for gaming. In addition, although there is little "proof" that participants learn more via gaming, there is no evidence that they learn less. Games seem to be at least as effective as other modes of teaching when tested with standard measurement devices; yet the types of learning that gaming seems to best promote require new tools of measurement.

Of course it is to be hoped that further research will be undertaken to discover just what games do, under what conditions, and for whom. Until such data are available, however, those willing to gamble and join the search for better, more involving approaches to teaching will find gaming filled with excitement, discovery, and fun.

NOTES

1. For elaboration, see Cathy S. Greenblat, "Simulations, Games, and the Sociologist," *American Sociologist* 6 (May 1971): 161-64; and John Raser, *Simulation and Society* (New York: Allyn & Bacon, 1969).

2. Guetzkow, Harold et al., INTER-NATION SIMULATION. Available from Science Research Associates, 259 East Erie Street, Chicago, Ill. 60611.

3. For a list of available games, see David Zuckerman and Robert E. Horn, *The Guide to Simulation Games for Education and Training* (Cambridge, Mass.: Information Resources, Inc , 1970); or Cathy S. Greenblat, "Partial Bibliography of Games and Simulations in the Social Sciences" (Rutgers University, 1971; mimeographed).

4. Cathy S. Greenblat, "Teaching with Simulation Games: A Review of Claims and Evidence" (Douglass College, Rutgers University, 1971; mimeographed); Sarane Boocock and Erling O. Schild, *Simulation Games in Learning* (Beverly Hills, Calif.: Sage Publications, 1968).

5. William Gamson, SIMSOC: *Simulated Society* (New York: Free Press, 1972 Edition).

6. These accounts present only a small piece of the work being done by these individuals. The selections have been chosen to accentuate different utilizations, but all are using games for several purposes, are in touch and share ideas, and use some of one another's games from time to time.

7. Allen Feldt, COMMUNITY LAND USE GAME. Available from Environmental Simulation Laboratory, 109 East Madison, Ann Arbor, Mich. 48104.

8. Larry McClellan et al., URBAN DYNAMICS. Available from Urbandyne, 5659 South Woodlawn Avenue, Chicago, Ill. 60637.

9. Urbandyne, EDGE CITY COLLEGE. Available from Urbandyne, 5659 South Woodlawn Avenue, Chicago, Ill. 60637.

10. Frederick Goodman, POLICY NEGOTIATIONS. Available from Environmental Simulation Laboratory, University of Michigan, 109 East Madison, Ann Arbor, Mich. 48104.

11. Frederick Goodman, THEY SHOOT MARBLES, DON'T THEY? Available from Environmental Simulation Laboratory, University of Michigan, 109 East Madison, Ann Arbor, Mich. 48104.

12. R. Garry Shirts, STARPOWER. Available from SIMILE II, 1150 Silverado, La Jolla, Calif.

13. SITTE. Available from SIMILE II, 1150 Silverado, La Jolla, Calif.

14. NAPOLI. Available from SIMILE II, 1150 Silverado, La Jolla, Calif.

15. Cathy S. Greenblat, Peter J. Stein, and Norman F. Washburne, *The Marriage Game: Understanding Marital Decision-Making* (New York: Random House, 1973).

GAMES AS VEHICLES FOR
SOCIAL THEORY

JAMES S. COLEMAN

Games and play have been examined by a number of authors, with attempts at identifying their relation to life activities and their distinctive character.[1] The importance of such an attempt lies in what it might tell us about the potential usefulness of games for the study of life in general, and in particular, social organization. However, the general absence of any success in these attempts lies, I believe, in failing to look carefully at the nature of life itself. If the sequence of activities that constitute life is seen as a game, as Bernard Suits (1967) has done,[2] then it appears possible to distinguish those activities which we call "play" and "games" from the remainder of this sequence of activities.

In describing life as a game, I mean to give it the formal characteristics of a game: (a) the players have goals toward which they act, although these goals may be changed by the course of the game; (b) their actions are governed by a set of rules that specify which actions are prescribed, which are permitted, and which are proscribed; (c) there is another set of rules, which may be either stated in advance or discovered only in the course of play, that specify the consequences of each action in aiding or inhibiting each player's movement toward his goal.

These are, perhaps, as good a set of defining properties of a game as any; yet at the same time, they define most of the activities in the sequence which constitutes life. Most, but not all. For if life is conceived as a game with these properties, then those activities we know as "play" and "games" do not fit. They are not actions of the player toward his goal in life, but actions quite irrelevant to the otherwise connected sequence.

Their relation to life can best be seen by examining a specific event that arises in all games: the "time out," that is, for a break in the sequence of play, which is not to be counted as part of the play, and during which the rules of the game no longer govern. The players may do anything during the time out, but when play

Reprinted from the **American Behavioral Scientist**, Volume 12, July/August 1969. ©1969 by Sage Publications, Inc.

AUTHOR'S NOTE: *This is a version of a paper presented at the 1968 American Sociological Association annual meetings, Boston, Massachusetts.*

begins again, it is wholly unaffected by the activities during "time out." From the point of view of the game, these activities did not exist. They were taken only because some other needs of the players, often personal physical needs, necessitated the time out.

My essential point is that if life is conceived as a game, it too has its times out, and the activity which takes place during these times out is either play or games—play if it does not proceed according to the criteria for a game set out above, and a game if it does. Play and games "don't count" in the normal sequence of life activities, just as activity during time out in a game "doesn't count" in the game. In this view of life, all else except play and games consists of a connected sequence of actions directed toward goals; play and games constitute the interruption or time out in this sequence. Games are more fully "time out," for they are more fully insulated from the normal "rules" of life by a set of explicit rules of their own.[3]

When one establishes the rules of a game, he in effect abrogates some of the rules of everyday life. For other purposes, players might want to continue to obey some of the everyday rules that do not conflict with play of the game. For example, players might continue to maintain the rule of not killing another person, although in a game of football they abrogate the rule of not hitting another person violently with one's body, because such violent aggression is allowed by the rules and helpful toward reaching the goal.

Why, then, do persons playing this large game of life take these times out which constitute play or games? The most reasonable explanation is that they do so for the same reason they take time out in a parlor game: because they have psychological needs and physiological needs which can only be satisfied by declaring a temporary moratorium—taking time out and attending to those needs.

It appears clearer, assuming all this is so, that players in the game of life should take time out for play than that they should take time out from life for a game. For this is a busman's holiday, playing games during the time out from the large game itself.

The puzzle of why they do so leads to important questions in socialization, for it is recognized that the playing of games is an important element in socialization. Thus the suggestion arises that in playing a game, a child or man is taking time out from a single sequential set of activities that constitute a complex game to establish a parallel set of activities, but with a beginning and an ending, which will aid him when he returns to the continuing sequence. One can see the possibility for a variety of types of socialization aids and psychological aids provided by this delimited and unconnected parallel set of activities; but it is not my intent to investigate these here. Rather, my aim is to indicate how the general category of activities that are described as games, these times out from life, can, because of their peculiar resemblance to life itself, be important elements in the construction of social theory.

For this purpose, it is useful to focus on one function that games appear to have for children. In learning to cope with the physical environment, the young child carries out a variety of playful and exploratory and experimental actions toward his environment: putting objects in his mouth, trying to grab a handful of water, putting his fingers in a fire, playing with clay or mudpies to make new shapes, rolling a ball downhill, and numerous similar actions. These actions occupy a large portion of time for a period during which the child learns certain rules of the physical environment. He does not learn physical theory, but a set of general laws or empirical regularities. He learns, in a qualitative way, the laws of mechanics and a few chemical facts.

When children begin to cope with a social environment, they find themselves subject to a more complex framework of action-and-response. Interaction with another person involves a double contingency: the other's response is contingent upon one's own action, just as one's own action is contingent upon his. Furthermore, the contingent action is not an automatic response governed by mechanical laws, but a purposive action, directed by the actor's goals, and constrained by the rules of the social organization within which he is acting.

This increased complexity brings enormous learning problems for a child, problems that require a learning environment comparable to that provided in an early period of development by exploration and experimentation with the physical environment. The social play and games of young children constitute, I believe, such a comparable environment. What the child learns in them is not social theory, but empirical regularities about the way other persons behave in particular situations, and in response to particular kinds of actions of his, when they have certain goals and are subject to certain rules or constraints. He learns, in a qualitative way, the laws of a system of behavior comparable to that of mechanics, that is, human purposive behavior.

In play, he learns about behavior; however, in games with explicit rules, he learns about a system within which purposive behavior takes place. The necessity for establishing games with rules separate from the normal sequence of life activities lies in the fact that this normal sequence fails to provide a wide enough range of experimentation and exploration of social organization. Piaget's observation of children playing marbles shows the extremely elaborate and detailed set of rules and procedures that children develop and learn, and the numerous variations in these rules—a far richer, more precise, and more directly enforced body of rules than the rules governing their current sequence of normal life activities.

These activities of young children suggest that just as casual exploration of the physical environment provides the experience which forms the basis for physical theory, exploration of the social environment through games may constitute a fruitful avenue toward social theory. In physical science, experimentation formalizes the practical investigation that each of us carries out upon his physical environment; similarly, games with explicit rules and structure may be the appropriate formalization of the practical investigation of our social environment that each of us carries out in childhood games. This methodology contrasts sharply to those that sociologists presently use as avenues toward social theory. In contrast to survey research and observations in natural settings, it depends on the creation of special environments, governed by rules that are designed precisely for the study of the particular form of organization. In contrast to experiments, with their experimental probe or stimulus and the consequent response, the principal element in game methodology is the construction of rules which can elicit a given form of social organization. The involvement of persons in a game is also different from the use of persons (or "subjects") in psychological experiment. In a game, the goals of each player and the incentive to play must be generated by the rules of the game itself. The players are not passive subjects, but active participants or players. As in any social subsystem, the players in a game find their rewards in the game itself, while an experiment ordinarily merely uses the services of its subjects for a period of time. There are, to be sure, a few sociological experiments that have many of the characteristics of games; but it is relatively unimportant whether these are called games or experiments.

GAMES AS EXPERIMENTS

If the potential of games for sociology is to be realized, then an appropriate methodology is required, a paradigm appropriate to the investigation of social structure in the same way that the experimental paradigm is appropriate to investigation of the physical or psychological structure.

The physical environment interests us as persons because of the regular responses it makes to our own actions toward it, regularities that can be described by physical laws. It is a property of the physical environment that the responses it makes depend only on the physical character of the actions taken upon it, independently of whether these actions are taken upon the initiative of a person, or derive from some other source. That is, the same physical laws govern the vertical velocity of a falling body, whether it is dropped by Galileo in a physical experiment or falls from a tree without human intervention. As a consequence, an experimental paradigm can be established in which the human experimenter, in order to learn the action-principles of the physical environment, acts upon the physical system by carrying out a particular physical intervention. He observes the response of the physical entity, and then if he has described both his action and the response in terms of the appropriate physical parameters, he can describe the regularity or lawfulness in the response of the physical body.

The experimental paradigm in physical science thus consists of: (1) human intervention, (2) description of the physical properties of that intervention, (3) measurement of the response of the physical system to that intervention, and (4) discovery of the regularity or relation between the physical properties of the intervention and the physical properties of the system's response.

The child's probings of his physical environment by attempting to grasp water, or by rolling a ball, or by putting his hand or a piece of paper in the fire constitute the early prototypes from which the paradigm is itself developed. The child, as the physical scientist, wants to learn about the behavior of his physical environment so that he can anticipate or predict its action in a future similar circumstance.

In constructing and playing a game, a child is engaged in a somewhat different endeavor. He is not merely probing a responsive environment composed of physical entities. He is studying a system of social behavior, and his own actions, governed by the rules of this system, are an intrinsic part of the system. Before any action takes place, the players must establish a set of rules to govern their actions. These rules limit the kinds of actions that players may carry out, and also provide their motivation, by defining their goals. Thus setting up a game means establishing a new and different set of relations between elements of a system, that is, players, and then observing the behavior of the players and the functioning of the system, The players in a game do not respond to the person who establishes the game unless he himself becomes a player, for otherwise he is outside the system; they respond to each other.

The child entering a game is entering a new social order; and he learns by observing both his own behavior and the behavior of others in that order. His necessity for entering a new social order to learn these things lies in the fact that he learns by the method of comparative observation, by the differences in behavior under different sets of rules. But if he were interested only in learning about behavior, about how people respond under different circumstances, he could do so through social play, in activity that follows the paradigm of investigation in physical sciences. He can learn the responses of people by teasing them, cajoling them, by obsequious actions, by anger and threats, by all sorts of probes that young children are wont to carry out, actions comparable in learning about his social environment to the action of grasping a handful of water in learning the properties of a liquid.

In playing games, he is doing something else. He is not learning about the responses of persons so much as he is learning about the functioning of systems of rules. The elements in these rules are not persons in the usual sense; they are actors-in-roles, utilizing some of the properties of persons, but not others. In a baseball game, a shortstop is an actor-in-a-role, utilizing some properties of appropriately skilled individuals (the ability to catch and throw a ball, the knowledge of where best to throw a ball once caught), but not utilizing others (his preference for dogs above cats, his belief in God, his childhood memories, the color of his hair). In a game of hide-and-seek, the elements of the game are not full-fledged persons, but rather players having those properties relevant to play: ability to run, to hide, to find another. The rules of the game take into account those specific properites of individuals that are relevant to performance of the player's role, but not others.

Sometimes the rules take specific note of physiological limitations of the players, as when an athletic game is divided into quarters, or chukkers, or halves, or rounds, with a designated rest period in between, or even a break for lunch and a break for tea, as specified in the rules of cricket. And in children's games, special rules are often established for a child much larger or much smaller than the others, to take account of his prowess or his limitations. But beyond this recognition in the rules of certain attributes of individuals that might interfere with the game if they are not attended to, the rules disregard other attributes of individuals. The game is a system of roles in relation, and play of the game shows how that particular system of roles in relation operates.

Thus the child's use of games in exploring his environment is much different from his physical probes of the physical objects around him or his emotional and behavioral probes of the human objects around him. It is an exploration of systems of roles, of social organization. This exploration has some features that indicate its nature and extent. For example, in observing the play of games among young children, an adult is often struck by the seemingly endless arguments and discussions over rules. The game often is stopped for long periods because of arguments about violations of the rules, and arguments about the rules themselves. The adult is often tempted to intervene to get the game going again, in the belief that nothing can be accomplished if the children can't even agree on enough rules to keep playing. But the adult may here be wrong, for it may be that the principal value of the game for the child is in learning about rules of a social system: their universality, their modifiability, their fairness, their enforceability and means of enforcement, and their justification. (As in the game of life, some rules in any game can be justified in terms of goals of the player, others in terms of maintaining a viable social order; still others are arbitrary rules, or "ultimates," that have no justification.)

An adult also observes that a young child of age three or four finds it difficult to accept the universal application of rules to himself and others in the same role. In playing hide-and-seek, he attempts when hiding to have different rules apply to himself than to others. Or in learning to play checkers, he refuses to accept the rules when they lead to his loss of the game. In these actions, he is apparently still in the process of learning to separate the idea of social organization, and rules governing role relations, from his particular position within the organization.

The conclusion I want to draw from all this is that the construction and observation of games constitute for the sociologist an activity analogous to the physical scientist's or psychologist's use of experiments, in that each constitutes a formalization of the means that children use in learning about their environment. The activity in the two cases is quite different. In the case of the physical or psychological experiment, the experimenter makes a specific and measurable probe or action or stimulus, followed by a measurement of the response of the physical or human object of the probe. In the case of the game, the sociologist establishes a social organization, a set of roles in relation and goals of the players, defined by the rules of the game, and then observes the way this "hypothetical" social organization functions.

AN ILLUSTRATION: A GAME ON COLLECTIVE ACTION

An extended example may make clearer how the sociologist may use games in this way. I will use as an example a game of collective decisions that I have worked with for the past several years. The game was devised because of the long-noted paradox (usually called Condorcet's paradox) that any decision rule to choose a collective action from among several alternatives can produce inconsistencies; for example, the selection of a different alternative may result if the order in which pairs of the alternatives are voted on is changed. I reasoned first that the problem is more fundamental yet: if there is only one collective action that binds the members of this collectivity, then why would any member participate in an action that was not his first preference? Why could any collective action be taken that was not unanimous? The answer appeared to lie in the fact that a set of individuals are seldom related through only a single collective action, but ordinarily through a whole sequence of actions, and that it is the possibility of benefits he might experience through some other action in this sequence which allows the individual to accept a collective decision that he sees as inimical to his interests.

To observe, then, how such collectivities function without breaking down (as they would be expected to do if only one action is considered in isolation), I constructed a game involving from six to eleven players and eight collective actions to be taken, with individuals' interests differing on any one action, and with the collective decision on each action to be made by majority vote of the players. The rules of the game in general followed parliamentary procedure.

In observing the play of this game, it quickly became evident not only that players took account of possible future actions when voting on any one action, but they used their interests in those future actions to mitigate their losses on this one. The principal (but not the only) means by which a player employed this strategy was to give up his vote on this issue in return for a promise of a vote from another player on a future action of more importance to him, or, if this issue was itself of great importance, to promise a vote on a future issue in return for a vote from another player on this one. Other means were used as well: since promises were not necessarily kept, a player would offer a vote on an action to the player who had control over determining which action was to be voted on next, in exchange for the right to determine what the next action would be. Since the likelihood of obtaining agreements depended upon one's reputation for keeping promises, a player would often forego a potential immediate gain if it meant breaking a promise, but more so early in the game than later.

Nevertheless, some players lost: Sometimes because they were intrinsically disadvantaged through the distribution of interests; sometimes because they failed to use their resources efficiently. What kept them playing? To answer this question, several games were played in which the players kept the same distribution of interests for each play of the game. After the first game, a coalition of players formed, all of whom could win by a given pattern of bloc voting (since it was possible for a bare majority of the players to win). The other players quickly lost interest, and the game broke down. Thus, between games as well as within a game, it became clear that what allowed the collectivity to continue to operate was the possibility of gains in the future; when that possibility was removed, then the collectivity broke down.

As a result of playing this game and observing its play, a possible conceptual framework for describing the system emerged. It was clear that each player was using his votes as generalized resources to realize his interests, recognizing that his vote on one action was valuable to others even if he had no interests in the action. Thus I conceived of each player having as resources his partial control over each action, and as interests his potential gains or losses resulting from each action. His behavior could then be described as an attempt to employ his resources to realize his interests. The crucial element of the social theory was not this action principle, which is merely a restatement of rational or purposive behavior, but the concepts of partial control over actions and interests in (or consequences of) each action for each player, and the emergent concepts of the *value* of control over an action (defined as the interests that powerful actors had in the action) and the *power* of actors (defined as control over valuable actions). These concepts then led into a formal mathematical theory for describing interdependent actions in any collectivity, work that I will not discuss here.

Returning to the game, it was evident that to realize their interests, players were exchanging the resources given to them by the rules of the game, i.e., by the constitution of this collectivity. The question arose: How was this exchange different from economic exchange in a barter economy? The most obvious difference is that the exchange was neither a physical exchange in which the resources actually came into a new owner's possession, nor an enforceable contract. As a consequence, it was not negotiable. Purely conjectural, or speculative, or theoretical activity could not carry very far, because of the absence of a well-developed conceptual framework. But the most obvious way to examine the difference was to change the rules of the game to make the vote a physical commodity, a piece of paper that could be transferred and voted by whoever held it at the time of the vote. In play of the game, this change led to an intensification of the market in votes, a much greater likelihood that two persons could make an exchange, since a vote came to have value in exchange to a prospective buyer even if he had no interest in that issue. It enabled players to maximize their interests more fully, because it facilitated the exchange of resources.

Again, since each player had equal control over each action (one vote), the question arose: Why not make exchange unnecessary by giving each player eight votes, any of which he could cast on any action? This change would allow each player to concentrate his resources directly on those actions that interested him most, and reduce the inefficiencies brought about by the exchange process.

In play, it quickly became clear that such a distribution of control changed a number of things. First of all, the vote could not be taken by open ballot sequentially, for the last players to vote found themselves in an especially advantageous position: They could vote only the precise number of votes necessary to win, and save others for a future action. But even when the vote was taken secretly, game strategy, in the sense of game-theoretic principles, came to be much more widespread. The game was no more a zero-sum game than before, but now that no joint action such as exchanging votes was necessary to realize one's interest, each player's activity came to be more concentrated upon the problem of the best deployment of forces. He no longer carried out marginal, or incremental, activity, as was previously necessary in gaining control of an action, and thus had little way of knowing what was the best action. The game was more nearly seen, and responded to, as a game of pure conflict of interests.

In this case, as in the case of the physically exchangeable votes, the change in rules did not lead to new conceptual development. It did, however, show the empirical consequence of this rule change, thus providing a stronger base for the development of a conceptual or abstract description of the variation.

Another variation studied by a change in the rules was the introduction of a two-stage decision process through the use of committees. The action could not be brought before the collectivity as a whole except by positive action of a smaller committee. It was quickly clear that this enriched greatly the amount and kinds of resources of the collectivity members. First, much of the bargaining, negotiation, and exchange was now directed to obtaining a positive action in the committee. Second, the smaller size of the committees, about one-third of the collectivity, made the committee action much more dependent upon specific individuals, and thus concentrated the control of particular actions much more in the hands of a few people. Third, this two-stage structure of decision-making resulted in many fewer positive actions than in the single-stage case, even though the distributions of interests for and against the actions remained balanced overall, as they were in the case of the single-stage decision.

It might well be argued that these same generalizations might even more easily have been stated from a casual acquaintance with the American Congress, or another legislative body with a committee structure. That may well be so. In the case of the first simple form of the game, naturally occurring social organization may have provided the necessary framework, and made unnecessary the construction of a game with special rules. But if so, it is merely a fortunate circumstance in this case; another variation of theoretical interest, such as physical transfer and full negotiability of votes, may not exist in society.

Some variations we have not been able to carry out, because the very expression of the rule requires a degree of theoretical sophistication beyond the present state. For example, in social organization generally, the future actions, which players balance off or negotiate against current ones, consist of an endless sequence arising in part through the action of individuals, but in part through external events. Such a structure must obviously change behavior very much, since explicit vote exchanges are not possible. Something like generalized political credit must come to exist, but we have not yet been able to establish the appropriate game structure and thus can only speculate.

A more important variation for which we have not yet been able to develop an appropriate set of rules is the use of resources from outside the actions of a collective body to affect those actions. If two members of a collectivity are also members of a second collectivity, an exchange can be made involving resources of both collectivities. We know from observation of naturally occurring social organization that these exchanges tend to be negatively sanctioned and defined as illegitimate by the collectivities involved; but exploration through establishing a range of such social organizations as games is necessary in order to gain a better idea of the processes involved: What determines the rates of exchange, what the effect is on the autonomy of each collective body, and so on.

These examples are sufficient to indicate the way in which the construction of games with various rules can be used toward the development of social theory. The pattern by which this can best occur is still only very roughly known, but it clearly involves a first step of abstraction in being able to set up appropriate rules and thus establish the game, and a second step of abstraction in drawing from the rules-and-behavior a conceptual scheme that constitutes social theory. The first stage of abstraction, establishing the rules of play, is part of the conceptual labor, for it often involves making explicit, in the information provided to the player, those considerations that remain implicit in actual social organization. For example, in the collective decision game, the interests of each player, which determine his winning or losing, are the votes of his constituents toward reelection. The game exposes this consideration, so to speak, by making it explicit in the rules rather than implicit. Thus the game is embodying a particular structure of events, control over events, and interests in events, which constitute the beginnings of a theory about the social organization of which legislatures consist. But formulation of the rules and play of the game is only a first step of abstraction, for it still involves the concrete playing-through of the game. The second step of abstraction is the development, from the game, of a fully abstract system of concepts that describes the functioning of a given form of social organization.

A part, and perhaps the most important part, of this methodology, is the study of types of rules in games and in social organization generally. It is clear that rules are of very different types. There are, for example, rules that define the required procedure, such as parliamentary rules in legislatures, rules that specify the obligations incumbent upon the player in a given role, and rules that concern only the punishment of behavior that breaks other rules.

Since rules are at the center of this methodology, such a typology of rules, or recipe, or theory about the types of rules necessary for a game representing a social organization, will allow the method to progress beyond the level of an art. For when such a theory of rules of social organization does exist, then it will become possible to create systematic variations in games, rather than merely ad hoc ones, and a methodology for sociology comparable to that of the physicist's or psychologist's experimentation will exist.

NOTES

1. See Michael Inbar (1969) for a review of the numerous attempts to capture the essential difference between play and games on the one hand, and all other activities on the other.

2. Suits (1967) argues that games are not distinguishable from other activities in life except by the explicitness of their rules and their goals.

3. This view of games runs immediately into the objection that some games are played as part of life itself, as an occupation: a professional athlete, or a professional card-player, does not take time out from life to play his games; his sequence of life activities includes these games as an intrinsic part.

This objection is quite valid, but it shows merely that games can be used in the sequence of life activities—they can be brought back into life, ordinarily by connecting success in the game to some consequence in the regular sequence of life, such as money reward, or prestige. But this "connecting up" to life in general requires an extrinsic operation; the game itself is by definition self-contained and unconnected to the normal life sequence.

REFERENCES

INBAR, MICHAEL (1969) "Toward a Sociology of Autotelic Behavior." La Critica Sociologica.
SUITS, BERNARD (1969) "What is a Game?" Philosophy of Science 34 (January): 48-56.

COLLECTIVE DECISIONS

JAMES S. COLEMAN

This paper constitutes an attempt to extend a style of theoretical activity in sociology which is quite new to our domain. It is theory which rests upon the central postulate of economic theory, that of rational man attempting to pursue his selfish interests. The introduction of this approach into sociological theorizing was carried out by Homans.[1] Homans has since then by degrees led himself away from such an approach into that of operant conditioning;[2] hence it is my hope that this paper will help to rescue the fresh direction in social theorizing which Homan initiated in 1958.[3]

THE CENTRAL PROBLEM OF SOCIOLOGY, AND THE FAILURE OF SOCIAL THEORY

Perhaps the central problem in sociological theory is that posed most succinctly by Thomas Hobbes.[4] Why is there not a war of all against all? Hobbes took as problematic what most contemporary sociologists take as given: That a society can exist at all, despite the fact that individuals are born into it wholly self-concerned, and in fact remain largely self-concerned throughout their existence.

Instead, sociologists have characteristically taken as their starting-point a social system in which norms exist, and individuals are largely governed by those norms. Such a strategy views norms as the governors of social behavior, and thus neatly bypasses the difficult problem that Hobbes posed.

Yet this approach has numerous defects. The image of man with which it begins is *homo sociologicus,* a socialized creature whose freedom has been stripped from him by the norms of society and by the processes that socialized him. One consequence of this is that sociological theory is unable to treat

Reprinted from Sociological Inquiry, Volume 34, Spring 1964, with the permission of the publishers.

AUTHOR'S NOTE: *Work reported in this paper was partially supported under a grant from the Carnegie Corporation for research in simulated social environments for use in education. I am indebted to Robert Peabody and Arthur Stinchcombe for comments on a draft of this paper.*

problems for which no norms exist. In particular, all collective decisions involve some matter for which no norms exist. It is this very fact which makes a collective decision necessary. A striking example of this can be seen in the recent actions of some communities concerning the behavior of adolescents. In highly stable communities, where parents are in frequent communication with one another, norms exist concerning the behavior of adolescents; and although these norms are not always obeyed, there is no question about the norms themselves— no question about what the parent has a right to expect or what the adolescent has the right to do. However, the normative system in some communities has so fully broken down that adolescents and parents have met in a kind of town meeting and established, by collective decision, the rights and responsibilities of adolescents.

Yet sociological theory has little to say about how such collective decisions are made. It is forced to introduce, ad hoc, such notions as "interpersonal influence," and "persuasion," using these as crutches.

Conflict with a communal group, which can occur in the process of collective decision-making, illustrates well the dilemma in which sociologists find themselves. Seeing conflict as a massive element in the functioning of society, but unable to say anything about it as social theorists, they split themselves as individuals off from themselves as sociologists: they form *Journals of Conflict Resolution,* join disarmament and peace committees, yet continue to purvey sociological theory which has nothing to say about conflict.

In this paper, I will proceed in precisely the opposite fashion to that taken by the advocates of homo sociologicus. I will make an opposite error, but one which may prove more fruitful. I want to begin the development of a theory of collective decisions, and in so doing I will start with an image of man as wholly free—unsocialized, entirely self-interested, not constrained by norms of a system, but only rationally calculating to further his own self-interest. This is much the image of man held by economists, and with it the economists have answered one part of Hobbes's question: How it is that although the men who make it up are wholly self-interested, the economic system can operate with one man's actions benefitting others. It was the genius of Adam Smith to pose an answer to this part of Hobbes's question—Adam Smith, who was first of all a religious Scotsman, a moral philosopher who was concerned with the question of how virtue could exist in society. His answer is indicated by the classic statement from *The Wealth of Nations:*

> it is not from the benevolence of the butcher, the brewer or the baker that we expect our dinner, but from their regard to their own interest.[5]

It is interesting to see the apparent effect on Adam Smith of Mandeville, an early moral philosopher who wrote in the first quarter of the eighteenth century. Mandeville wrote a tract called *The Fable of the Bees,* in the form of a poem. In this, Mandeville described a hive of bees, each intent on crassly furthering his own interests without regard for his fellows. But certain members of the hive grumbled because there was no virtue, no concern for the common good. Finally, Jove granted their wishes and made the members of the hive virtuous and altruistic. At this, sloth overtook the hive, and with its members no longer concerned with their own interests, the hive fell to pieces.

Adam Smith, in his *Theory of Moral Sentiments*,[6] reserved special reproach for Mandeville, whose theories he called mischievous, attempting to make virtue out of vice. Yet less than twenty years later, *The Wealth of Nations* contained the essence of Mandeville's thesis.

The aim of this paper will be to start with much the same premises as those on which the theory of a free market is based, and to begin the development of a theory of collective decisions. Thus it will not begin with socialized man, but with rational man, wholly concerned with pursuit of his own interests.

THE ELEMENTS OF A SYSTEM

In order for collective decisions to be in any way necessary, a situation must exist in which actions of some individuals have consequences for others. If each actor had control over only actions that affected himself, then no one would be affected by any other, and no social system would exist. But whenever one actor's actions can have consequences for another, the bases for a social system arise. If actor B is affected by A's action, then it comes to be to his interest to have some effect on this action, so that its consequences for him will be predictable, and not wholly harmful.

Thus we will assume to begin with a set of actors (wholly self-interested, as indicated before), whose actions have some effects for one another. Their actions are interdependent, in that each actor engages in some action whose consequences are not only for himself, but for other actors as well.

In establishing this perspective, several elements have been set down as basic: *actions* of individual actors or sets of actors; *actors* themselves, who we here consider as individuals; and for each actor a certain amount of *power* over each action, and a certain amount of *effect* that the action has for him. The effects of the action upon an individual constitute the utility of the action for the individual. This utility is in fact a utility difference: the various possible outcomes of the action will have different utilities for him, and it is the difference between these utilities which constitutes the utility or importance of the action for him. This utility difference we will call the *interests* of the individual in the action.

In considering collective decisions, we will narrow the area of investigation to those decisions over which at least several actors have some power, and for which at least several actors will feel the consequences.

In such a situation, and with wholly self-interested actors, then how can there be anything but a war of all against all? To be sure, under certain conditions something approaching such a war occurs in actual communities, when the community erupts into conflict. Thus any theoretical approach must not only show how a collective decision is made, but also show the conditions under which the community instead will break out into conflict.

The answer which the present theory will give to the war of all against all question is that only under special conditions can, in fact, a set of such self-interested actors reach collective decisions. But before examining those conditions, it is useful to state the postulates of the theory at the level of the actors. The central postulate about behavior is this: Each actor will attempt to extend his power over those actions in which he has most interest. This is a very simple application of the economist's postulate of maximization of utility. The actor will attempt to carry out this power extension in such a way as to maximize his interests, though the specific way he will go about it is not determined. (Indeed, as will be evident in the illustration, it is not clear to the actor even in a relatively simple situation what is the best way to proceed, since the strategy is so complex).[7]

The kinds of power held by each actor can be grouped into two, for purposes of the theory. First, he has power over actions within the system, including those which are of little or no interest to him. Certain of this power is of little or no direct value to him; however it may be of value to others. Second, he has power which lies outside the interdependent actions of this set of actors. Physical force is the best such example, though there are many others, involving the use of power from any external source to force the other actors into the action he desires.

For the present, let us consider the first of these. The second, since it employs power from outside the system at hand, cannot be dealt with as completely as the first.

Most collectivities are extremely rich in the kinds of actions they or subgroups within them carry out. Thus it will be necessary, in the discussion which follows, to consider a set of actors from which a great part of this richness is deleted. This is important not merely for purposes of analysis, but also because part of this richness includes investment of the actors in one another and in the collectivity itself, which generates the kind of behavior that sociologists are most quick to seize upon as being "governed by norms." These investments can be dealt with elsewhere,[8] but here we are concerned with the primitive state of a set of actors engaged in decisions in which power and interests are distributed among them—a set of actors not bound together by a wealth of internal investments.

EXCHANGES OF POWER

Faced with a situation of a lack of power over actions which interest him, together with a surplus of power over actions which interest him little or not at all, the rational man will make an exchange of power. Thus he will engage in a special kind of economic transaction. He has resources which are of little use to him. For example, in the case of a group's leader, he may have the power to determine what activity the group will carry out, though it may be unimportant to him. He will exchange these for resources which are of use to him. He may wish the group to expand, but not have control over the decisions about composition of the group. Thus he will lead the group toward an activity desired by those members whose power over adding new members to the group is greatest. Thus he will be making an economic exchange, explicit or not, in which he creates an obligation on the part of certain group members by taking an action that satisfies their interests. They may not pay off the obligation of course, but they fail to do so under the risk of not obtaining their way on a future issue that may be of importance to them.

Such exchanges may be more or less explicit and acknowledged by both sides; but whether they are explicit or not, they nevertheless exist; and I suggest they are the essence of political behavior, that is, of collective decisions.

The very first deduction, then, that can be made concerning collective decisions among a set of self-interested actors, is that there must be more than one issue upon which decisions are to be made. It is evident what occurs if there is only one action. Actors have power only over the one action, and each will employ whatever external sources of power he has (so long as their cost to him is less than his interest in this action) to gain full power over the action. This can lead, if the action is of sufficient interest to the actors, to a war of all against all. [9]

Thus the existence of more than one collective action is necessary if the decision is not to lead to the use of external power, and thus to intense and even violent conflict in a collectivity; but before discussing that point, I would like to consider a particular illustration of collective decisions at some length.

A LEGISLATIVE GAME[10]

A legislature is an extremely important case of collective decision-making. As representatives of their constituents, legislators are embodiments of the interests of society. It is for that reason that I propose to focus attention upon an abstract representation of a legislature as a means of studying collective decisions. This representation is in the form of a game, in which the players are legislators, involved in eight collective actions. [11] These eight collective actions are decisions on eight issues, with each legislator having approximately equal power on each. Each legislator has a single vote on each issue; this plus the power to raise an issue to the floor, a prerogative which is rotated, and other actions of parliamentary procedure, constitute his power. His interests, in turn, are interests in getting reelected with as large a majority as possible. His reelection, however, depends on the outcome of those issues in which his constituents are interested. If he manages to get an issue passed, those constituents who favor it will vote for him, those who oppose it, against him. If it is defeated, those who oppose it will vote for him, those who favor it will vote against him. His interests are clear, then, in such a situation. To receive constituents' votes (which constitute his interests), he must get certain issues passed, others defeated. For other issues, the outcome is unimportant for the votes he receives from constituents.[13]

Yet in many respects, the "unimportant" issues are of major importance for him. His votes on these issues are commodities which he can use, exchanging them to further his control over those issues which do matter. Typically, in the playing of this game, the principal behavior of the players is exchange of votes: Each player giving up votes on issues which are of little interest to him for votes on those which are of much interest. Other types of exchange exist also, involving the subsidiary types of powers held by each of the actors over the collective decisions: the order in which issues come to the floor, and the vote on tabling an issue.[13]

To illustrate how this game functions (and thus how collective decisions can occur when each acts in his own interest), it is useful to consider a particular configuration of interests. Cards representing constituencies and their interests are distributed to players. In a game of five, the distribution of interests might be set up as in Table 1.

TABLE 1. Net Votes for Each of Five Legislators on Eight Issues

Issue	A	B	C	D	E
Civil Rights	100+	0	100−	100+	120−
Aid to Education	80−	0	60+	60−	40+
Defense Appropriation	20−	0	160+	60+	100−
Medical Care for Aged	20−	0	120+	0	100−
Offshore oil to states	300+	100−	0	100−	100−
Federal seashore park	50+	0	50+	100+	200−
Retaining Military Base	50−	100−	200+	0	50−
Federal Dam	0	200+	100−	100−	0

In this distribution, many agreements to exchange votes can be made in the hope of gain. For example, B is interested in only three issues, and he has five votes which he can use to gain passage or defeat of those issues. If he talked first to A, he might propose an exchange of his civil rights vote in return for A's federal dam vote. (This might ensure passage of civil rights, while if he had talked to E first, a similar agreement would have defeated civil rights.) Then he could agree with E to vote for the defense appropriation in return for another federal dam vote. If A and E keep their agreements, this would ensure passage of the dam bill for B. He could then go ahead to attempt to gain defeat of either the military base bill or the offshore oil bill, using any of the other three votes on which he is uninterested. B is obviously, because of his narrow range on interests, in a position of great flexibility to gain control of legislation that does interest him.

Legislator C, in contrast, has much less latitude. Because his interests are widespread, he will very likely find it necessary to sacrifice the less strong ones in order to gain votes on those most important to him.

As is evident, the rational strategy for a legislator would require a great amount of rapid calculating ability, including estimates of the agreements that others will make and of their likelihood of honoring agreements.

The distribution of interests in this example illustrates certain important kinds of variations in the distribution of interests in collective actions such as legislation. Issues 5-8 have been termed "pork-barrel legislation," because for each, there is a single constituency with very strong interests in one direction (in favor of passage, except in the case of the seashore park), while the opposing interests are more widely distributed through several constituencies. In the game, such issues characteristically generate more vote exchanges than do others, because of the intensity of interests of the constituency most directly affected.

Other issues show different variations. For example, some of the issues are "balanced," with as many constituents' votes for as against; but defense appropriation has an excess of 100 votes for passage; civil rights and aid to education have excesses of 20 and 40 against; and medicare has an excess of 40 for.

A third variation is the difference between constituencies which are internally split, and those which are solidly for or against. In civil rights, nearly all precincts are strongly for or against. (Each card is a precinct with a total of 100 votes on a single issue, except in the case of the affected precincts on pork-barrel issues, which have a total of 300 votes. A legislator may have several precinct cards on the same issue, which can either reinforce or cancel each other.) In aid to education, on the other hand, most precincts are internally divided. As a result, the net interests of legislators tend to be strongly one way or the other on civil rights, but less strongly so on aid to education.

The effects of these variations on the bargaining and on the outcome of legislation have not yet been studied systematically. (The difficulties in analysis—and even data collection—in a system with as many internal transactions as this one become apparent when one attempts to study this game quantitatively. Though it has been played hundreds of times, modes of analysis are only beginning to be developed by several students of the game.) It is evident, however, that these variations—which mirror variations found in collective decisions in society—do have some effect, both on the behavior of the legislators and on the outcome of the legislation.

The question which can be in part answered by this game is the question with which we started: How can there be other than a war of all against all in such collective decisions? Obviously, so long as an actor can exchange power over actions which have little or no effects on his interests, in return for power over actions which interest him, this is the best means of gaining his interests. Introduction of external power usually involves a sacrifice of other interests. Thus under many conditions it is quite irrational for him to introduce such external power, and quite rational for him to act within the existing framework of power.

Observation of the playing of this game gives some insight into this question. Players do not quit the game in the middle; seldom are they uninterested in playing subsequent sessions. They do not attempt to use external power.[14] Instead, they avidly seek to gain useful power by exchanging power that is useless to them. However, some sessions have been played by keeping the same constituencies for several sessions. Those actors who hold constituencies which consistently defeat them are not highly motivated to continue in this way. Thus after a few such sessions, there is always a strain toward the redistribution of interests by redistributing constituencies.

One can treat the tendency to withdraw or stop playing as a weak manifestation of the tendency to use external power, and thus the tendency for collective decisions to break down. The condition described above, of a permanent disadvantage (in this case, through a particular distribution of interests, but in many empirical collectivities, through a particular distribution of power), is the major condition in this game that generates such a tendency to stop playing. There are other tendencies, however, which indicate the conditions under which collective decisions can break down. The most interesting of these that has been observed in this game is the tendency to break an agreement after another actor has paid off his part of the agreement with his vote. When the issue in which an actor is interested comes up early in the session, those who have agreements with him must make their payment first. Then he may without immediate cost feel free to break his side of the agreement. Such behavior is not prohibited by the rules of the game, and it does occur.

However, the following behavior has been observed, by a group of players who have continued playing together over a number of sessions: In the first sessions, such agreements were broken with some frequency. Certain actors gained reputations of being unreliable, with the consequence that others made agreements with them less frequently. As the sessions continued, however, all actors became quite trustworthy, and in particular, these "unreliable" actors made direct attempts to restore the confidence of others in them, by keeping agreements and not attempting to extract the last ounce of gain in making their exchanges.

Thus the short-range rationality predominated when these actors were engaged in a very temporary set of collective decisions—one or two sessions of the legislature. But when the actors became engaged in long-term collective action, a long-term rationality came to dominate. This long-term rationality necessitated, purely in the actor's own interests, developing trustworthiness (or in economists' terms, a high credit rating), often at the cost of immediate gain.

CONDITIONS FOR COLLECTIVE DECISIONS

Beyond answering the major question of how a set of rational actors can engage in collective decisions, we can begin to study the conditions under which such collective decisions can occur without the use of external power. A major one of these conditions has already been stated: That there be more than one collective action. This is obvious in the present context; it is not so obvious in empirical situations. For example, one would predict that as fringe benefits, vacations, shorter hours, automation adjustments, come to be added to the single issue of wages, collective bargaining could be more successful.[15] This prediction is not at all "commonsense," and in fact common sense would have led me to make the opposite predication: As the number of issues increases, the difficulty of reaching a successful negotiation would increase.

Other predictions would be similarly non-commonsense: That in "bedroom" suburbs community decision-making would much more quickly erupt into conflict than in socially and economically self-contained towns. In the former communities, there are far fewer decisions and far fewer interests, which could provide possibilities of exchange. Empirically, in the United States, the major decisions of importance in bedroom communities are schools; and it is true that the most frequent and intense community conflicts over schools have been in such communities. In another decision-making area also, water fluoridation, these communities show the greatest intensity of controversy and the greatest frequency of upsetting a community council's decision (ordinarily by defeating fluoridation).

Still another prediction has to do with the viability of democractic political systems in voluntary associations and social movements: Since voluntary associations are ordinarily segmental, involving only a single interest of most members, democratic politics would be particularly liable to erupt into conflict. This is most relevant in associations in which strong interests are involved, such as labor unions, or political or religious movements. In confirmation of the prediction, labor unions, and political and religious movements are particularly subject to division and uncontrolled strife. The best example of a viable democratic system in a labor union is the printer's union, which differs sharply from most unions in the range of a member's interests that are associated with the union. Political movements stand in contrast to political parties (which incorporate a wide range of interests) in their much greater tendency to fragmentation.[16]

The second deduction which is corollary of the first is that if a collective decision is to be reached without the use of external power, there must not be one action which interests each actor more than all the other actions in which they are collectively engaged. If such focus of interest did exist, this would mean that no actor is willing to make an exchange of his power over this action in return for power over others. Thus just as in the case of a single action, the concentration of interests on this one action prevents exchanges of power.

Again one would make predictions from this: When in a community an issue arises that is of great interest to all members, this issue would more likely break out of the bounds of normal decision-making than would an issue which is of interest to a smaller fraction of the population. For if there is great interest in the issue on the part of all, the community divides into a majority and minority, each holding this issue to be most important; and thus the minority is motivated to seek external power.

A third deduction that can be made concerning the conditions for collective decisions is that the various interests be distributed across persons in several different ways. For if the distribution of interests on one issue is the same as that on all others, the various different issues in effect collapse into one. No or almost no exchanges are possible, as illustrated by the following example from the game which approximates this situation.

| | | | Actor | | |
Issue	A	B	C	D	E
1	+100	+ 50	+ 75	−100	− 50
2	+ 50	+100	+ 75	− 50	−100
3	−100	− 50	− 75	+100	+ 50
4	− 50	−100	− 75	+ 50	+100

In this case, actor A could exchange votes with actor E, issue 2 for issue 1, and issue 4 for issue 3; but he would not be wise if he did so, because without exchanges, a natural majority exists of A, B, and C on all issues. By making exchanges, he would lose at least on his issues of minor interest, 2 and 4. Thus in this case the distribution of interests coincide, and no exchanges will be profitable to both sides. The minority has no recourse except to external power.

Under certain conditions in society, such coincidences in the distribution of interests on important issues often arise. In Canada at present, for example, there are major social divisions which generate interests on the part of members: language, religion, region, economic status. Unfortunately for decision-making in Canada, these divisions cut the society in much the same way: French-speaking, Catholic, Quebec, economically depressed, vs. English-speaking, Protestant, non-Quebec, prosperous. In the United States, there is a similar coincidence of division: Negro, economically depressed, and culturally deprived vs. white, prosperous, culturally advantaged. In the United States, however, the lack of regional coincidence with this division has led not to legislative breakdowns, but to the use of different kinds of popular power through direct action by and on behalf of Negroes.

In communities in which conflict erupts with particular intensity, there appears also to be, with great frequency, such a coincidence of interest-generating divisions in the community.[17]

These predictions obviously cover only a small subset of the total deductions which might be made when this theory is well developed. One set of deductions would have to do with the distribution of power over various collective decisions in the society. In the game discussed here, the distribution was a very simple one: A single vote for each actor; and all collective decisions are made by the whole set of actors. However, in any existing collectivity, different sets of actors control different actions of the collectivity, which makes matters enormously complex. In a legislative body such as the U.S. House of Representatives, for example, different committees hold power over different types of legislation, and over different stages in the passage of legislation. In turn, a body of the political party determines membership and position on the committees. In any representative assembly, there is another alternation of power: The legislators' votes control the passage of laws affecting the people; and the citizens' votes control the fate of the legislature's members.

Another variation is exemplified by the difference between a Congress and a Parliament. In a parliamentary system, the executive and legislative functions of government are more nearly merged. Members of a parliament in the controlling majority are responsible to the executive as well as to their constituents, for their reelection is contingent on two decisions: The party's decision to put them up for reelection, under control of the central party (i.e., the executive), and their reelection itself, under control of the citizens. In the U.S. Congress, in contrast, reelection does not depend upon support by the national party, except in minor ways, such as speeches and patronage.[18]

NORMS AND COLLECTIVE DECISIONS

Let us return now to the question of norms and the necessity for introducing homo sociologicus, governed by norms, as an intrinsic element in theories of collective decision-making. A sociologist might well point out that the players in such a game as described above are all well-socialized persons, and thus homo sociologicus has entered the game in the form of socialized players. In one sense these players are socialized. They are socialized as rational beings, who can estimate the consequences for themselves of their actions, in particular the consequences at several removes in time. They may be socialized in other ways as well, but if so, that has no relevance to the game. For one can conceive of totally unsocialized persons (except as he can calculate the consequences of his actions, and anticipate the behavior of others) as players of this game. (One could even program on a computer the behavior of such persons in a game like this, as one player, Clinton Herrick, is presently doing.) These players, in order to satisfy their interests, would behave just as do the persons who do play the game, with only minor variations. It would require ingenuity beyond my own to discover strategies they could employ which would be more likely to gain them their interests than to engage in the exchange of power that takes place when "socialized" persons are the players.

I would go one step further, and say that a major aspect of socialization is not "the internalization of norms," but rather coming to see the long-term consequences to oneself of particular strategies of action, thus becoming more completely a rational, calculating man. It is likely that this socialization in fact comes about through the games that children play, and through the extended game they play with their principal childhood authorities, their parents.

A sociologist might also suggest that the rules of this game, or any game, are the analog of social norms. This, I believe, is not so in the conventional sense of norms (as rules which, if followed, can sometimes cause him to act against his selfish interests). For a game is an activity which an individual enters voluntarily. Thus he stays in the game only so long as it is in his interests to do so. If there are rules which cause him to act against his interests in the game, then he will change the rules or leave the game, unless he feels that the rules are compatible with his long-range interests.

This is not to say, of course, that norms do not develop in a social system. It is to say that norms require acquiescence from all actors insofar as they have freedom to withdraw from the system; and such acquiescence is given in terms of the actors' own interests. It is to say further that it is not norms, and individuals socialized to them, which are the starting points of a fruitful theory of social systems, but instead, collective actions and rational actors, each with interests and power relative to these actions.

CONCLUSION

This paper has attempted to show the way in which rational self-interested actors can engage in collective decisions without engaging in a war of all against all. It has attempted as well to indicate some of the structural conditions under which such collective decisions can be made without recourse to external power. It should be recognized, of course, that there are other processes operative in such collective decisions, because empirical collectivities are rich in internal investments of members: Investments of self in one another, in sub-groups of the collectivity, and in the collectivity itself. Such investments, and their impact on collective decisions, offer an important realm of investigation, along with further investigation of structures of power and exchanges of power discussed in the present paper. Beyond this lies the investigation of these processes in organizations such as bureaucracies, which have very different structures of control over various decisions, and different distributions of interest, than do collectivities of the sort discussed in this paper.

NOTES

1. George C. Homans, "Social Behavior as Exchange," *American Journal of Sociology,* 63 (May, 1958), pp. 597-606. Of historical interest is the fact that in the first journal devoted to sociology in the United States, *The Sociologist,* Albert Chavannes, its founder, propounded a "law of exchange" which was strikingly like that of Homans. See John Knox, "The Concept of Exchange in Sociological Theory," *Social Forces,* 41 (1963), pp. 341-345. This work, however, apparently had no influence on the subsequent development of social theory.

2. George C. Homans, *Social Behavior: Its Elementary Forms,* New York: Harcourt, Brace and World, Inc., 1961.

3. My own interests in this direction were intensified by a paper by Talcott Parsons, "On the Concept of Influence," *Public Opinion Quarterly,* 27 (Spring, 1963), pp. 37-62. In the course of a critical commentary on this paper–James Coleman, "Comment on 'On the Concept of Influence,' " *Public Opinion Quarterly,* 27 (Spring, 1963), pp. 63-82–I began to see for the first time ways in which the general strategy of Homans could be applied to more macroscopic social structures. This paper of Parsons represents a considerable departure from his earlier approaches to theory in sociology, a departure I consider very profitable indeed.

4. Thomas Hobbes, *Leviathan,* Oxford: Blackwell, 1960; first published in 1651. Hobbes said (p. 82), "Hereby it is manifest, that during the time men live without a common power to keep them all in awe, they are in that condition which is called war; and such a war, as is of every man, against every man."

5. Adam Smith, *The Wealth of Nations,* Modern Library Edition, New York: Random House, 1937, p. liv.

6. Adam Smith, *The Theory of Moral Sentiments,* published in abridged form in H. W. Schneider, ed., *Adam Smith's Political and Moral Philosophy,* New York: Hafner Publishing Co., 1948.

7. Several political scientists have argued that decision-makers do not calculate complex rational strategy, but solve one problem at a time. For example, Simon proposes that decision-makers "satisfice," Lindblom proposes a theory of "incremental" decision-making, and March suggests that there is successive attention to problems of urgency. My intent is not to argue that "maximizing" behavior takes place, rather than one of these alternatives. I propose only that the actor will attempt to gain more power over those actions which interest him most. Obviously, different actors may employ different strategies in doing so. The focus, however, should not be on the type of strategy, but on the structure of power and interests which generates activity in this direction.

8. I have discussed them in two papers, though the subject is only opened for examination by these papers: op. cit., and "Individual Autonomy in Theories of Action," a paper presented at meetings of the International Sociological Association, Washington, D. C. September, 1962.

9. The failure of welfare economists to consider more than one action in their attempts to devise an aggregating function that will maximize social welfare—see, e.g., Kenneth J. Arrow, "Social Choice and Individual Values," Cowles Commission Monograph 12, New York: John Wiley, 1951, and Jerome Rothenberg, *The Measurement of Social Welfare,* Englewood Cliffs, N. J.: Prentice-Hall, 1961–has led to the present impasse in welfare economics. It is possible to show that with a number of issues, such exchanges as discussed above lead toward a maximazation of the social welfare, relative to the existing power distribution.

10. I am indebted to those at Johns Hopkins with whom I developed this game: James Diltz, James Kuethe, Edward McDill, Anthony Neville, Robert Peabody, and Sarane Boocock. Peabody's studies of the House Rules Committee–R. L. Peabody and N. W. Polsby, eds., *New Perspectives on the House of Representatives,* Chicago: Rand McNally and Company, 1963, papers by Peabody, and Cummings and Peabody–were particularly useful in stimulating the game's construction. Most of the insights into the functioning of the game have come from a continuing group of players who are faculty members and graduate students at Johns Hopkins. This group consists of Louis Goldberg, Clinton Herrick, James McPartland, Robert Peabody, Leo Rigsby, Seymour Spilerman, Arthur Stinchcombe, Donald Von Eschen, Benjamin Zablocki, and myself. In addition, I am grateful to members of seminars at McGill University and Carleton University for comments.

11. The game begins by a distribution of constituencies to each of the players. All 52 cards are dealt (even when the number of cards per player is not equal), each card representing the votes in a given precinct. Alternatively, constituencies may be chosen by players, each taking a card in turn from the set of cards placed face up. Each precinct is interested in only one issue, though in most precincts, voters are divided on the issue.

Each player calculates the net vote for or against each issue in his constituency, and in turn the players make short statements to inform others of their constituents' interests in particular issues. Following the speeches, a period for initial negotiation occurs, halted at the end of three minutes by the session chairman (the dealer). The player to the left of the chairman moves an issue to the floor, either for passage or defeat. After a second to the motion, and in the absence of a successful tabling motion, the negotiation period on this bill begins, and continues for two minutes. A vote on this bill is then taken by roll call, and the floor is given to the next man to the left, who may move another issue. Defeat of a bill is final for the session, but a tabled bill may be brought up again.

Scoring occurs as indicated in the body of the paper. Constituents give their votes to legislators not on the basis of the legislator's vote, but on the basis of the success or failure of the bill in which they are interested.

The precincts are as indicated below:

Civil rights	Pro	75	0	75	0	100	100	70	0	0	70
	Con	25	100	25	100	0	0	30	100	100	30
Aid to education	Pro	70	100	0	30	70	0	100	70	20	20
	Con	30	0	100	70	30	100	0	30	80	80
Defense appropriation	Pro	90	90	90	90	90	0	0	0		
	Con	10	10	10	10	10	100	100	100		
Medical care for aged	Pro	70	30	0	80	80	100	30	30		
	Con	30	70	100	20	20	0	70	70		
Offshore oil to states	Pro	0	0	0	300						
	Con	100	100	100	0						
Federal seashore park	Pro	75	100	75	50						
	Con	25	0	25	250						
Retaining military base	Pro	25	0	25	250						
	Con	75	100	75	50						
Federal Dam Con.	Pro	0	0	0	300						
	Con	100	100	100	0						

12. For purposes of exposition, it may be confusing that the actor's interests are the votes of constituents. This is distinct from his own vote, which constitutes his power over the legislative actions. Such a representative system is not necessary to illustrate the processes under discussion here. A "direct democracy" game in which each player is a citizen with specified interests in particular collective actions has been developed, and that would more simply illustrate these processes. The legislative game is used here as illustration because there has been more experience in playing it.

13. I played this game once with a group of political scientists highly skilled in the use of parliamentary procedure. We play by full parliamentary rules rather than the restricted set of rules otherwise used in this game, and it became quickly evident that there are many auxiliary types of power held by an actor who knows these rules.

14. For example, in a game with graduate students and faculty members, faculty members do not use the external power they hold over graduate students to force exchanges which are not in the student's best interest. This is not because of any altruism on the part of faculty members, but because the use of such power would cost them more in respect than it would gain them in the game. They see also that introduction of such power would so unbalance the power structure that those without power would withdraw from subsequent play, and thus the advantage gained would be wholly a temporary one.

15. This predication from the above theory was not evident to me until it was pointed out by Albert Breton of the University of Montreal.

16. I am grateful to Maurice Pinard of McGill University for drawing my attention to the empirical confirmation of the prediction, in the fragmentation of ideologically focused political movements.

17. For examples and discussion see James S. Coleman, *Community Conflict*, New York: The Free Press, 1957.

18. Each of these systems shows its defects under particular conditions. In the U.S. Congress, the lack of responsibility to the executive produces great difficulties in the passage

of legislation, except that which is in the direct interest of a majority of Congressmen. In the Canadian Parliament, the defect shows up in current political strife over separatism: the provincial governments declare themselves, rather than members of Parliament, as the true representatives of their province, and hold much more power than do state governments in the U.S., even though the British North America Act (the Canadian Constitution) grants them less power than the U.S. Constitution grants the states. As a generalization, one could say that in the U.S., state and regional power is exerted through the Congress; in Canada, it is exerted through the provincial governments.

THE SIMULATION OF AN ARTISTIC SYSTEM

PHILIP H. ENNIS

Wesleyan University

The sociology of art, though quickening in recent years, is still in an undeveloped state. Its practitioners are few, and fewer still spend full time on the job. It has been more often than not a field to dabble in, to begin one's sociological career, and then go on to more "basic" areas or about which to make windy pronouncements. There has been, consequently, little cumulation of theory, method, and findings. Among the many things the field needs, one of the most important is a theoretical apparatus of sufficient power to identify the basic components of artistic systems, to specify the range and types of interaction among the components, and to locate the artistic subsystem within the larger society, or, more accurately, develop concepts to account for the different and varying ways artistic subsystems are articulated with the larger society.

The second thing necessary for a developing sociology of art is the monograph—indeed, a chain of monographs—intensive studies analysing one moment in the life of an art form. There are good reasons, I suppose, why the monographic study of artistic systems has been largely the preserve of art historians and to a lesser extent of psychologists. Perhaps it is imperatives of historical detail, the necessity to weave a mass of concrete individual and institutional stories into a coherent sequential pattern that is beyond the interest or skill of many sociologists.

It was that oppressive historicity that led me to the experiment to be described below. I wanted to see how an artistic system worked without having to master the burdens of three different kinds of historical material. First, I didn't want to sort out tangles of institutional interconnections such as I have had to do in my work in the development of rock music. Second, I didn't want to learn the particular language of a different art form. This history of the *content* of painting, music, theatre, dance, etc., all require great knowledge of the specific languages of those forms. It takes half a lifetime of specialized work to fully understand these languages. Third, I did not want to traffic with the Zeitgeist, the total cultural atmosphere that drenches a given artistic subsystem and shapes its growth in subtle and mysterious ways.

I wanted an artistic system composed of the barest structural components. I wanted control over the social processes that made these components work, and I wanted an instant tradition of the art language so that I could understand how it would change over time. If these three conditions could be met, an artistic system could be simulated.

With respect to the instantly created artistic language, several critieria had to be met. First, the basic unit (equivalent to the single painting, poem, etc.) of the language has to be brief enough so that many of them can be processed through the system in a short time; that is, the simulation has to be able to do the centuries of the "history of Western painting" in a few moments. Second, the unit has to be rich enough and complex enough to carry messages equivalent to the compressed and meaning-laden content of real music, art, theatre, etc. Third, the unit has to be a part of a stream—that is, to be constructed from a formal language, with a vocabulary, a grammar, and a set of artistic and normative conventions as to form and content. Fourth, there must be a sufficient supply of these units to make that formal language discernible, and to make it capable of variation and development. There must be, in short, the "tradition" which supplies both the boundaries of the genre and the potentialities for further creative use of the tradition for contemporary additions to it. Finally, and most critically, the unit must be capable of being produced anew by the actors in the simulation. We have to be able to make new contributions to the tradition.

The ordinary television commercial and video tape equipment with an editing capacity fulfills these conditions. Accordingly, about one and one-half hours' worth of commercials was video taped directly from the air in early July of this year. The tradition, instantly collected and permanently closed, consisted of 132 commercials and some brief segments of programming taken from the three major networks on a Saturday from about 9 a.m. to 11 p.m.

The eight or nine participants in the simulation were four college students, one of whom had worked the previous semester on a pilot version of the simulation. The rest, high school students, were then subjected to hours of exposure to the tradition. There were several purposes for this sustained viewing. One was to get them sufficiently familiar with the TV commercial in the special context of viewing it as "the history of Western painting" so that they could break out of their previous set of responding to a commercial as a persuasive—that is, an *instrumental*—message, and to learn to experience it as an *expressive* symbol. Second, sustained exposure to the tradition as a group provided almost the only explicit *socialization* into their future roles they were to receive. We talked about types of commercials, what are liked and disliked, and, most importantly, which ones we wanted to work with subsequently as actors in the system. To facilitate this, each student was given his or her own video tape and allowed to record the specific commercials that were of interest. These work tapes were used later by the actors in the course of their role requirements. Finally, prolonged exposure to the full tradition was intended to bind all the participants together in a shared experience which their future audiences would not have had, which would differentiate them from the audience and put, so to speak, the burden of proof on them to communicate with that audience.

I realized only later how important this extended viewing was. There is much to learn about television commercials. They are created by a complex and highly differentiated industry. The cost of a single commercial ranges from a few thousand to over sixty thousand dollars. One commercial, described in *The Anatomy of a Television Commercial* by L. Diamont, took 100 professionals 15 months—10,000 man hours—to make. They involve the efforts of teams of writers, directors, still and film photographers, designers, actors, and, not least important, market researchers. Their studies are generally embodied in the characteristics of the people portrayed in the commercials, mirroring the target audiences sought and in the appeals to be made to those audiences. Commercials range from about 10 seconds to 2 minutes, though most today are 30 seconds in duration. There is thus a great compression of message and form in the commercial. Much slips by the ordinary viewer unnoticed.

This compression reproduces, unintentionally perhaps, one of the basic processes in all art—that is, to reveal some things and to hide some things. In all art systems, there is a changing balance of what is hidden and what is revealed to the various participants in the system by each art object which flows by. This is one way to define the change of style in an art form. Television commercials do this in fully as many ways as does any other medium. An example: Many of you have seen commercials for laundry products that promise the housewife she will be able to wash out stubborn stains. For example, college boy and girl meet in cafeteria—they bump trays—he gets mustard on his shirt—she gets ketchup—the announcer's voice promises that the product can remove mustard and ketchup stains. A subsequent and more revealing commercial for another similar product tells the housewife she can remove stains of chocolate pudding and blood from the family's soiled laundry. The hidden is becoming clearer but not so much that the proprieties of television can be breached to say out loud that these products are aimed at the recalcitrance of excrement and menstrual blood stains. This is on the instrumental side.

On the artistic side, there has not been a film technique that has not been used in commercials. You are all now quite familiar with the freeze, the fade, the speed up, slow motion, or the yellow submarine psychedelic animation on commercials. Yet it takes sustained viewing to see the varieties of zoom shots and the endless jokes in casting actors who while waxing, walking, eating, dieting, look suspiciously like Norman Mailer, Lady Bird Johnson, or Ted Kennedy. But commercials are an acquired taste. I do not make any claim they are an art form in their natural state just because they borrow art-like techniques and personnel from the more acknowledged art media.

Now for the social process. Eight students were assigned to the three basic "interior" roles of the systems and given their normative guidelines. Four students chose or were selected to be artists. Two were to be distributors, and two were to be critics. The *artists* were instructed as to their basic norms. The first master norm is, in I. A. Richard's term: The artist's job is to get it right. He or she must work out that picture in his head until it is within his own boundaries of acceptability. The second master norm is to communicate that completed picture to someone somehow: Behavior guided by these two directives has complex, sometimes contradictory, outcomes.

The artists were also given the boundary conditions of their product. The commercials they were to make had only three restrictions. First, they had to be sixty seconds or less in duration. Second, they had to include a product, and, third, only the tradition could be used. In one way or another, all three of these norms were violated in the course of this simulation as they were in the earlier pilot project.

The critic's basic role requirement was to develop the standards of evaluation of the tradition and the new commercials. They were to study, explain, and persuade artists, distributors, and audience of their view of the tradition. Their operative response was to say which commercials were good and which were bad, and why.

The distributor's job was to create a show, teach an audience to understand and like it, and try to win the competitive battle among themselves for the biggest share of that audience.

The audience was a group of about sixty high school students enrolled in a summer school enrichment program. Their job was to do what audiences all over the world do, to say yea or nay, they like it or they don't. All these people, sixty audience members, four artists, two distributors, and two critics, went through three generations; this was the major socialization of the experiment.

The procedure began with the artists creating their first commercials. I instructed them (rather, we taught each other) singly and as a group in the techniques of the Sony 3650 and 3600 video tape recorders. These machines have a limited editing capacity. They allow the artist to separate the sound track from the video track, to put the sound of one commercial with the picture from another. They allow him to place a video sequence from one commercial next to the video sequence from another. A still shot—that is, a single frame—or a set of still shots can be inserted into a video sequence. The speed of the video part can be slowed down or speeded up. The artists, not surprisingly, varied in the time they put in, their skill on the machines, their ideas, and their responses to the other participants in the simulation.

The artists produced nine commercials for the first generation. The two distributors selected the ones they liked. They then each selected about 35 commercials from the tradition to surround the newly created ones and developed a logo—a distinctive video mark that identified their own shows. The audience was assembled and told they were to look at the show as if it was a film parley—a succession of little movies. On a sheet of paper they were to score each commercial on a scale from 1 to 5. If they loved it, 1; if they hated it, 5; 2, 3, 4 for intermediate degrees of liking. The audience could and did this scoring with no difficulty. They clearly discriminated among the commercials. Some were almost totally hated, others were enjoyed by almost everyone, and some were responded to neutrally. The critics too scored the shows, also on a scale from 1 to 5. A really good commercial was scored 1, a very bad one, 5. Most of the artists attended all the shows. It is my judgment that the collective response of the audience during the show—laughter, groans, tension, release—was more important to artists and distributors than the scores, which I summarized and showed to all the participants the next day.

The two critics wrote reviews of this first show and circulated them to the artists and distributors. Everyone digested everyone else's responses. One artist dropped out; another took her place. The artists then began to create commercials for the second generation. In two weeks, they had prepared ten new ones. The distributors again chose those they wanted to use. They selected additional ones from the tradition to make a show. The critics screened the show the night before the performance and distributed a pre-review to the audience. The audience was assembled and again scored each of the commercials in the show. The critics did likewise and wrote their reviews. The artists once more took stock of how their work was received, and what their brother and sister artists had done. They proceeded to make the third generation of commercials. They made nine which were given to the distributors, who produced and presented a new show to the audience. The simulation was stopped at this point.

Now, what did the commercials look like and how did they change over the three generations? For the first generation, the artists had as a guide only a brief view of the commercials made during the pilot project and our discussions of the tradition. Those discussions produced a threefold categorization of commercials. It is unclear how much this categorization determined their creative style and how much it was a post hoc rationalization, making sense of a very mixed bag. Most of the commercials in the tradition were what we called the *ironic mode.* That is, they sold a product around a funny story that had a punch line or a humorous theme. A second type was the *serious*, straightforward appeal to do or buy something. These were largely public service commercials, against pollution, for the Peace Corps, Red Cross, and the like. They usually had a mixture of strongly lyrical music and visual images. The third type was the lively, action-packed commercial for children's products. That was more direct than the ironic ones, but more light-hearted than the lyrical ones.

The artists stayed very close to the style of the tradition in their first generation's commercials. All nine commercials used a single stylistic device, and eight of the nine used a single technical method. The stylistic device was an extension of the ironic mode. All the artists made a joke about the commercials. The technical method was to use the audio part of the commercial over the video of another. Again, all the artists used this approach in all nine of the commercials. Whether this restriction of creative form was due to a lack of technical skill in using the video tape equipment or whether it was due to the artists' defensive, distance-providing rejection of the medium, or both is difficult to say. The critics' response to these first-generation commercials was to complain of their technical qualities. There was too much break up, roll-over, junk. It was difficult to get to the message because of this. The audience was also clearly upset with the distortion. "Technical excellence" in an art form is a vital necessity to audiences. They are ready to give themselves up to a situation of release. Every day roles are dropped. But the audience role includes, just as do other roles, its repertoire of vigilance and defense mechanisms. The main direction of these mechanisms is to ensure cognitive safety and intelligibility. The artist and the distributor violate these expectations of safety and intelligence at their peril. As with other boundary maintaining norms, technical excellence is subject both to rigidification on the one hand and deliberate "creative" flouting on the other.

In the preparation for the second generation of commercials, most of the artists concentrated on technical mastery and, in so doing, expanded their artistic vocabulary considerably. They also responded to each other's work more attentively. They stole and borrowed extensively.

The developing concern with technique revealed, if only incipiently, the "two language" problem. In addition to artists' hiding and revealing, there is another process that separates artists from the other participants in the system. It is that they become absorbed in the techniques of their medium and develop a set of appreciative standards that is more or less invisible to outsiders. More important, their more intense involvement with the medium, compared to audiences and even critics and distributors, means they burn out forms more quickly. They are always ahead of the audience in both languages.

The ten second-generation commercials showed considerable advance technically and substantively. Only four of the ten (compared to *all* nine of the first generation) used the two-commercial technique. Three of this generation used three commercials, one used six, one used nine, and one used twenty-two different commercials. The still shot, the still-into action sequence, and the audio loop were other technical advances.

In terms of content, there was a clear movement away from the ironic distance stance into a greater acceptance of the medium. But the ironic style still had audience appeal, and some of the artists responded accordingly. Finally, there was a peculiar focus on automobiles in this generation. I can't explain it; but I can almost hear an art historian of the simulation saying, "Ah, the Zeitgeist, it was in the air." There was also in some of the second generation's new commercials a continuation of social commentary. Critics' response to the generation struck a balance between interpretation of the possible meanings of the social commentary commercials and praise, if puzzlement, for the artistic ones.

The third generation expanded the artistic direction even further, producing in fact two important normative violations of the artist's guidelines.

The first was the transformation of a 30-second commerical which showed a service station attendant changing a car's muffler in the speeded-up style of early movies. The artist slowed the tape into an excruciatingly long four minutes. The audience fought its restlessness and puzzlement, eyes glued to the screen. Thus, the narrative guideline of 60 seconds was creatively breached.

The second violation was more subtle. The artist compressed the entire hour and a half's tradition into about 40 seconds by recording and recording the tradition tape played at the "fast forward" speed. This process, of course, eliminated ordinary visual images leaving only a moving set of shaded patterns on the screen. He titled it "The Tradition" and in so doing forced the distributor to go beyond his normatively implicit code of silence into a verbal explanation to the audience.

The simulation was, in my view, partially successful, especially because of its surprises. It was successful first because we got untrained young people to enact the basic social roles of an artistic system, transforming thereby a randomly collected and inert collection of objects into a living and developing tradition. There was one surprise that is still mysterious. I had expected that as each generation proceeded the artists would gradually increase their use of the newly made commercials and decrease their use of the tradition. Yet not a single artist directly used any of the commercials made by another artist. They borrowed techniques, subject matter, and even whole concepts from each other. And there was a heavy use and reuse of a limited number of commericals from the tradition. Second, as the simulation unfolded, I observed the growing separation and even conflict among artists, critics, distributors, and, to a lesser extent, the audience. Yet I simultaneously observed the tendency for critic and distributor not only to influence the artists, but to become them. Both distributors actually created partial commercials which they put on their shows, and at least one of the critics displayed consistent interest in the aesthetic possibilities of the medium.

The greatest failure of the simulation was that there did not develop any really discernible *selective* impact of the audience on the shows as a whole or on the new commercials. The artists and distributors were *generally* constrained and excited by the presence of the audience, and in one case they were powerfully directed. The first show was too long and too repetitious. The distributors were quite careful subsequently to keep it varied and briefer.

We also moved the show to an air conditioned auditorium and provided the audience with free Cokes. But there was not very much audience impact on the specific content of the shows or as far as I could discern, on the creative process. I even attempted to alter the distributors' behavior by instructing one of them to be an "art" distributor by emphasizing the most advanced art commercials and the other to be a "commercial" distributor by emphasizing the units that were most liked by the audience. Not even the critics noticed any difference in the programming. There were two reasons for this failure, I think. One was that the audience was a captive one. They could neither vote with their feet and stay away nor could they differentially respond to the two shows, since they were presented back to back and were not very clearly differentiated in content. The second reason is that there were too few artists. By the end of the first generation's production, the artists appeared to have developed an informal norm of productivity—at least one commercial, but not more than three or four. The distributors used all the new commercials produced, destroying thereby a powerful weapon that might have influenced the artists. The ability of the distributor's selective power on the artist and subsequently on audiences depends upon an abundance of material to choose from. Inadvertently, the artists' restriction of production protected them from audience and distributor influence.

Two other problems arose, both dealing with socialization into an artistic system. The audience, you will recall, had no way to make sense of the first-generation commercials unless they had seen enough of the tradition to understand the transformations in the original commercials. The only way the audience could be persuaded to see the tradition was if it was packaged in the form of a show. But a show and an educational experience tend to have different requirements. The distributor and the educator have different interests. In this case, the distributors began to develop an interest in presenting sequences of commercials which would exemplify certain themes—e.g., the female distributor strung a series of commercials in a row underscoring the sexist-female as object items. This attempt clashed both with the showcasing of the new commercials and with the presentation of enough of the tradition to make them intelligible. Moreover, the very presentation of the tradition interfered with the distributor concerns for high audience liking and their dislike of repetition. There are a number of procedures that can obviate these difficulties, but they need testing and experimentation.

The other problem stems from the absence of formal socialization of the artists, distributors, and critics and the more or less arbitrary assignment of students to these roles. One of the persistent problems in all art is the natural and social selection of individuals into the role of artist, critic, etc. With such a truncated procedure as used in this simulation, it was not possible to study this process, other than to record the incipient social labeling that took place: this artist was creative but slovenly in technique, that artist was a fine technician, and so on.

Finally, a word about one basic social process in artistic systems that was amenable to study. This is the effect of repeated exposure to an art object. There is satiation on the one hand, which tends to exhaust the audience's liking for a play, movie, song, etc. But on the other there is the nostalgia effect which makes an old favorite more liked after some period of non-exposure. The simulation repeated many commercials once or twice, and a few up to five times across the three shows, while in general there was great stability of audience response—if they liked it once, they liked it three times. There was sufficient variation of response to invite analysis of the net direction of change and the individual components of that aggregate shift. It is still premature, but it does appear that the extent of initial liking and initial intelligibility of the commercial are important variables in the struggle between satiation and nostalgia.

APPENDICES

APPENDIX A
Bibliography

I. THE NATURE AND RATIONALE OF GAMING-SIMULATION.

Abelson, Robert P.
   1968 "Simulation of Social Behavior." In Gardner Lindzey (ed.), Handbook of Social Psychology, Vol. II Cambridge, Massachusetts: Addison Wesley.
Back, K. W.
   1963 "The Game and the Myth as Two Languages of Social Science." Behavioral Science 8 (January): 67-71.
Becker, H. A. and H. M. Goudappel (eds.)
   1972 Developments in Simulation and Gaming. Utrecht, Netherlands: Boom Meppel.
Boocook, Sarane
   1971 "Instructional Games." In Encyclopedia of Education. New York: Macmillan Company.
Callois, Roger
   1961 Man, Play and Games. New York: The Free Press.
Cangelosi, Vincent E. and William R. Dill
   1965 "Organizational Learning: Observations Toward a Theory." Administrative Science Quarterly 10 (September): 175-203.
Coleman, James S.
   1964 "Collective Decisions." Sociological Inquiry 34 (Spring): 166-181.
   1966 "Introduction: In Defense of Games." American Behavioral Scientist 10 (October): 3-4.
   1967 "Game Models of Economic and Political Systems." Pp. 30-34 in Samuel Klausner (ed.), The Study of Total Societies. New York: Anchor Books.
   1968 "Social Processes and Social Simulation Games." In Boocock and Schild (eds.). Simulation Games in Learning. Beverly Hills, Calif.: Sage Publications, Inc.
   1969 "Games as Vehicles for Social Theory." American Behavioral Scientist, XII (July-August): 2-6.
Coplin, William
   1966 "Inter-Nation Simulation and Contemporary Theories of International Relations." American Political Science Review 60 (September): 562-578.
   1970 "Approaches to Social Sciences through Man-Computer Simulations." Simulation and Games 1 (December): 391-410.
Druckman, Daniel
   1968 "Ethnocentrism in the Inter-Nation Simulation." Journal of Conflict Resolution 12, (March): 45-68.
   1971 "Understanding the Operation of Complex Social Systems: Some Uses of Simulation Design." Simulation and Games 2 (June): 173-195.
Duke, Richard D.
   1964 Gaming Simulation in Urban Research. Institute for Community Development and Service, Continuing Education Service, Michigan State University, East Lansing, Michigan.
   1974 Gaming: A Future's Language. Beverly Hills: Sage Publications, Inc.

Feldt, Allan G.
    1966 "Operational Gaming in Planning Education." Journal of the American Institute of Planners 32 (January): 17-23.
Gagne, Robert *et al.*
    1962 Psychological Principles in System Development. New York: Holt, Rinehart and Winston.
Goffman, Erving
    1961 Encounters: Two Studies in the Sociology of Interaction. Indianapolis: Bobbs-Merrill Co.
    1969 Strategic Interaction. Philadelphia: University of Pennsylvania Press.
Goodman, Fred L.
    1972 "Games and Simulations." 2nd ed. In Robert Travers (ed.) Handbook of Research on Teaching. Chicago: Rand McNally.
Greenblat, Cathy S.
    1971a "Le Developpement des Jeux-Simulations a l'usage du Sociologue." Revue Francaise de Sociologie 12 (avril-juin): 206-210.
    1971b "Simulations, Games, and the Sociologist." The American Sociologist 6 (May): 161-164.
    1974 "Gaming and Gaming-Simulation. An Overview for Teachers, Trainers, and Community Workers." In Marshall Whithed and Robert Sarly, Urban Simulation Design and Analysis. Netherlands: Sitjhoff International Publishing Company.
House, Peter
    1974 The Urban Environmental System: Modeling for Research, Policy-Making and Education. Beverly Hills: Sage Publications.
Huizinga, Johan
    1955 Homo Ludens. Boston: Beacon Press.
Inbar, Michael
    1969 "Development and Educational Use of Simulations: An Example 'The Community Response Game.' " International Journal of Experimental Research in Education 6 (January): 5-44.
Kinley, Holly J.
    1966 "Development of Strategies in a Simulation of Internal Revolutionary Conflict." American Behavioral Scientist 10 (November): 5-9.
Klietsch, Ronald G.
    1969 An Introduction to Learning Games and Instructional Simulations: A Curriculum Guideline. Saint Paul: Instructional Simulations, Inc.
Lauffer, Armand
    1973 The Aim of the Game. New York: Gamed Simulations Inc.
Livingston, Samuel A. and Clarice S. Stoll
    1973 Simulation Games for the Social Studies Teacher. New York: Free Press.
Long, Norton E.
    1958 "The Local Community as an Ecology of Games." American Journal of Sociology 64 (3): 251-261.
McLaughlin, B.
    1971 Learning in Social Behavior. New York: The Free Press.
McLuhan, Marshall
    1964 "Games: The Extensions of Man." In Marshall McLuhan, Understanding Media. New York: McGraw-Hill.
Mead, George H.
    1934 "Play, the Game, and the Generalized Other." In G. H. Mead, Mind, Self and Society. Chicago: University of Chicago Press.

Meier, Richard L.
    1963 "Game Procedure in the Simulation of Cities." Pp. 348-354 in L. J. Duhl (ed.) The Urban Condition: People and Policy in the Metropolis, New York: Basic Books.
    1967 "Simulations for Transmitting Concepts of Social Organization." Pp. 156-175 in Werner Z. Hirsch et al., Inventing Education for the Future. San Francisco: Chandler Publishing Co.

Modelski, George
    1970 "Simulations, 'Realities', and International Relations Theory." Simulation and Games 1 (June): 111-134.

Moore, Omar Khayyam and Alan Ross Anderson
    1969 "Some Principles for the Design of Clarifying Educational Environments." In David Goslin (ed.) Handbook of Socialization Theory. Chicago: Rand McNally.

Prud'homme, Remy, Jean de la Brunetiere, and Gabriel Dupuy
    1972 Les Jeux de Simulation Urbanistiques. Paris: Tema-Editions.

Raser, John R. L.
    1969 Simulation and Society. Boston: Allyn and Bacon, Inc.

Ray, Paul and Richard D. Duke
    1968 "The Environment of Decision-Makers in Urban Gaming Simulations." In William Coplin (ed.) Simulation in the Study of Politics. Chicago: Markham.

Rhyne, R. F.
    1972 "Communicating Holistic Insights." Fields Within Fields . . . Within Fields 5 (1).

Russell, Constance J.
    1972 "Simulating the Adolescent Society: A Validity Study." Simulation and Games 3 (June): 165-188.

Shubik, Martin
    1971a "On Gaming and Game Theory." Technical Report P-4609. Santa Monica: The RAND Corporation.
    1971b "On the Scope of Gaming." Report P-4608. Santa Monica: The RAND Corporation.

Suits, Bernard
    1967 "What is a Game?" Philosophy of Science 34 (June): 148-156.

Tansey, P. J.
    1971 Educational Aspects of Simulation. New York: McGraw-Hill.

Taylor, John L. and Rex Walford
    1972 Simulation in the Classroom. Baltimore: Penguin.

Verba, S.
    1964 "Simulation, Reality and Theory in International Relations." World Politics 16 (April): 491-519.

II. ELEMENTS OF DESIGN AND CONSTRUCTION

Adair, Charles H. and John T. Foster, Jr.
    1972 A Guide for Simulation Design. Tallahassee, Florida: Instructional Simulation Design Inc.

Berger, Edward, Harvey Boulay and Betty Zisk
    1970 "Simulation and the City: A Critical Overview." Simulation and Games 1 (December): 411-428.

Boocock, Sarane S.
    1972 "Validity-Testing of an Intergenerational Relations Game." Simulation and Games 3 (March): 29-40.

Burgess, Philip
  1969 "Organizing Simulated Environments." Social Education 33 (February): 185-192.
Druckman, Daniel
  1971 "Understanding the Operation of Complex Social Systems: Some Uses of Simulation Design." Simulation and Games 2 (June): 173-195.
Duke, Richard D.
  1974 Gaming: A Future's Language. Beverly Hills: Sage Publications.
Environmetrics
  1971 The State of the Art in Urban Gaming Models. Springfield, Virginia: Clearinghouse for Federal Scientific and Technical Information.
France, William and John McClure
  1972 "Building a Child Care Staff Learning Game." Simulation and Games 3 (June): 189-202.
Gamson, William A.
  1971 "SIMSOC: Establishing Social Order in a Simulated Society." Simulation and Games 2 (September): 287-308.
Glazier, Ray
  n.d. "How to Design Educational Games." Cambridge, Massachusetts: Abt Associates.
Greenblat, Cathy S.
  1974 "Sociological Theory and the 'Multiple Realities Game.'" Simulation and Games 5 (March).
Helmar, Olaf
  1972 "Cross-Impact Gaming." Futures (June): 149-167.
Inbar, Michael and Clarice S. Stoll
  1972 Simulation and Gaming in Social Science. New York: Free Press.
Lauffer, Armand
  1973 The Aim of the Game. New York: Gamed Simulations, Inc.
Livingston, Samuel and Clarice S. Stoll
  1973 Simulation Games for the Social Studies Teacher. New York: Free Press.
Raser, John R.
  1969 "Simulation and Society." Boston: Allyn and Bacon, Inc.
Twelker, Paul A.
  1969 "Designing Simulation Systems." Educational Technology (October): 64-70.

III. GAMING-SIMULATION FOR TEACHING AND TRAINING

Abt, Clark
  1967 "Education is Child's Play." Pp. 123-155 in Werner Z. Hirsch et al., Inventing Education for the Future. San Francisco: Chandler Publishing Co.
  1970 Serious Games. New York: The Viking Press.
Armstrong, R. H. R. and John L. Taylor (eds.)
  1970 Instructional Simulation Systems in Higher Education. Cambridge, England: Cambridge Institute of Education.
Attig, J. C.
  1967 "Use of Games as a Teaching Technique." Social Studies 58 (January): 25-29.
Baldwin, John D.
  1969 "Influences Detrimental to Simulation Gaming." American Behavioral Scientist 11 (July-August): 14-20.
Benjamin, Stanley
  1968 "Operational Gaming in Architecture." Ekistics, 26, 157 (December): 525-529.

Bloomfield, L. P. and N. J. Padelford
    1959 "Three Experiments in Political Gaming." American Political Science Review
        53 (Dec.): 1105-1115.

Boocock, Sarane S.
    1966a "An Experimental Study of the Learning Effects of Two Games with Simu-
        lated Environments." American Behavioral Scientist 10 (October): 8-17.
    1966b "Games with Simulated Environments in Learning." Sociology of Education
        39 (Summer): 215-236.
    1966c "Toward a Sociology of Learning: A Selective Review of Existing Research."
        Sociology of Education 39 (Winter): 1-45.
    1967a "Games Change What Goes on in the Classroom." Nation's Schools 80
        (October): 94-95.
    1967b "Life Career Game." Personnel and Guidance Journal 46 (December):
        328-334.
    1970 "An Innovative Course in Urban Sociology." The American Sociologist 5
        (February): 38-42.
    1971 "Instructional Games." In Encyclopedia of Education. New York: Macmillan
        Company.
    1972 An Introduction to the Sociology of Learning. Boston: Houghton Mifflin.

Boocock, Sarane S. and E. O. Schild
    1968 Simulation Games in Learning. Beverly Hills, California: Sage Publications.

Brodbelt, Samuel
    1969 "Simulation in the Social Studies: An Overview." Social Education 33 (Febru-
        ary): 176-178.

Bruner, Jerome S.
    1966 Toward a Theory of Instruction. Cambridge, Mass: Harvard University Press.

Burgess, Philip
    1966 "Political Science Gaming in Teaching and Research." Ohio State University.
        Fall WOSU Faculty Lecture Series, mimeographed.

Carlson, Elliot
    1967 "Games in the Classroom." Saturday Review. (April 15): 62-64, 82-83.

Charles, Cheryl L. and Ronald Stadsklev
    1973 Learning with Games. Boulder, Colorado. Social Science Education Con-
        sortium, Inc.

Chartier, Myron R.
    1972 "Learning Effect. An Experimental Study of a Simulation Game and Instru-
        mented Discussion." Simulation and Games 3 (June): 203-218.

Cherryholmes, Cleo H.
    1966 "Some Current Research on Effectiveness of Educational Simulations: Impli-
        cations for Alternative Strategies." American Behavioral Scientist 10
        (October): 4-7.

Cohen, Kalman J. and Eric Rhenman
    1961 "The Role Of Management Games in Education and Research." Management
        Science 7 (January): 1-17.

Cohen, B. C.
    1962 "Political Gaming in the Classroom." Journal of Politics 24 (May): 367-381.

Coleman, James S.
    1967a "Academic Games and Learning." Invitational Conference on Testing Prob-
        lems, Educational Testing Service, Princeton, New Jersey.
    1967b "Learning through Games." NEA Journal 56 (January): 69-70.

Cruickshank, D. R. and F. W. Broadbent
    1969 "An Investigation to Determine Effects of Simulation Training on Student
        Teaching Behavior." Educational Technology 9 (October).

Cruickshank, D. R.
    1970 "The Use of Simulation in Teacher Education: A Developing Phenomenon." Journal of Teacher Education 20 (Spring): 23-26.

Daniellan, Jack
    1967 "Live Simulation of Affect-Laden Cultural Cognitions." Journal of Conflict Resolution 11 (October): 312-324.

Degnan, Daniel A. and Charles H. Harr
    1971 "Computer Simulation in Urban Legal Studies." Journal of Legal Studies 23 (2): 353-365.

DeKock, Paul
    1969 "Simulations and Changes in Racial Attitudes." Social Education 33 (February): 181-183.

Edwards, Keith J.
    1971 "The Effect of Ability, Achievement, and Number of Plays on Learning from a Simulation Game." Baltimore: Center for Social Organization of Schools, The Johns Hopkins University Report No. 115.

Fennessey, Gail M., Samuel A. Livingston, Keith J. Edwards, Steven J. Kidder, Alyce W. Nafziger
    1972 "Simulation, Gaming and Conventional Instruction: An Experimental Comparison." Baltimore: Center for Social Organization of Schools, The Johns Hopkins University, Report No. 128.

Fletcher, Jerry L.
    1971a "The Effectiveness of Simulation Games as Learning Environments: A Proposed Program of Research." Simulation and Games 2 (December): 425-454.
    1971b "Evaluation of Learning in Two Social Studies Simulation Games." Simulation and Games 2 (September): 259-287.

Garvey, Dale M.
    1967 Simulation, Role Playing and Socio-Drama in the Social Studies. Emporia, Kansas: Emporia State Research Studies.

Giffin, S. S.
    1965 The Crisis Game: Simulating International Conflict. New York: Doubleday.

Goldhamer, Herbert and Hans Speier
    1959 "Some Observations on Political Gaming." World Politics 12 (October): 71-83.

Gordon, Alice Kaplan
    1970 Games for Growth. Palo Alto, California: Science Research Associates.

Greenblat, Cathy S.
    1973 "Teaching with Simulation Games: A Review of Claims and Evidence." Teaching Sociology 1 (October).

Guetzkow, Harold
    1959 "A Use of Simulation in the Study of International Relations." Behavioral Science 4 (July): 183-191.
    1962 Simulation in Social Sciences: Readings. Englewood Cliffs, New Jersey: Prentice-Hall.
    1963 Simulation in International Relations: Developments for Research and Teaching. Englewood Cliffs, New Jersey: Prentice-Hall.

Heap, James L.
    1971 "The Student as Resource: Use of the Minimum Structure Simulation Game in Teaching." Simulation and Games 2 (December): 473-487.

Henderson, Bob G. and George Gaines
    1971 "Assessment of Selected Simulation Games for the Social Studies." Social Education (May): 508-513.

Inbar, Michael
    1966 "The Differential Impact of a Game Simulating a Community Disaster." American Behavioral Scientist 10 (October): 18-27.

Ramey, James W.
  1967 "Simulation in Library Administration." Journal of Education for Librarianship, VIII (Fall): 85-93.
Raser, John R.
  1969 "Games for Teaching." In Simulation and Society. Boston: Allyn and Bacon Inc.
Robinson, James A.
  1966 "Simulation and Games." In Peter Rossi and Bruce Biddle (eds.), The New Media and Education. Chicago: Aldine Publishing Co.
Robinson, James A., Lee F. Anderson, Margaret G. Hermann, and Richard C. Snyder
  1966 "Teaching with Internation Simulation and Case Studies." American Political Science Review 60 (May): 53-64.
Shirts, R. Garry
  1970 "Games Students Play." Saturday Review (May 16): 81-82.
Sprague, Hall
  n.d. "Using Simulations to Teach International Relations." La Jolla, California: SIMILE II, Western Behavioral Sciences Institute, mimeographed.
Sprague, Hall T. and R. Garry Shirts
  1966 "Exploring Uses of Classroom Simulations." La Jolla, California: SIMILE II, Western Behavioral Sciences Institute (mimeographed).
Stewart, Edward C.
  1967 "The Simulation of Cross-Cultural Communication." Washington, D.C.: Human Resources Research Office.
Stoll, Clarice S. and Michael Inbar
  1970 "Games and Socialization: An Exploratory Study of Race Differences." Sociological Quarterly 2 (Summer): 374-380.
Stoll, Clarice S. and P. T. McFarlane
  1969 "Player Characteristics and Interaction in a Parent-Child Simulation Game." Sociometry 32 (September).
Tansey, P. J.
  1970 "Simulation Techniques in the Training of Teachers." Simulation and Games 1 (September): 281-303.
  1971 Educational Aspects of Simulation, New York: McGraw Hill.
Tansey, P. J. and D. Unwin
  1969 Simulation and Gaming in Education. London: Methuen, and New York: Barnes and Noble.
Taylor, John L.
  1971 Instructional Planning Systems. New York: Cambridge, University Press.
Taylor, John L. and R. N. Madison
  1958 "A Land Use Gaming Simulation." Urban Affairs Quarterly (June): 37-51.
Taylor, John L. and K. R. Carter
  1967 "Instructional Simulation of Urban Development: A Preliminary Report." Journal of Town Planning Institute 53 (December): 443-447.
Taylor, John L. and Rex Walford
  1972 Simulation in the Classroom. Baltimore: Penguin.
Twelker, Paul A. (ed.)
  1969 Instructional Simulation: A Research Development and Dissemination Activity. Corvallis, Oregon: Teaching Research, 1969.
  1972 "Some Reflections on Instructional Simulation and Gaming." Simulation and Games 3 (June): 147-153.
Walford, R.
  1969 Games in Geography. London: Longmans.

IV. OTHER ARENAS AND APPLICATIONS

A. Public Policy

Abt, Clark C.
  1970 Serious Games, New York: The Viking Press.
Armstrong, R. H. R. and Margaret Hobson
  1969a "Games and Urban Planning." Surveyor 31 (October): 32-34.
  1969b "Models for Life." Education 5 (September).
  1969c "Planning Games Are More than Just Fun." Municipal and Public Services
     Journal 2089 (November).
Duke, Richard D. and Barton R. Burkhalter
  1966 The Application of Heuristic Gaming to Urban Problems. East Lansing:
     Institute for Community Development.
Feldt, Allan
  1966 "Potential Relationships between Economic Models and Heuristic Gaming
     Devices." In Vincent P. Rock (ed.), Policymakers and Modelbuilders: Cases
     and Concepts. New York: Gordon and Breach, Science Publishers.
Fuller, Buckminster
  1969 "The World Game." Ekistics, 28 (October): 286-291.
Greenblat, Cathy S.
  1974 "Gaming as Applied Sociology." In Arthur Shostak (ed.). Putting Sociology to
     Work. New York: David McKay.
Helmar, Olaf
  1972 On the State of the Union. Middletown, Connecticut: Institute for the Future.
House, Peter and Philip D. Paterson, Jr.
  1969 "An Environmental Gaming Simulation Laboratory." Journal of the American
     Institute of Planners 35 (November): 383-388.
House, Peter
  1974 The Urban Environmental System: Modelling for Research, Policy-Making and
     Education. Beverly Hills: Sage Publications.
Laska, Richard M.
  1972 "Games People Play Help Solve Urban Ills." Computer Decisions 4
     (February): 6-10.
Lauffer, Armand
  1973 The Aim of the Game. New York: Gamed Simulations, Inc.
Lindsay, Sally
  1972 "APEX" Saturday Review (May 13): 55-57.
Little, Dennis
  1972 "Social Indicators, Policy Analysis and Simulation." Futures (September):
     220-231.
Meier, Richard L. and Richard D. Duke
  1966 "Gaming Simulation for Urban Planning." Journal of American Institute of
     Planners 32 (January): 3-17.
Ray, Paul and Richard D. Duke
  1968 "The Environment of Decision-Makers in Urban Gaming Simulations." In
     William Coplin (ed.) Simulation in the Study of Politics. Chicago: Markham,
     1968.
Steinwachs, Barbara
  1971 "The Urbanarium: A Museum Responds." The Museologist 121 (December):
     5-9.
Tansey, P. J.
  1971 Educational Aspects of Simulation. New York: McGraw-Hill.

Taylor, John L.
  1971 Instructional Planning Systems. Cambridge: University Press.
Thomas, Clayton J. and Walter L. Deemer, Jr.
  1957 "The Role of Operational Gaming in Operations Research." Operations Research 5, 1 (February): 1-27.

B. Research and Theory Construction

Boguslaw, Robert, Robert H. Davis, and Edward B. Glick
  1966 "A Simulation Vehicle for Studying National Policy Formation in a Less Armed World." Behavioral Science 11 (January): 43-61.
Drabek, Thomas E. and J. Eugene Haas
  1967 "Realism in Laboratory Simulation: Myth or Method." Social Forces 45 (March): 337-346.
Dukes, Richard
  1973a "Learning Tools to Research Instruments: A Research Package for STAR-POWER." Boulder: University of California Sociology Department, Mimeographed.
  1973b "Symbolic Models and Simulation Games for Theory Construction." A paper delivered at the 1973 Annual Meetings of the American Sociological Association.
Guetzkow, Harold
  1972 "Simulations in the Consolidation and Utilization of Knowledge about International Relations." Pp. 674-690 in Randall L. Schultz, Simulation in Social and Administrative Science: Overviews and Case-Examples. Englewood Cliffs, New Jersey: Prentice-Hall, Inc.
McFarlane, Paul
  1971 "Simulation Games as Social Psychological Research Sites: Methodological Advantages." Simulation and Games 2 (June): 194-161.
McGuire, William J.
  1969 "Theory-Oriented Research in Natural Settings: The Best of Both Worlds for Social Psychology." In Muzafer Sherif and Carolyn W. Sherif (eds.), Interdisciplinary Relationships in the Social Sciences. Chicago: Aldine Publishing Co.
Phillips, Bernard S.
  1971 "Simulation." Pp. 171-190 in Social Research: Strategy and Tactics. New York: The Macmillan Co.
Raser, John and Wayman J. Crow
  n.d.   "A Simulation Study of Deterrence Theories." La Jolla, California: Western Behavioral Sciences Institute.
Raser, John R.
  1969 Simulation and Society. Boston: Allyn and Bacon Inc.
Smoker, Paul
  1969 "Social Research for Social Anticipation." American Behavioral Scientist 12 (July-August): 7-13.
Straus, Murray A.
  1970 "Methodology of a Laboratory Experimental Study of Families in Three Societies." Pp. 552-577 in Reuben Hill and Rene Konig, Families in East and West. Paris: Mouton.
Terhune, K. W. and J. H. Firestone
  1970 "Global War, Limited War and Peace: Hypotheses from Three Experimental Worlds." International Studies Quarterly 14 (June): 195-218.

V. RESOURCES

A. Bibliographies and Directories

Belch, Jean
> 1973 Contemporary Games: Volume 1 Directory. Detroit: Gale Research Company.

Charles, Cheryl L. and Ronald Stadsklev
> 1973 Learning with Games. Boulder, Colorado: Social Science Education Consortium, Inc.

Duke, Richard D.
> 1969 Operational Gaming and Simulation In Urban Research: An Annotated Bibliography. Ann Arbor: Environmental Simulation Laboratory.

Gibbs, G. I.
> 1974 Handbook of Games and Simulation Exercises. Beverly Hills: Sage Publications.

Joint Council on Economic Education
> 1968 Bibliography of Games–Simulations for Teaching Economics, and Related Subjects. New York: Joint Council on Economic Education.

Kidder, Steven J.
> 1971 Simulation Games: Practical References, Potential Use, Selected Bibliography. Baltimore: Center for Social Organization of Schools, The Johns Hopkins University.

Klietsch, Ronald and Fred Wiegman
> 1969 Directory of Educational Simulations, Learning Games, and Didactic Units. St. Paul: Instructional Simulations, Inc.

Lowenstein, Louis K.
> 1971 An Annotated Bibliography on Urban Games (Exchange Bibliography #204). Monticello, Illinois: Council of Planning Libraries.

Nagelberg, Mark and Dennis Little
> 1970 "Selected Urban Simulations and Games." Simulation and Games 1 (December): 459-481.

Nagelberg, Mark
> 1970 Simulation of Urban Systems–A Selected Bibliography. Middletown, Connecticut: Institute for the Future.

Shubik, Martin and Garry D. Brewer
> 1972 Reviews of Selected Books and Articles on Gaming and Simulation. Santa Monica, California: The RAND Corporation.

Thornton, Barbara
> 1971 Gaming Techniques for City Planning: A Bibliography. Exchange Bibliography #181. Monticello, Illinois: Council of Planning Libraries.

Twelker, Paul
> 1969 Instructional Simulation Systems: An Annotated Bibliography. Corvallis, Oregon: Continuing Education Publications, Teaching Research.

Werner, Roland and Joan T. Werner
> 1969 Bibliography of Simulations: Social Systems and Education. La Jolla, California: Western Behavioral Sciences Institute.

Zieler, Richard
> Games for School Use: An Annotated List. Yorktown Heights, New York: Board of Cooperative Educational Services.

Zuckerman, David W. and Robert E. Horn
> 1973 The Guide to Simulation Games for Education and Training. Cambridge, Massachusetts: Information Resources, Inc. (2nd ed.).

B. Periodicals and Newsletters

American Behavioral Scientist has devoted several entire issues (October 1966, November 1966, July-August 1969) to simulation and games. From Sage Publications, 275 South Beverly Drive, Beverly Hills, California 90212.

Simulation and Games: An International Journal of Theory, Design, and Research is intended to provide a forum for theoretical and empirical papers related to man, man-machine, and machine simulations of social processes. The journal publishes theoretical papers about simulations in research and teaching, empirical studies, and technical papers about new gaming techniques. Each issue includes book reviews, listings of newly available simulations, and 'simulation reviews'." Published quarterly by Sage Publications, 275 South Beverly Drive, Beverly Hills, California 90212.

Simulation/Gaming/News. Published five times during each school year, beginning February 1972. "Simulation/Gaming/News will facilitate communication between people involved, or interested in getting involved, with exciting new alternatives to traditional instruction. . . . It will be published in tabloid newspaper format . . . it will tend toward the informal and will be access-oriented, putting people in touch with each other and with other information sources." Subscriptions from Simulation/Gaming News, Box 3039, University Station, Moscow, Idaho.

Simulation Sharing Service . . . designed to "promote and facilitate the use of simulations in the church's ministry. It will deal with existing and emerging games, adaptations and modifications, trends in gaming, uses of gaming, information about designing, films about games, and other happenings in simulations as applicable to the church." Published by Simulation Sharing Service, Box 1176, Richmond, Virginia 23209.

Strategy and Tactics. "This is what you get in each issue: a ready-to-play conflict-simulation game, complete with a 22 x 28 inch playing surface, die-cut playing pieces, and complete rules. An historical magazine 32-40 pages long containing one feature article on the same subject as the game in that issue as well as other feature articles of the same length on different subjects. Also included are game and book reviews, commentary on existing games and discussions of subscribers' questions on the gaming field." Subscriptions available from Simulations Publications, Inc., 44 East 23rd St., New York, New York 10010.

APPENDIX B

Guidelines For Recording Game Information

Users of gaming-simulations often have difficulty communicating with one another about the critical components and characteristics of the materials of interest to them. The "Gaming-Simulation Record Sheet" presented here is meant to serve as a form for keeping records of new and old gaming-simulations: their major characteristics and the evaluations of the user.

GAMING-SIMULATION RECORD SHEET*

TITLE: _____

Designer:
Date of construction: Present stage of development:
Subject matter:
Purpose:
Intended use:

Related games: ancestors or descendants

PRAGMATICS
  Availability:
    Cost:
    Source:

  Space and paraphernalia:
    Space requirements:
    Computer requirements:
    Other media required (not provided in kit):

    Kit paraphernalia:
      Materials and quantities needed:

      Standard/custom-made:
      Documentation:

  Personnel requirements:
    Number of operators needed:
    Operator roles and skills:

  Player characteristics:
    Number of players:
    Age range:
    Prior knowledge or sophistication required
      or recommended:
    Desired degree of player homogeneity:

* © 1973, R. D. Duke and C. S. Greenblat

Time parameters:
    Preparation time:
        Operator training:
        Player preparation:
    Duration of play:
        Introduction:
        Flying time:
        Critique:

DESIGN AND OPERATING CHARACTERISTICS:

Steps of play and plot outline:

Player organization:
    Individual/team/coalition:

    Number of players/role or team:

OTHER NOTES:

APPENDIX C
A General Framework For Evaluation

In addition to recording his evaluation of gaming-simulations a user may wish to elicit the evaluations of participants and to collect information on what transpired. Appendix C presents a very skeletal evaluation form that might be employed for such purposes. It is deliberately sufficiently general to be used with almost any gaming-simulation. Specific questions relevant to a particular gaming-simulation can be added by the user who wishes more detailed feedback from players.

GAMING-SIMULATION EVALUATION FORM
FOR PARTICIPANTS

Name of simulation: _____
Date of simulation: _____
Group (if more than one session run simultaneously): _____

1. What was your individual or group identity in the game? (e.g., city politician, "circle," "green")

2. If your role changed during the session, please give later roles played:

3. If there were formal leaders in your group, please indicate the position you had by circling the appropriate number:
   1 leader
   2 participant
   0 not applicable

4. If there were no formal leaders, please indicate how you would describe your participation in the decision-making process. Would you say you were:
   1 a "leader"
   2 an active participant (i.e., vocal in expressing your ideas)
   3 a passive participant (i.e., followed the deliberations but contributed little
   4 physically present but uninvolved
   0 not applicable

5. Please describe *your group* in terms of each of the dimensions listed. These are obviously subjective evaluations, so yours may differ from that of other group members. In each case 1 = low, 2 = medium, and 3 = high.

	Low	Medium	High	Inapplicable
a. degree of activity of group	1	2	3	0
b. broad participation by group members (e.g., everyone participating	1	2	3	0
c. group cohesion	1	2	3	0
d. degree of cooperation	1	2	3	0
e. ability to reach concensus despite disagreements	1	2	3	0
f. quality of leadership exercised by formal or informal leaders	1	2	3	0

6. Which three people do you feel had the *most effect,* for better or for worse, on what happened in the course of play? If you think they had a positive effect, put a + sign after the name; if you think they had a negative effect, put a - sign after the name.

_____   _____   _____

7. Please describe each of the following dimensions of your personal participation by using the scale below. Again, 1 = low, 2 = medium, and 3 = high.

		Low	*Medium*	*High*	*Inapplicable*
a.	interest in the session(s)	1	2	3	0
b.	enjoyment of the session(s)	1	2	3	0
c.	learning about the system or topic simulated	1	2	3	0
d.	development of empathy for those who play your role in real life (i.e., understanding of their problems, tasks, etc.)	1	2	3	0
e.	development of empathy for those who in real life play other game roles	1	2	3	0
f.	self-awareness, self-understanding	1	2	3	0
g.	getting to know others in the group or class	1	2	3	0

8. How would you rate the simulation in general as a learning/communication experience? Would you rate it:
   1   Very good or excellent
   2   Good
   3   Fair
   4   Poor
   or 5   Terrible

9. What did you find the most valuable about this experience?

10. What did you find the least valuable about participating?

Please use the back of this page to give us any general comments.
Thank you for your assistance.